Climatology in Cold Regions

Climatology in Cold Regions

Chenghai Wang
School of Atmospheric Sciences
Lanzhou University
Lanzhou, China

WILEY Blackwell

This edition first published 2023

Registered Office
John Wiley & Sons, Inc., 111 River Street, Hoboken, NJ 07030, USA

Editorial Office
9600 Garsington Road, Oxford, OX4 2DQ, UK

For details of our global editorial offices, customer services, and more information about Wiley products visit us at www.wiley.com.

Wiley also publishes its books in a variety of electronic formats and by print-on-demand. Some content that appears in standard print versions of this book may not be available in other formats.

Library of Congress Cataloging-in-Publication Data applied for
Hardback ISBN: 9781119702658

Cover Design: Wiley
Cover Image: © Kirill Skorobogatko/Shutterstock

Set in 9.5/12.5pt STIXTwoText by Straive, Pondicherry, India

Contents

List of Figures

List of Tables

Foreword

It is a difficult task to put fragments together. Because I have been lived and worked in cold and arid regions, my research interests have focused on climate in cold and arid regions for a long time. Witnessing the wealth of exceptional research results in this fields in the past decades, I have dreamed of assembling the results of research in cold regions into a book to promote the formation of a systematic discipline.

This dream came true with the assistance of my students. Some former students have now become my colleagues. They follow me to work and study in this field and their insightful opinions and contributions are the basis of this book. I am extremely grateful for all of their contributions to the quality of this book. I would like to thank specifically Dr. Kai Yang and Dr. Feimin Zhang. They sacrificed a lot of spare time for the completion of this book.

I sincerely thank the teachers who guided and unselfishly helped me enter this research field, Professors Youxi Gao, Dahe Qin, Guodong Cheng, and Professor Wenjie Dong. I was afraid that the book might not live up to their expectations, so I dropped the idea of inviting one of them to write the foreword at the last minute. I am also grateful to many experts and friends who previously supported me and gave me attention.

As this is a new subject, I am only the first visitor to this particular scene. I was almost lost in the splendid view, that was created by many scientists. I am aware that I have omitted a lot of important works, and have overlooked some citations. I hope that publishing this book reflects the Chinese adage: "throw a brick to meet jade". I also expect feedbacks and corrections from peers.

Finally, I thank Dr. Collison Mandy and Ms. Rosemary Morlin for their careful and professional assistance with this book, one of the important reasons why I chose Wiley to publish it. I also thank Dr. Frank Weinreich for designing an eye-catching book cover.

September, 2022

Chenghai Wang
Lanzhou University, China

1

Introduction

Cold regions are special geographical units. Climate characteristics on a global scale have become well known, and regional climate features have attracted more and more attention. This chapter will define cold regions and types of cold region, to illustrate climate characteristics in cold regions.

1.1 Scope of the Subject and Recent Highlights

The climate in cold regions is a new, interdisciplinary, subject that is concerned with the climate and its changing in cold regions over the earth's atmosphere system. It focuses mainly on the high-altitude and middle-high latitude geosciences and their physics, chemistry, and fluid dynamics.

Many large-scale climate characteristics and processes have been understood, particularly entering the twentieth century. The climate in limited and specific regions has recently attracted more attention.

There is mounting evidence that the human activity has been influenced on climate, ecology, and biological systems. For instance, researches revealed that temperatures in cold region rise twice times faster than in middle-low latitude; the minimum temperature rises faster than the maximum temperature. One more concern is the glacial retreat and frozen ground degeneration. Particularly, more and more attention has been paid to Arctic ice melting and relevant Arctic "amplification."

Cold is a relative concept to warm. The climate can be roughly divided according to latitude into tropical, sub-tropical, and polar regions. Cold regions can be found in high-altitude areas and high mountains. Cold means a region lacks of solar radiation, in other words, where surface-energy negative balance, and absorbing less radiation than what is being released.

Owing to latitudinal differences in the distribution of solar radiation, there are latitudinal zones in which the climate varies regularly according to latitude. The globe is divided into multiple climate zones according to the characteristics of fundamental climate indicators (e.g. surface air temperature and precipitation).

1.2 Some Definitions and Terms of Reference

Cold regions are generally regarded as the cold environment in both polar regions and more temperate locations, which are closely related to the cryosphere. The areas covered by stable seasonal snow is usually considered to be cold regions. In that case, cold regions consist of glaciers, ice caps,

Climatology in Cold Regions, First Edition. Chenghai Wang.

stable snow, permafrost, sea ice, river ice, and lake ice. Based on this definition, the global area of cold region is about 0.6×10^8 km^2, covering 40% of land area. This does not include unstable snow areas and parts of the seasonally frozen ground (SFG) area.

There is no universally accepted definition of cold regions. In the early stage, Köppen (1936) defined the cold region in Canada by the criterion that the near-surface air temperature in the coldest month of a year is lower or equal to −3 °C and there are only four months with a monthly average temperature greater than 10 °C. Gerdel (1969) considered the annual mean temperature zero isothere for the division of cold regions and other regions. Wilson (1967) suggested a definition that cold regions with temperature and precipitation as essential elements. Yang (1997) used Köppen's criterion and defined the cold region in China as a places where rivers and lakes were frozen for more than 100 days. Based on these definitions, Chen et al. (2005) used the criterion that the near-surface air temperature in the coldest month of the year is lower or equal to −3°C, that the number of months with a monthly mean temperature greater than 10°C is not more than 5 months, and that the annual mean temperature is lower or equal to 5°C to define the cold region in China, along with the distribution and boundary of permafrost, glaciers, stably seasonal snow and vegetation zone divisions.

Large amounts of permafrost, SFG, and snow are distributed in cold regions. A distinct feature of land surface in cold regions is that the soil freezes and then thaws. The seasonal variation of the thermal regime in soil is a representative indicator of the climate in a cold region. In climatology, similar to Gerdel's (1969) definition, the area north of the zero isothere of the January mean temperature is defined as a cold region (Figure 1.1). This region not only includes the area defined by Köppen as the circumpolar cold region, but also contains the areas with the coldest monthly mean temperature of nearly 0°C.

The circumpolar cold region generally covers the permafrost area and the cold region in mid-latitudes covers the SFG. The freeze–thaw cycle in frozen ground can influence the soil water–heat

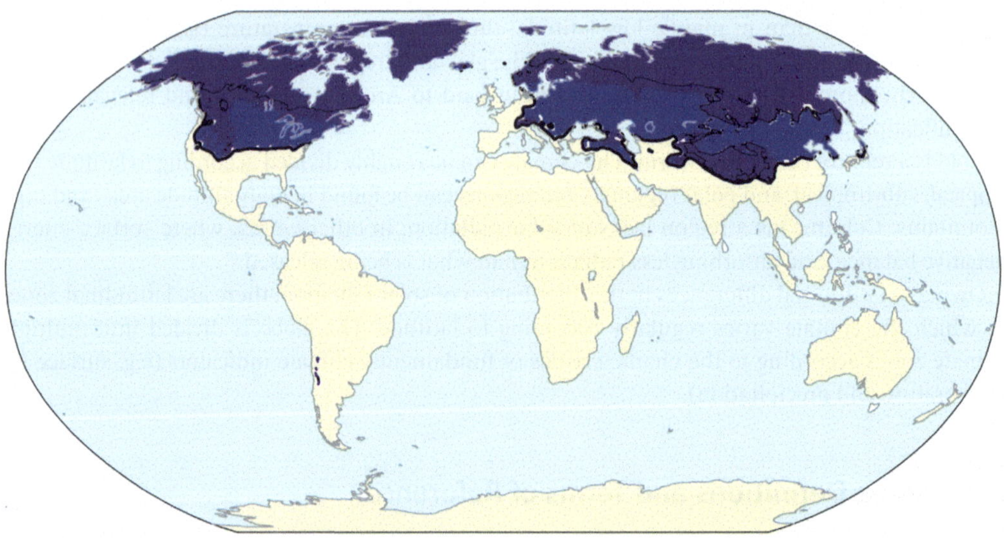

Figure 1.1 Definition of cold regions. The areas colored purple are the cold regions defined by Köppen's (1936) criterion. The areas where soil undergoes seasonally freeze–thaw is also regarded as the cold regions (slate blue). The thick and thin lines are the zero isotherms of the January temperature and annual mean temperature, respectively, which define the SFG and permafrost regions.

transport and variations of surface diabatic heating, which makes the climate in this region unique. Consistent with essential awareness, this region is also covered with snow, while the maximum extent of the snow-cover basically coincides with the zero isothere line of the January mean temperature. In the next two sections, the characteristics of climatology in cold regions and their discrepancies will be introduced.

1.3 Characteristics of Cold Region Climatology

According to the world map of the Köppen-Geiger climate classification, cold regions cover the snow zone, and polar zone, while the characteristics of precipitation and temperature have distinct spatial differences. The climate in a region is closely related to the variations of precipitation and temperature, which decides the formation of the climate. The following sections will analyze the temporospatial characteristics of precipitation and near-surface temperature.

Compared with the tropics or summer monsoon regions, precipitation in winter and summer in cold regions is obviously less, especially in winter. The annual mean precipitation in winter is only 100–150 mm; in the western region of North America, eastern Siberia, and northern China, the annual mean winter precipitation is less than 50 mm. In summer, the mean precipitation is about 200–300 mm (Figure 1.2). The distribution of precipitation has a distinct spatial difference in cloud regions; precipitation in high latitudes is greater than mid-latitudes in both winter and summer, and winter precipitation in northern Europe is greater than summer precipitation.

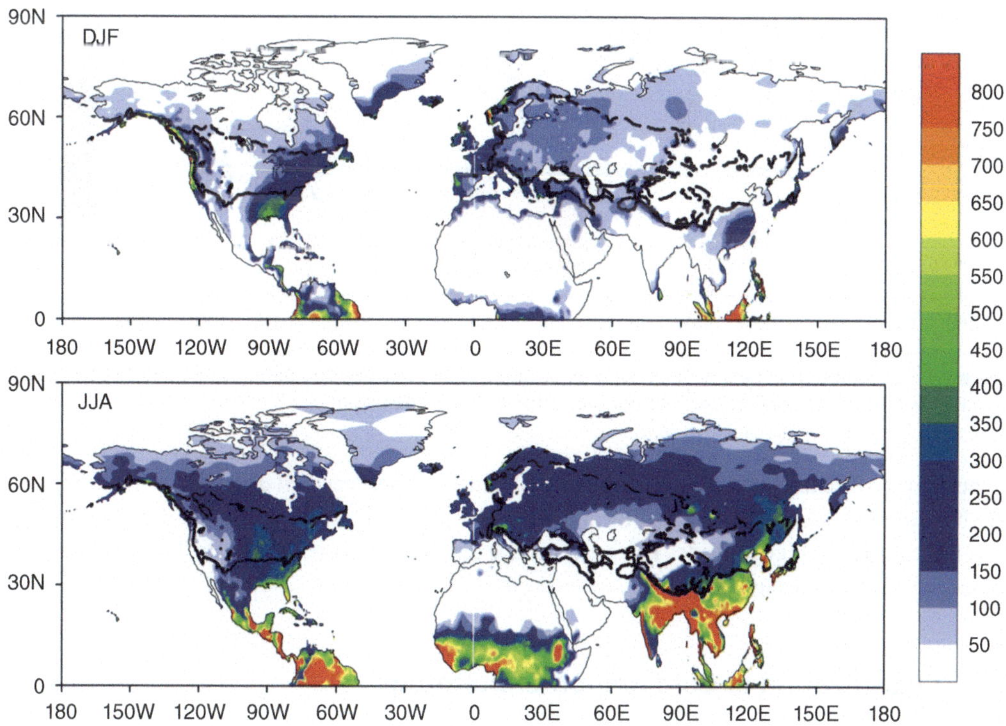

Figure 1.2 Distribution of precipitation (mm) for winter (December–January–February, DJF) (upper panel) and summer (June–July–August, JJA) (lower panel) in cold regions in the northern hemisphere.

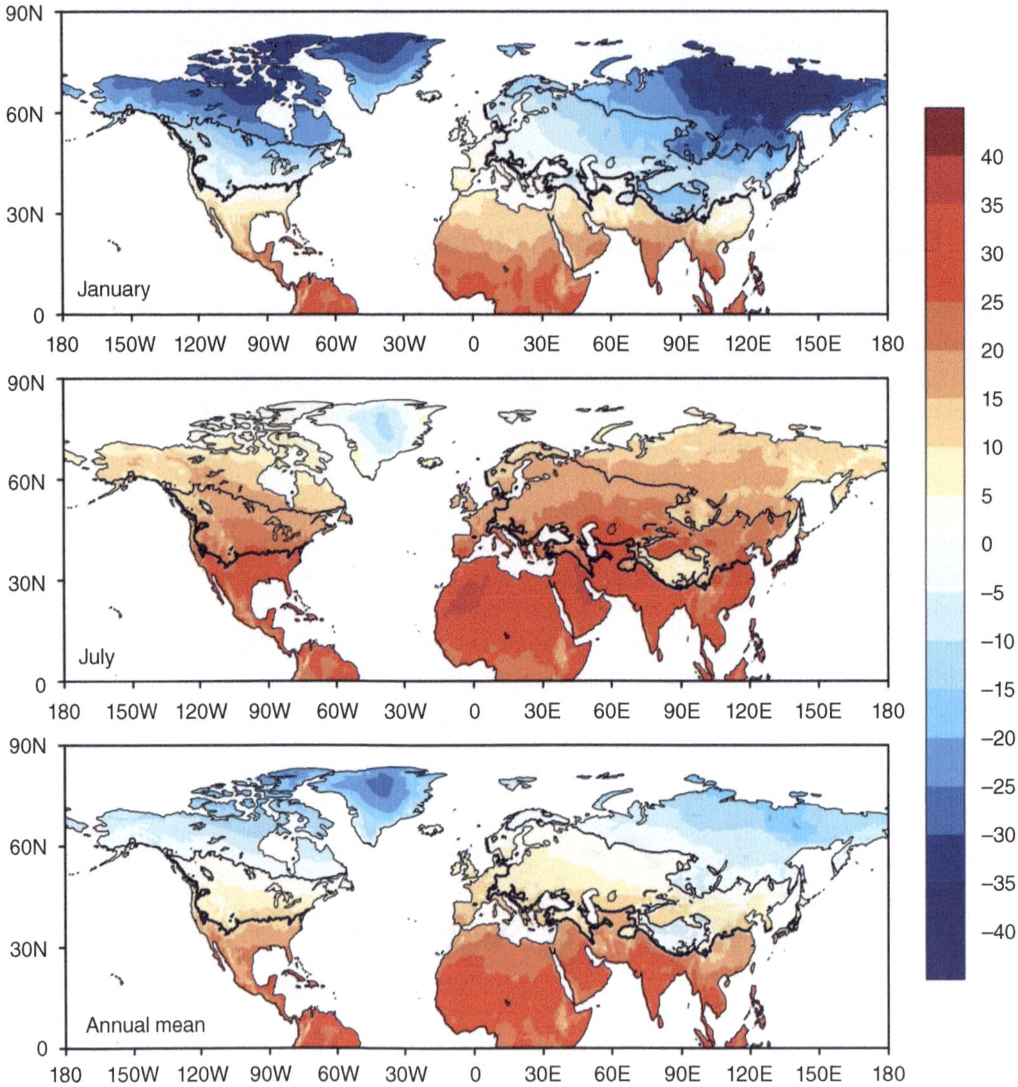

Figure 1.3 Distribution of near-surface air temperature (°C) for January (top panel), July (middle panel), and annual mean (bottom panel) in the cold regions of the northern hemisphere.

The near-surface temperature in the cold regions has distinct seasonal differences, and temperature has zonal variations, decreasing from low latitudes to high latitudes (Figure 1.3), decreasing from low attitude to high altitude. The near-surface temperature is related to the land surface processes (e.g. snow, frozen ground).

1.4 Classification of Cold Region Climatology

The climate in different parts of cold regions has heterogeneity. Global cold regions are mainly located in high latitudes and high altitudes. According to the types of frozen ground, cold regions can

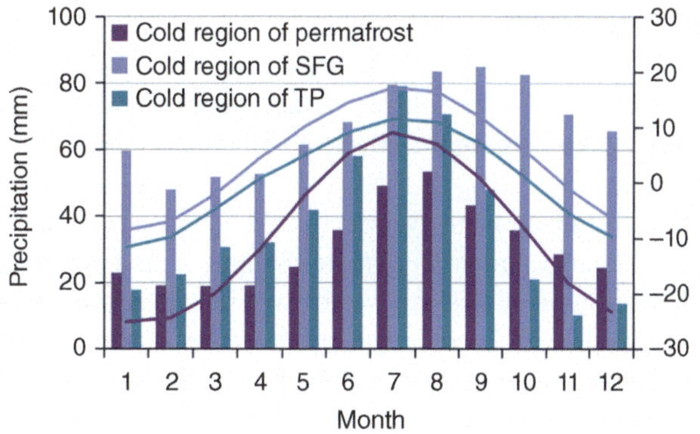

Figure 1.4 Comparisons of precipitation (mm, bar) and near-surface air temperature (°C, lines) between high latitudes (permafrost and seasonally frozen ground regions, SFG) and high altitudes (e.g. the Tibetan Plateau (TP) cold regions.)

be divided into two categories, the region of permafrost, the region of SFG. According to geographical location, it can be divided into polar regions, middle and high latitudes and the Tibetan Plateau (TP).

Comparing the precipitation and near-surface temperature between high latitudes (permafrost) and SFG and high altitudes (TP) in cold regions (Figure 1.3), it is clear that near-surface monthly temperature in the permafrost region is generally lower than that in the SFG region, and the near-surface temperature in the TP is higher than that in the permafrost region, but lower than the SFG. The variations of near-surface temperature have distinct dependence on the variation of latitude and altitude. Monthly precipitation in the SFG region is generally greater than that in the permafrost region, which is concentrated between August and October. In contrast, the precipitation in the permafrost region is concentrated between July and September. The annual variations of precipitation in the TP are different from the cold regions in high latitudes (Figure 1.4), which is concentrated between June and August with 60–80 mm, and its amount is more than that of the permafrost regi9on, but less than that in the SFG region. Comparing the annual variations of precipitation and near-surface temperature in three different cold regions, these have distinct differences.

1.5 Impacts of Cold Regions on Global Climate

On average, the temporal scale of energy (heat), moisture, and matter exchange between land and atmosphere is shorter in tropical regions than in cold regions. In tropical regions, the interaction between oceanic surface and atmosphere has quick and short features. In tropical areas, convective precipitation and tropical cyclones (typhoons or hurricanes) frequently occur, with smaller monthly change features than in sub-tropical regions. These probably result as an energy export to tropical regions. The natural function of the cold region is energy accumulation. According to this theory, cold regions are more likely to be energy hoarders for seasonal climate change.

1.6 Contribution of Cold Regions to Energy Balance

In addition to the energy from solar radiation, there is an inter-conversion of energy in the climate system. The heat absorbed from surface radiation balance in colds region would be stored in soil, and adjusted by soil moisture and soil heat capacity of different soil kinds. Energy will be released to the atmosphere in summer when evaporation becomes significant. This process corresponds to ground freezing and thawing.

1.7 Summary

- Cold regions play the role of "storage/memory" and "repeater" of energy in the climate system.
- Cold regions contain the effective and crucial signals (e.g. soil moisture, snow, etc.) for climate prediction.
- The moisture and energy cycle in cold regions are promotors for climate change.
- Global changes intensify climate changes in cold regions.
- Changes in the biology and vegetation in cold regions are more evident and vulnerable.
- Climate changes in cold regions result from dynamics and hydrothermal processes.
- Cold regions are the main and essential water sources of global fresh water, and is also the main source of global rivers.
- Engineering and construction risks in cold regions are increasing.
- Whether global warming has changed carbon and nitrogen cycles in cold regions.
- What are the hazards and risks in cold regions?

2

Climate in Cold Regions

The climate in cold regions is unique. Due to low temperature, the weather, climate, and their influence factors have obvious differences compared with subtropical and tropical regions. This chapter will present the characteristics of general circulation, the primary weather systems (including permanence and semi-permanence), meteorological elements, land–surface (including the distribution and changes in frozen ground, hydrology, and ecology), and winter monsoon in cold regions.

2.1 General Circulation

There are two basic factors that determine the general circulation of Earth's atmosphere: the latitudinal change of energy received by the Earth–atmosphere system from the sun, and the global uneven distribution of angular momentum of the atmosphere. General circulation on Earth is mainly due to the uneven heating of Earth's surface. Earth's energy balance involves incoming solar radiation and outgoing thermal infrared radiation, which together make up the net radiation (NR). The global general circulation, pressure distribution, and wind patterns play an integral role in the heat balance of Earth and the formation of global ocean currents. In the global general circulation, warm air is transported from lower latitudes to higher latitudes and cold air from high latitudes to low latitudes. This heat exchange keeps global temperatures relatively stable. Low latitude region is net incoming energy throughout the year as a result of continuous heating, and high latitudes are colder as a result of continuous cooling due to net energy loss.

Cold region is a product of regional stationary sate in general circulation evolution. According to the results of geopotential height and horizontal wind at 500 hPa in January and July (Figure 2.1), the general circulation shows obvious seasonal variations. The position of the troughs and ridges in winter and summer is essentially stable or gradually changing, and occupies a considerable part of the year, while it is relatively short in the two transitional seasons (spring and autumn). In midlatitudes, the general circulation behaves as a flat westerly wind; both the geopotential height and the horizontal wind in winter (January) are larger than that in summer (July). That is, there are three troughs and ridges in the middle-high latitude westerly in winter, four trough ridges in summer, the so-called "winter three summer four".

Due to the spatial discontinuity of global cold regions, some cold regions have regional and local scales, which induce local circulation/cells, for instance, the valley wind. A prominent regional scale cell is the monsoon cell on the southern Tibetan Plateau (TP), which is an indicator for the eastern Asian summer monsoon (Figure 2.2). Also, the high altitude of TP induces many regional cells in the west, east, north, and south directions, which influence the climate regime in eastern Asia. While wind passes across the underlying surface in cold regions, such as glaciers, snow, and

Climatology in Cold Regions, First Edition. Chenghai Wang.

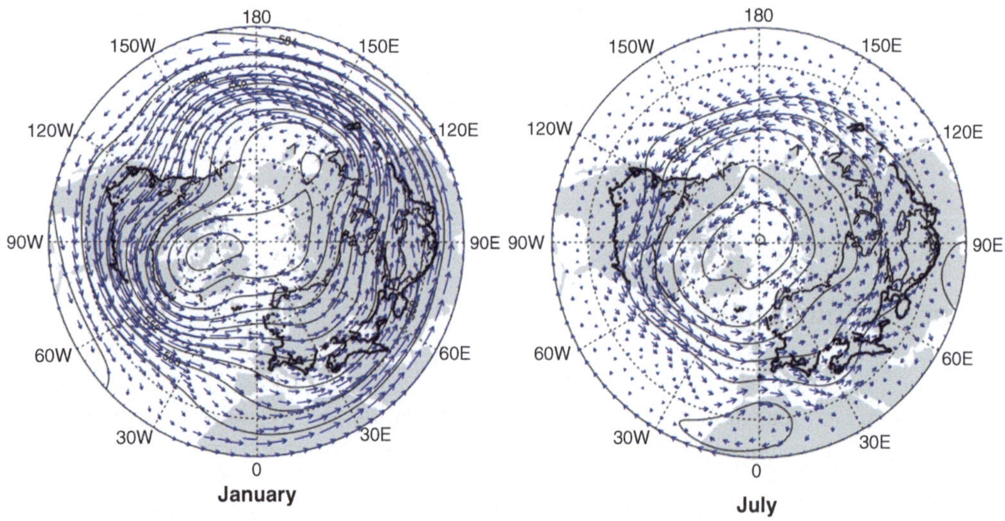

Figure 2.1 The geopotential height (contour) and horizontal wind (vector) at 500 hPa level in January (left panel) and July (right panel).

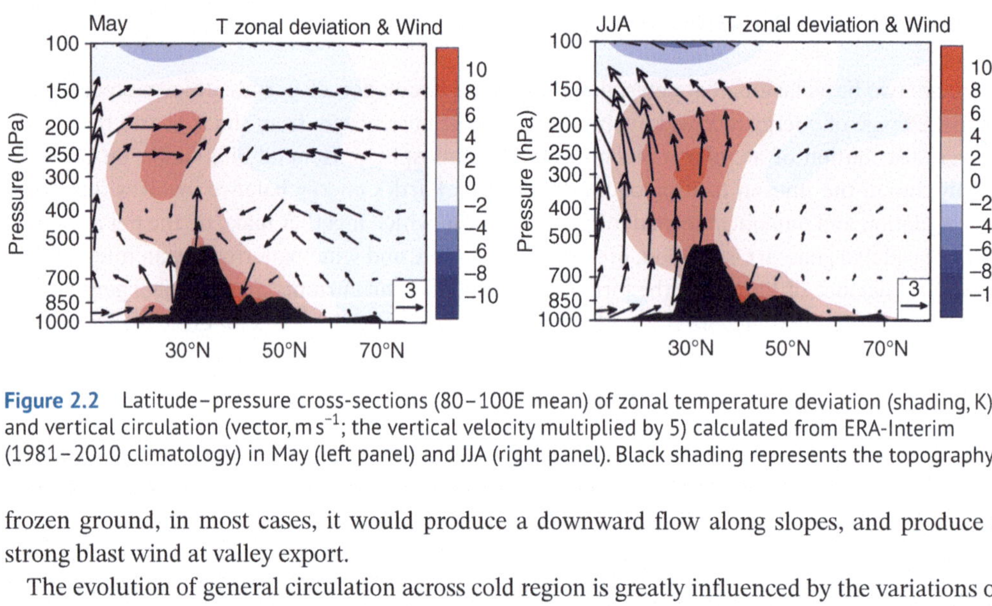

Figure 2.2 Latitude–pressure cross-sections (80–100E mean) of zonal temperature deviation (shading, K) and vertical circulation (vector, m s^{-1}; the vertical velocity multiplied by 5) calculated from ERA-Interim (1981–2010 climatology) in May (left panel) and JJA (right panel). Black shading represents the topography.

frozen ground, in most cases, it would produce a downward flow along slopes, and produce a strong blast wind at valley export.

The evolution of general circulation across cold region is greatly influenced by the variations of surface diabatic heating. As a typical cold region, TP is a huge, elevated heating source in summer, which can result in convergence in the lower atmosphere and divergence in the higher atmosphere in the eastern TP and the reversed structure in the western TP.

2.2 Low and High Frequency of Atmosphere Variability

Generally, the atmospheric motion can be divided into low- and high-frequency variability according to its temporal scales. Comparing the variance of unfiltered, 2.5–6-day filtered and 10–90-day filtered geopotential height in winter and spring (Figure 2.3), the three centers are the North

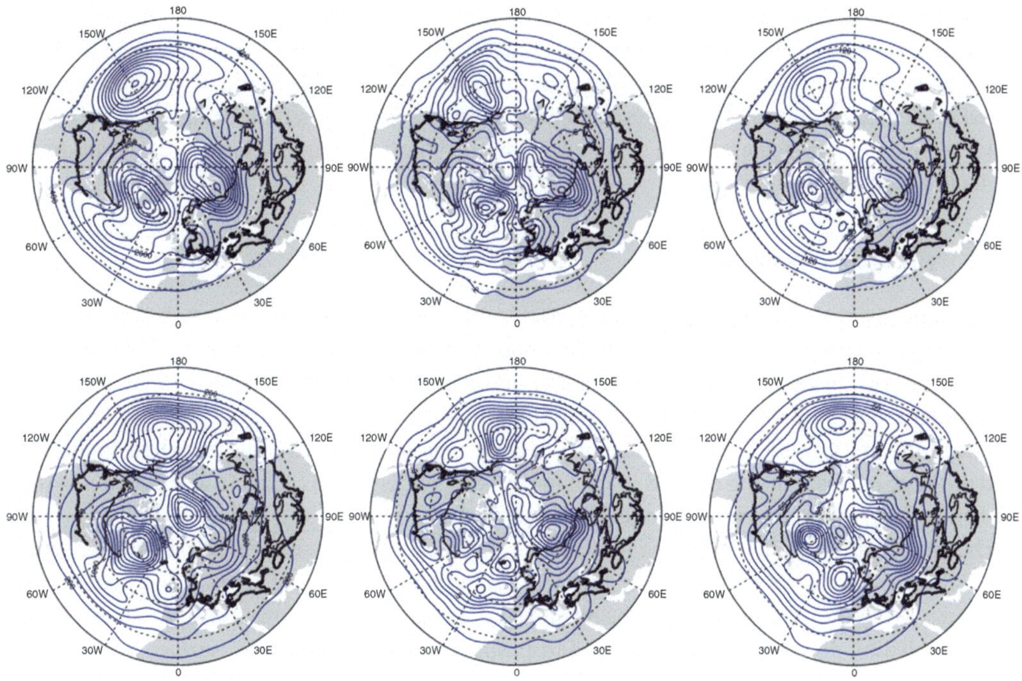

Figure 2.3 The variance of the unfiltered (left), 2.5–6-day filtered (middle), and 10–90-day filtered (right) geopotential height at 100 hPa in winter (upper) and spring (down).

Pacific, North Atlantic, and Siberia, respectively. The evolution pattern of the 10–90-day filtered geopotential height is more similar to the evolution pattern of the unfiltered geopotential height than that of the 2.5–6-day filtered geopotential height. Meanwhile, the evolution of the 10–90-day filtered geopotential height is much greater than that of the 2.5–6-day filtered geopotential height. It means that the interannual variation of geopotential height in winter and spring is mainly dominated by the low-frequency variation. It is also clear that the evolution of the 10–90-day filtered geopotential height is the main contributor over the land around North Atlantic and Siberia. This suggests that land surface in cold regions plays an important role in the interannual variation of geopotential height in winter. This effect can persist from winter to spring, though the distribution of geopotential height evolution has some changes due to land-sea thermodynamic differences.

The 2.5–6-day eddy is very active in the middle-latitude North Pacific and North Atlantic, which is called the storm track. The 2.5–6-day eddy involves strong momentum and heat flux, which causes the interaction between wave and flow in the storm track. The low-frequency Rossby wave train triggered by the cold region in spring can persist during the summer and influence the summer climate.

2.3 Weather Systems

General circulation usually refers to the mean state of large-scale and long-term atmospheric movement or the change processes of a specific period. The horizontal scale is larger than several thousand kilometers; the vertical scale is larger than 10 km, and the temporal scale is longer than 106 hours. This large-scale atmospheric movement not only restricts large-scale weather changes, but is also one of the basic factors of climate formation. Abnormal changes in general circulation will inevitably lead to abnormal weather and climate. The so-called weather system is a large part of the general circulation, and a particular weather process is the background of a specific general circulation. The typical weather systems include polar vortex, polar jet, blocking high, polar low, polar high, and related phenomena (e.g. sudden stratosphere warming). These will be introduced in detail as follows.

2.3.1 The Polar Circulation

Geographically, the regions north of 66.5°N and south of 66.5°S are defined as polar regions. The Arctic region is basically the ocean except Greenland, and the Antarctic region is mainly land. The Arctic Ocean is an ocean of icebergs, but it is not completely frozen even in winter. The mean atmospheric heat in polar region is net spending, thus the polar is the atmospheric cold source, medium and low latitudes of heat transports to polar regions via the average meridian circulation and the large eddies, causing the loss of heat in the polar region, forms cold air, and affects weather and climate in middle and low latitudes by heat exchange.

The mean winter circulation at sea level is dominated by three "centers of action": the Icelandic Low off the east coast of southern Greenland, the Aleutian Low south of Alaska, and the Siberian High in central Eurasia (Figure 2.4). The mean Icelandic and Aleutian lows are maintained largely by their position downstream of major mid-tropospheric stationary troughs, surface heating contrasts and regional cyclone development. The winter Siberian High is a low-level system driven by radiative cooling. During summer, the Icelandic and Aleutian lows are much weaker and the Siberian High is replaced by mean low pressure. Weak low pressure also characterizes in the central Arctic Ocean in summer, compared with a pronounced high over the Beaufort Sea in spring. Winter cyclone activity on the Pacific side of the Arctic is most common in the vicinity of the AL.

The polar vortex is a large area of low pressure and cold air surrounding both poles and it commonly exists near them. It is weaker in summer but stronger in winter. The term "polar vortex" dates back to at least 1853, but was revised in 1950, to become the "circumpolar vortex" and reverted to "polar vortex" in1959. The term "vortex" refers to the counter-clockwise flow of air that helps to keep the colder air near the poles. Commonly, during winter in the northern hemisphere, the polar vortex will expand, sending cold air southward with the jet stream. The polar vortex is the strong, midlatitude, quasi-zonal, stratospheric wind system that develops during winter as a result of the latitudinal gradient of ozone heating. In the literature of atmospheric science, the term "polar vortex" is mostly used as an abbreviation for circumpolar vortex and refers to a planetary-scale westerly (west to east) flow that encircles the pole in middle or high latitudes. The latitude at which the zonal wind reaches its hemispheric maximum can be considered as the approximate border of a polar vortex. Vortex in the troposphere is much larger than the vortex in the stratosphere. Furthermore, the tropospheric vortex exists for the whole year, but the stratospheric polar vortex exists only from autumn to spring (Waugh et al. 2016).

2.3.2 Polar Vortex

In winter, the height field points to strong westerlies in the middle and upper stratosphere flowing around a deep cold vortex (Figures 2.5 and 2.6). Above 40 km (3 hPa, not shown), there is a polar westerly jet stream in midlatitudes. Peak mean zonal winds of 80 m s^{-1} actually occur around 60–70 km altitude (at 0.3–0.1 hPa level). The winter mean stratospheric vortex is broadly symmetric, but is less symmetric than its Antarctic counterpart. There are troughs over eastern Asia and

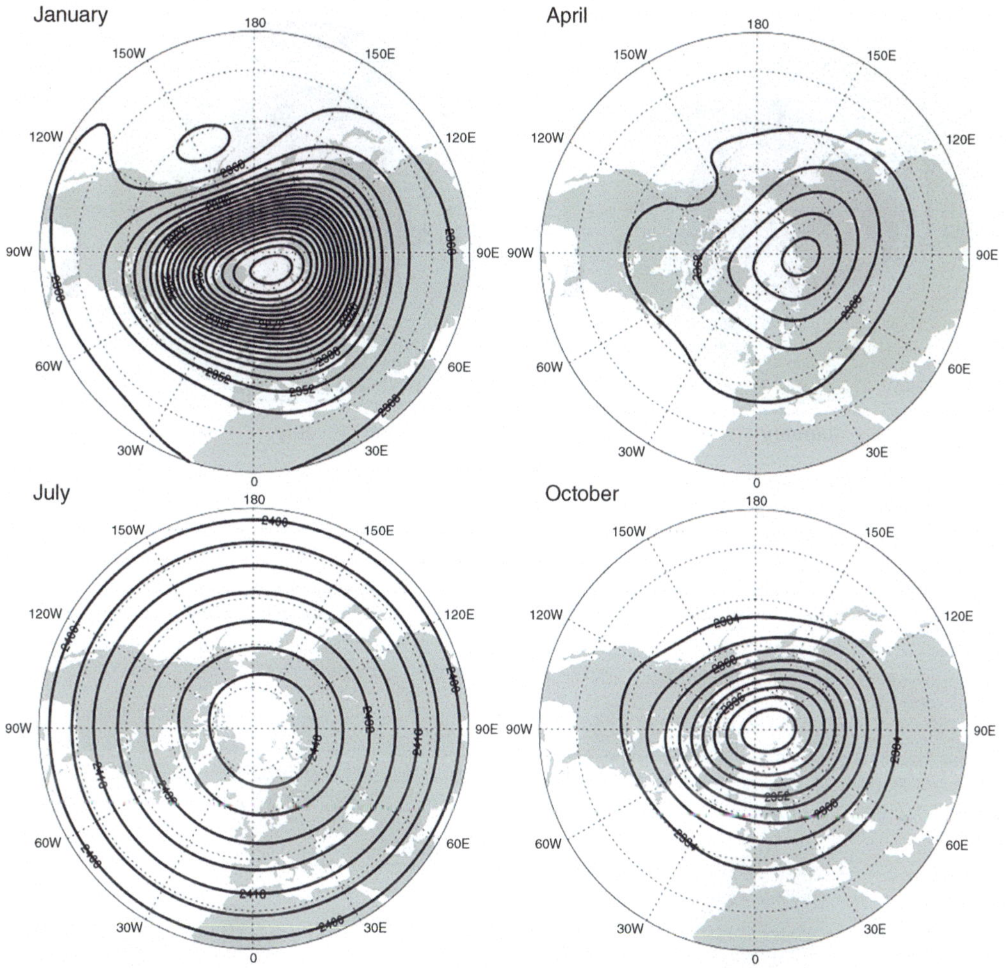

Figure 2.4 The mean geopotential height (gpm) at 30 hPa for the four mid-season months over 1981–2010, based on ERA5 data.

eastern North America and a weak ridge located in western North America. In spring, the vortex center has shifted well off the North Pole to north-central Eurasia.

In summer and above about 20 km, the cyclonic vortex has broken down. There is a polar easterly flow around a highly symmetric warm polar anticyclone. In summer, an easterly jet of about 60 m s^{-1} at 0.1 hPa is located at around 50–60°N. The October wind field illustrates the transition back toward the winter situation.

As discussed by Holton (2004) and Andrews et al. (1987), if there were no transports arising from the breaking of atmospheric waves propagating upwards from the troposphere, the zonal-mean temperature of the stratosphere would become to a radiatively determined state, with the temperature distribution corresponding to an annually varying thermal equilibrium that follows the annual cycle in solar heating. The circulation would hence represent a zonal-mean flow in balance with the meridional temperature gradient (a thermal wind balance), with essentially no meridional or vertical circulation and no stratosphere–troposphere exchange.

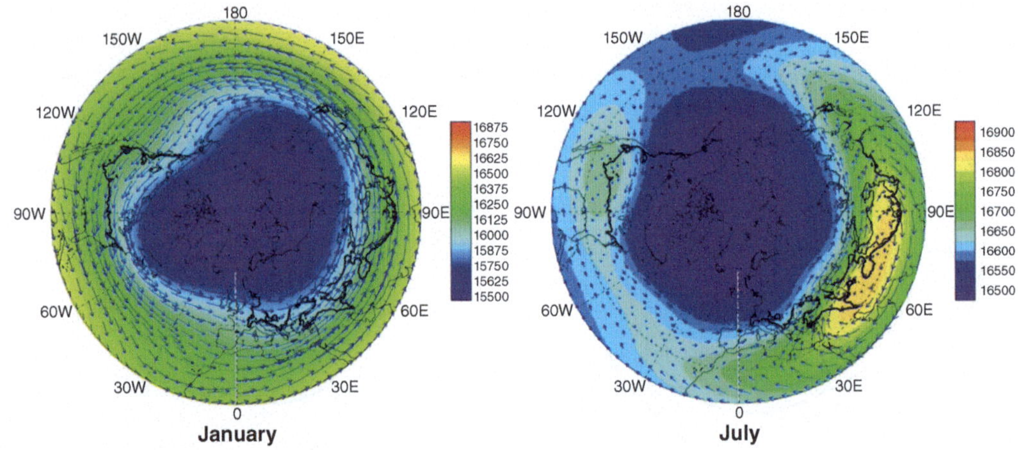

Figure 2.5 The geopotential height (gpm, color) and horizontal wind ($m\,s^{-1}$, vector) at 100 hPa in January (left panel) and July (right panel).

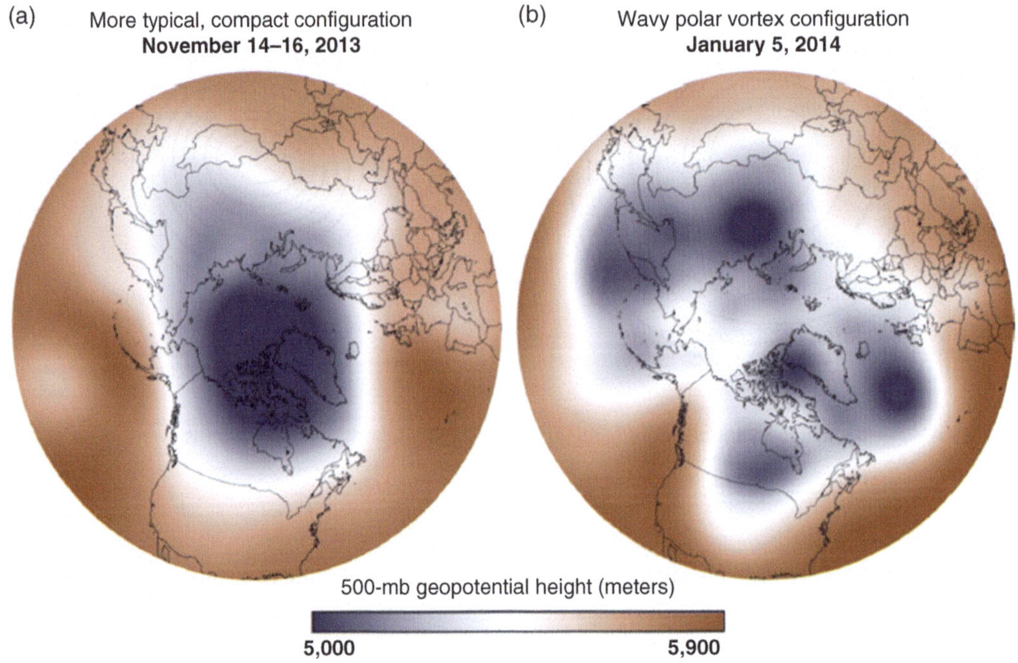

Figure 2.6 Comparison of geopotential height fields at the lower jet stream level (500 hPa) with low values in purple and the jet stream in white. (a) A single and more west to east jet stream encircling the pole contrasts with a more undulating configuration (b) with multiple low height centers (dark purple). *Source:* NOAA / United States Department of Commerce / Public Domain.

The existence of a westerly vortex in winter and an easterly vortex in summer is qualitatively that expected from radiative equilibrium, the climate in winter actually shows considerable departures from the radiatively determined state. In the 30–60 km region, the change in temperature from the winter pole to the summer pole is much smaller than the radiatively determined gradient. This is due to eddy transports that drive the flow away from a state of radiative balance. By contrast, the departure from radiative equilibrium in summer is small, implying a reduction of transport.

In the troposphere, these eddy transports are largely associated with moving synoptic-scale waves (they are associated with migrating cyclones and anticyclones at the surface). By contrast, winter transports in the stratosphere are associated with the long planetary waves, especially for wavenumbers 1 and 2, which can penetrate into the stratosphere under specific conditions. Wavenumber refers to the number of atmospheric waves around a latitude circle. For any given day, Fourier transform methods can be used to break down the total circulation at a given level in the atmosphere with respect to the relative contributions of long planetary waves (lower wavenumbers), which move slowly or remain stationary with respect to the Earth's surface, and shorter waves (higher wavenumbers). A strong wavenumber two component, for example, would have a pronounced expression of two ridges and two troughs, each separated by 180° longitude. If we look at the circulation on a typical winter day, we can see that while the troposphere has a strong contribution from the higher wavenumbers, the circulation of the stratosphere is more symmetric, indicating that the shorter waves are not readily penetrating into the stratosphere. For long-time averages, the effects of the shorter waves are marginal. But even with such time averaging, the winter circulation of the stratosphere is more symmetric than that of the troposphere. This can be related to the impacts of orography and land-sea thermal contrasts (Pawson and Kubitz 1996).

Waves will penetrate (vertically propagate) into the stratosphere provided that the zonal-mean wind is positive (the wind averaged around a latitude circle blows from west to east) but less than a critical value that depends strongly on the length of the waves. The longer the wave (that is, the lower the wavenumber), the stronger the zonal wind can be and still allow for vertical propagation. Since the zonal wind tends to be at a maximum near the tropopause in middle and high latitudes (much lower in altitude than the winter stratospheric jet), the shorter waves in winter tend to be "trapped" in the troposphere. In summer, the zonal-mean zonal winds in the stratosphere are easterly. There can be no vertical propagation of waves in these situations. This is consistent with the fact that the summer stratospheric circulation is highly symmetric with a thermal structure close to that expected from radiative equilibrium.

The typical thermal structure of the winter stratosphere features low temperatures in the vortex core in the lower stratosphere and increases outward from the axis of rotation. Conversely, there are high temperatures in the vortex core of the upper stratosphere and lower mesosphere that decrease outward. The vertical change in the temperature structure is attributed to warming through large-scale subsidence in the mesosphere and radiative cooling in the lower stratosphere during the polar night (Gerrard et al. 2002). The stratospheric jet maximum is located at the level of the temperature transition region in the vertical (10–1 hPa). Positive potential vorticity (PV) anomalies are found in the vortex core at the level of the thermal transition zone, with a cold trough below and a warm ridge above. The vortex winds, which are formed between the two thermal regimes, circulate around the core. PV represents the specific volume (volume per unit mass of air) times the scalar product of the absolute vorticity (the sum of relative and planetary vorticity) and the gradient of potential temperature. The PV maximum arises from the low density of the air at stratospheric levels (specific volume is large), the strong increase of potential temperature with height (strong stability), and the positive absolute vorticity.

Climatology of the stratospheric polar vortex in both hemispheres has been discussed based on areal extent (Baldwin and Holton 1988) and on elliptical diagnostics (Waugh 1997; Waugh and Randel 1999). Fitting an ellipse to a specified contour allows one to summarize various measures of the vortex. These include the equivalent latitude of the vortex edge (defining its area), the offset of the vortex from the pole, the longitude of its center and its elongation (aspect ratio or ellipticity).

Waugh and Randel (1999) defined these parameters on the basis of selected PV contours for the 440 K (lower stratosphere) to 1300 K (upper stratosphere) potential temperature, average from October 1978 to April 1998.

The equivalent latitude of the vortex edge is lower (i.e. the vortex covers a larger area) in winter compared to autumn and spring. Area increases with height for January, the equivalent latitude is around 70°N in the lower stratosphere, compared to about 60°N in the upper stratosphere. Depending on height and season, the vortex is centered 6–18° off the pole. The vortex tilts westward with a height between 500 and 1300 K by about 60°. The vortex also tends to be elongated in shape. In November–December, there may be rapid eastward and then westward shifts of the vortex center in the lower stratosphere between 60 and 120°E and 60–90°W, apparently related to changes in the planetary waves. Similar but fewer shifts occur in late January to early February.

The Arctic vortex is smaller and breaks down earlier in summer than its Antarctic counterpart. It is also highly variable throughout its lifecycle, especially in late winter. This variability is partly caused by extreme events distorting the vortex. A distorted vortex was present in about half of the winters from January 1990 to August 1997 (Waugh and Randel 1999). There were also other periods when the vortex was symmetrical and quiescent. Interannual variability in the stratospheric polar vortex has been linked to the phase of the equatorial quasi-biennial oscillation (QBO), as studied by Labitzke (1982). The QBO refers to an oscillation in the zonal winds of the equatorial stratosphere having a period that fluctuates between about 24 and 30 months. Stratospheric equatorial easterlies associated with the QBO favor a weaker, warmer vortex than equatorial westerlies, but the relationship has broken down in some winters.

The primary feature of the mid-tropospheric (500 hPa) circulation in northern high latitudes is a well-developed cyclonic vortex. The winter vortex is strongly asymmetrical. At 60°N, the mean flow is characterized by major troughs over eastern North America and eastern Asia and a weaker trough over western Asia (the Urals trough). A strong ridge is located over western North America, with much weaker ridges over the eastern Atlantic and central Asia. The lowest geopotential heights are found in the Canadian Arctic Archipelago. These features are related to dynamical forcing by orography (most importantly, the Rocky Mountains and the Himalayas which disturb the circulation), thermal forcing associated with the large-scale land-ocean distribution (warm oceans versus cold land) and radiative forcing: the stronger winter asymmetry at 500 hPa as compared to the stratospheric vortex, and the large relative shift in the vortex center. The more symmetrical nature of the stratospheric winter vortex relates to the fact that only the long planetary waves can propagate up to high levels. The three troughs and three ridges that are fairly well expressed at about 60°N indicate that the total circulation at this latitude has a strong wavenumber three component. In summer, the 500 hPa vortex is much weaker in response to the more even latitudinal distribution of radiation and temperature. It is also more symmetric than its winter counterpart, with the lowest pressure heights centered over the pole.

The mean thickness of the tropopause layer is only about 600 m over the Kara-Laptev seas in January, whereas in summer it exceeds 1.6 km in a belt from northern Norway to northern Novaya Zemlya and eastward to Wrangel Island (Makhover 1983). There is a secondary minimum (maximum) in tropopause height in October (December–January). By contrast, the tropopause temperature shows a simple annual cycle with a minimum of −62 °C in January and a maximum of −49 °C in July, apparently following the annual cycle of temperature in the lower stratosphere as identified for the 100 hPa (Wilson and Godson 1962). Highwood et al. (2000) also suggested that tropopause temperatures in winter are lower in the Eurasian sector and higher in the Canadian-Atlantic sector.

The existence of a double maximum/minimum in tropopause height, is attributed to several factors. The primary August maximum is attributed to the lag in the heating of the surrounding landmasses, while the second one in January is related to enhanced meridional heat transport by the planetary waves. The spring minimum is associated with the seasonal weakening of this transport. The autumn minimum relates to the cooling of the landmasses and a decrease in cyclone activity compared with late summer. Two patterns of the annual cycle in tropopause height are identified in different regions. Over North America and subarctic Siberia there is a simple annual wave (winter/summer–tropopause pressure maximum/minimum) while in the High Arctic, northern Europe, and western Siberia there is a double wave (spring/autumn – maxima, summer/winter – minima).

The Arctic stratosphere extends upward from about 10 to 40–50 km altitude. It is essentially a very dry, statically stable region where ozone dominates the absorption of solar ultraviolet radiation and emission of infrared radiation. In the troposphere (the region between the stratosphere and the surface, which contains about 80% of the atmospheric mass), the mean thermal structure is maintained by an approximate balance between infrared radiative cooling, surface radiative heating and the subsequent vertical transport of sensible and latent heat by turbulent processes, and large-scale horizontal heat transport by synoptic-scale disturbances. The dominance of surface radiative heating has the result that surface and tropospheric temperatures are highest in the equatorial regions and decrease toward the poles. As the troposphere is primarily heated from surface, there is a rapid decrease in temperature with altitude, however, the potential temperature (the temperature that an air parcel would have if raised or lowered to the 1000 hPa surface), generally increases with height. There are prominent seasonal and regional departures, such as strong low-level temperature inversions in the Arctic during winter. By contrast, for mean conditions in the stratosphere, the infrared cooling to space is balanced primarily by radiative heating owing to the absorption of ultraviolet radiation by ozone. As a result of ozone absorption, the mean temperature increases upward in the stratosphere (hence potential temperature increases strongly with height, meaning very stable conditions): above roughly 50 km, that is, in the mesosphere, where temperature decreases with height again due to reduced ozone absorption. In the stratosphere above about 30 hPa (roughly 25 km), the temperature decreases from the summer pole to the winter pole, which is qualitatively consistent with radiative equilibrium conditions (in June, total solar radiation at the top of the atmosphere is largest at the North Pole, and is least at the South Pole). Winter temperatures at around 30 km in the Arctic are below −70 °C while in midsummer with 24-hour sunlight they exceed −40 °C. In the lower stratosphere, the temperature is strongly influenced by upper-tropospheric processes. This region is characterized by a mean temperature minimum at the equator, and maximum at the summer pole, and in the midlatitudes of the winter hemisphere. The first order pattern is that the Arctic tropopause is encountered lower in the atmosphere (i.e. at higher pressure) than in lower latitudes. The annual mean height field is nevertheless strongly eccentric about the pole with a center of 280 hPa located in the Devon–Ellesmere islands. Weather forecasters examine the polar vortex by looking at conditions tens of thousands of feet high in the atmosphere; when we feel the extremely cold air from the Arctic regions at Earth's surface, it is sometimes associated with the polar vortex. Some confusion stems from the fact that these polar vortices are neither unusual nor extreme; they are simply basic features of Earth's climatology. Some extreme weather events at some locations are related to transient displacements of the edge of the tropospheric polar vortex.

During winter in the northern hemisphere, a stratospheric polar vortex develops over the North Pole, with strong zonal westerlies circulating around low temperatures at the center of the vortex. Impacts of stratospheric polar vortex events are not confined to the stratosphere. Dynamical coupling with the troposphere can cause an eddy-driven jet stream to shift southward and result in extreme weather in parts of Eurasia and North America (Kidston et al. 2015). Although the stratospheric polar vortex is typically thought of as being centered on the Arctic, in fact, its center is often found in the subarctic, in northern Canada or northern Eurasia. During January–February 1979–2018, stratospheric polar vortex at 100 hPa was found approximately 40% of the time on the North American side and 25% on the Siberian side (Kretschmer et al. 2018). Although continental cold air outbreaks can occur without a stratospheric polar vortex connection, a southward-shifted stratospheric polar vortex is often associated with cold temperature anomalies, snow storms, and rainfall that arise from the tropospheric polar jet moving equatorward and a weakening jet stream (Cohen et al. 2018; King et al. 2019; Overland and Wang 2019). A stratospheric polar vortex located over North America in February contributes to extremely low temperatures in the northern Great Plains, and allows high surface pressure with low temperatures to develop over Central Europe that moves westward over the United Kingdom and as far south as Italy (Greening and Hodgson 2019). In March, the center of the stratospheric polar vortex relocated over Eurasia and was associated with cold air advection over Europe. All European events have been referred to as "the Beast from the East" but they were associated with different synoptic situations. For example, the movement of the stratospheric polar vortex over North America in February 2018 set up a tropospheric ridge/trough height field with cold temperatures over Central and Eastern Canada and the United States (Overland and Wang 2019). This persisted for most of February and was particularly evident in the second half of the month. The 850 hPa tropospheric vortex is co-located under the SPV center near Hudson Bay. Cold temperature anomalies of less than −8 °C reached well into the Midwestern United States and were associated with strong cold air advection from the Arctic between the ridge over Alaska and the tropospheric vortex center.

2.3.3 Polar Jet

The stratospheric circulation during winter is dominated by a strong polar vortex; along its margin is the polar jet. The polar jet decays in spring, to be replaced in summer by a weak circumpolar easterly flow. The winter vortex sometimes rapidly breaks down due to sudden stratospheric warming, the effects of which can impact on the troposphere and surface. At 500 hPa, a circumpolar vortex is present year round. It is strongly asymmetric during winter, with major troughs over eastern North America and eastern Asia and a weaker trough over western Asia. A prominent ridge is located over western North America. The lowest winter potential heights at 500 hPa are found over the Canadian Arctic Archipelago. The asymmetry is related to orography, large-scale land–ocean distributions and radiative forcing. The mid-tropospheric vortex weakens during spring and summer and becomes more symmetric than its winter counterpart.

Weather connectivity between Arctic and midlatitude depends not only on the magnitude of Arctic temperature amplification, but also on the location, amplitude, and movement of meanders in the polar jet stream. The jet stream (a band of stronger winds (shown in white in Figure 2.6) halfway up in the atmosphere) can have more than one manifestation such as an elliptical zonal (west–east) configuration or a wavier shape. The wavy configuration moves warm air to the north and cold air to the south, depending on the longitudinal locations of the wavy structure. Excursions to the north are ridges as they are associated with high pressure, and those to the south are troughs as they are associated with low pressure, and often severe weather. While Arctic amplification (AA) is ongoing every year, potential linkages between the Arctic and subarctic/midlatitude weather depend on the variability and configuration of the jet stream in the mid/upper troposphere (5–10 km altitude) and the stratospheric polar vortex (at 15–40 km altitude).

That jet stream excursions shift both temporally and in terms of their longitudinal location argues for the potential weather linkages occurring irregularly, i.e. lack of trend or cycle on an annual or seasonal basis, despite continued Arctic amplification. Inspection of year-to-year variability (e.g. Kug et al. 2015) shows such irregularity. Amplified Arctic warming in early winter may not in itself initiate midlatitude connections, but instead intensify such interactions by enhancing the amplitude of existing large-scale jet stream excursions and therefore contribute to the establishment of stationary atmospheric wind patterns known as blocking (Woollings et al. 2018; Tachibana et al. 2019; Luo et al. 2019; Overland et al. 2021).

The wavy jet stream is not the only meteorological pattern that can influence midlatitude weather extremes. The winter–spring (January–April) 2020 Siberian heat wave (record Arctic temperature of 38 °C) with extensive wildfires resulted from the more southerly zonal location of the jet stream that reduced the potential penetration of cold air from the north. Another example of a southern location of a strong zonal type jet stream impact was the cold temperatures and snow in Texas, during mid-February 2021. Due to poor weather and loss of power, 80 lives were lost during this event. The Texas event was short-lived as the severe weather moved east-northeast rapidly with the strong jet stream. These extreme weather events are noteworthy, with Arctic conditions reaching into the subarctic and midlatitudes, but are not necessarily caused by Arctic amplification and often have historical analogs.

2.3.4 Polar Low

Polar lows are intense maritime mesocyclones of typically 100–500 km in diameter (Figure 2.7). They may intensify rapidly, and surface wind speeds can sometimes reach hurricane intensity (Businger and Reed 1989). They tend to be short lived, generally lasting only 3–36 hours. Polar lows are the most intense category of the family of mesoscale cyclonic vortices found poleward of the main polar front, which can present significant hazards to shipping, are also known as Arctic Instability Lows, comma clouds and, in the southern hemisphere, Antarctic coastal vortices. In the

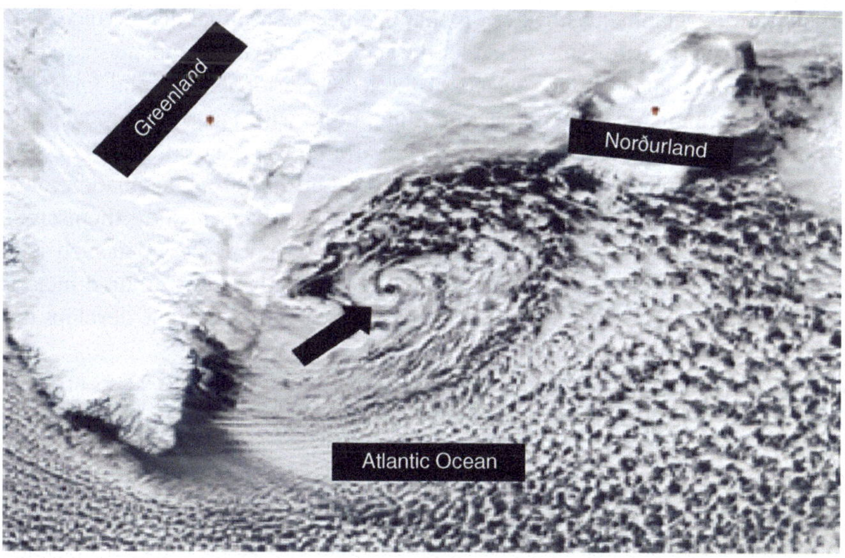

Figure 2.7 MODIS satellite image 06 April 2007 showing the polar low which forms around the central dark eye (marked by arrow). Photo from Earth data: https://worldview.earthdata.nasa.gov.

north polar regions, polar lows are particularly common in the Nordic seas, the Labrador Sea, the Bering Sea, the Gulf of Alaska, and the Sea of Japan. Hundreds of mesoscale lows can develop in these regions on an annual basis (Renfrew 2003).

Polar lows develop over open water. When moving over land or sea-ice cover, they tend to dissipate rapidly. They can be thought of as a “hybrid” system, typically having both baroclinic features and features convective in nature. A common feature of polar lows seen in satellite imagery is a spiral cloud (comma cloud) signature.

Two related mechanisms have been proposed to explain polar lows. The first is CISK (conditional instability of the second kind), which emphasizes the organization of cumulonimbus convection. An initial disturbance causes low-level convergence and ascent, which in a conditionally unstable atmosphere results in latent heat release through convection. The latent heat release leads to a drop in surface pressure, favoring more cyclonic relative vorticity, more low-level convergence and, through continuity, high-level divergence. The low-level convergence provides for a continued source of moisture and convective latent heat release a feedback process that can lead to rapid development. It seems that cold polar outbreaks over open water in high latitudes can provide for the deep layer of conditional instability associated with the CISK process.

The second mechanism is known as WISHE (wind-induced surface heat exchange). WISHE is feedback between the circulation and surface sensible and latent heat fluxes from the sea surface, with a stronger circulation giving rise to large surface fluxes of heat, which are then redistributed aloft by convection, in turn strengthening the circulation. The feedback arises in that the surface heat fluxes are proportional to the wind speed turning on the winds, and the heat flux strengthens. The redistribution by convection results in a pressure drop, further strengthening the winds and hence the surface heat fluxes. WISHE can explain the growth of both tropical cyclones and polar lows.

In WISHE, the emphasis is placed on the surface fluxes as the primary growth process, and the convection only serves to redistribute the heat. This contrasts with CISK, which emphasizes that the circulations amplify through their interaction with the convection itself. However, in the real world, the two mechanisms can be difficult to distinguish from each other. The common feature of both processes in polar low development, however, is that one needs an initial disturbance. The presence of some sort of baroclinic disturbance provides the key. Off East Greenland, mesocyclone development also appears to be forced by the effects of the ice sheet topography and low-level katabatic winds. Klein and Heinemann (2002) show from simulations with a mesoscale model that channeling of the flow in the large valleys of East Greenland around Angmagssalik leads to convergence and cyclonic vorticity generation. This can be enhanced during strong katabatic events that supply cold air, augmenting low-level baroclinicity. These katabatic winds are themselves favored by the presence of a synoptic-scale low in the Greenland Sea. They propose that such synoptic support is important because the pure katabatic wind layer is only a few hundred meters deep. The positive feedback on the katabatic system can occur when a mesocyclone develops off the coast.

In recent years, the Japan Sea is thought to be one of the most frequent occurrence areas of polar lows (Ninomiya 1989, 1991, 1994; Yanase et al. 2002, 2004; Fu et al. 2004a,b; Tsuboki and Asai 2004). The Japan Sea, located in the east of the Eurasian continent and covered the area of 35°–45°N, 130°–140°E, is an almost enclosed basin. In winter, the outbreak of the cold polar air mass that formed over the Eurasian continent often strongly influences the Japan Sea basin wherein the warm Tsushima Current flows, which makes the air–sea interaction very active (Fu et al. 2004a,b) and provides favorable conditions for the development of polar lows. Polar lows may cause severe weather such as strong winds and heavy snowfall.

2.3.5 Blocking High

During the development and evolution of the ridge of the longwave trough in the westerly belt, when the ridge extended northward, the connection between the south and the warm air in the south would be cut off by the cold air, and the closed circulation appeared in the north of the ridge, forming the warm high center. It is a deep warm high in the middle and upper troposphere at middle and high latitudes. This is a quasi-stationary ridge of long waves containing a closed center of high pressure formed by the increase in the amplitude of the long wave, which moves infrequently and for longer periods of time, impeding the eastward movement of upstream westerlies and weather systems. Blocking high occurs mainly in the northern hemisphere and is often associated with cut-off low.

The formation and maintenance of blocking high and cut-off low block the upstream fluctuation from spreading to the downstream, destroying the sinking westerly circulation, so that the movement of cyclones and anticyclones on the ground is blocked. This circulation situation is also called a blocking situation. This is a stable situation that can be maintained for quite a long time and has an important influence on the circulation weather process in a large area of the upper and lower reaches under its control. Under the direct control of blocking high pressure, the weather is generally sunny and less cloud, cold advection and subsidence movement are common in the east, and the weather is mainly cold and sunny. In the west, it is generally warm advection and ascending movement, warmer and cloudy.

The general characteristics of blocking high are defined as follows: the closed isolines must appear simultaneously on the surface diagram and 500 hPa, and the blocking high divides the westerly jet into two branches. The center of the blocking high is located north of 30 °N. The duration is not less than five days. Blocking high is most common over the Atlantic Ocean, Europe, and western Alaska, and occurs more frequently over the Atlantic than the Pacific. In Asia, blocking high often occurs over the Ural Mountains and the Sea of Okhotsk. In terms of latitude, most blocking occurred in the 55–59°N latitude band, and the least in the 40–50°N latitude band. The blocking high in the northern hemisphere is most likely to occur over Alaska in the eastern Pacific and over the eastern Atlantic to northwestern Europe. In Europe, blocking high can generally stay for about 20 days, the shortest being more than five days; In Asia, the average is eight days and the shortest three to five days. In Asia, the most occurs in May, June, and July, and the least in March and November. Europe, on the other hand, is two months ahead of Asia, with November, March, April, and May with most and January and July with least.

The establishment and collapse of a blocking high is often accompanied by a dramatic shift in circulation patterns over a large or even hemispheric scale. Its establishment marks the transition from zonal circulation to meridional circulation. Its persistence indicates that the meridional circulation is at a strong stage. Its collapse marked the transition from meridional to zonal circulation. Blocking high can affect the weather and climate in a large area, and its long duration can bring drought and continuous rain to a large area, resulting in climate anomalies. The breakdown of the blocking situation in winter is closely related to the outbreak of cold waves. The collapse of the Urals blocking high often causes cold waves in East Asia, and the maintenance of Okhotsk blocking high is an important large-scale circulation condition for the occurrence of rainy weather in China. Therefore, the establishment, maintenance and destruction of blocking situations play a very important role in China's weather and climate. The blocking high is usually quasi-stationary and moves slowly, sometimes even receding westward. The average duration of blocking high pressure is five to seven days, sometimes more than 20 days. The region is often formed in persistent single weather, the eastern part has long sunny weather, while the western part has precipitation.

The Ural Mountains and the Sea of Okhotsk are often blocked by high pressure. When they are stable, southern China is prone to continuous rain and freezing weather in winter. When the Ural blocking high weakens and collapses, it often causes cold snaps in China.

2.3.6 Polar High

Polar high refers to the cut-off, deep, and warm high pressure maintained for more than three days in the north of 70°N (Cai 1978; Tedesco et al. 2016). Due to the support from the warm and humid airflow in the south, the ridge continues to extend northward. Since the simultaneous existence of the cut-off low pressure in its upstream and downstream, the connection between the north extension and the south heating flow is cut off, and a closed circulation appears in the north of the ridge. In this way, the strong warm air from the south will be isolated in the north and form a warm high-pressure center, called polar high. Observations have shown that the average lifetime of polar high is about 10 days, and the occurrence areas are concentrated in the vast ice-ocean area from the east of the New Siberian Islands to northern Alaska and Greenland (Cai 1978). However, there are also polar highs with abnormally long duration, for example, the polar high was maintained for 35 days from January 9 to February 13, 1977.

The polar high is a very deep system, extending from the ground to 300 hPa. During the maintenance of polar high, the troposphere near the polar region is unusually warm, so the temperature distribution increases from south to north in high latitude, and the frontal area tilts southward with the height, which is very different from the structure of tropospheric temperature and pressure field in general (Liu 1986). The polar high will raise the polar tropopause to the same height as the tropopause near 40°N. The tropopause height above the partial polar vortex is the lowest, and the tropopause breaks. The existence of the polar high makes the front area tilt southward with the height, and the position corresponds to the polar high jet, which is called the polar high front area (Xu and Guo 1978). There is an obvious easterly jet on the south side of the polar high (70–80°N) located at 300–400 hPa, and the maximum wind speed is 35 m s^{-1}, called polar high easterly jet (Liu 1986).

As an important weather system in the Arctic, the maintenance of the polar high will significantly change the circulation in the northern hemisphere. After its establishment, development and collapse, there will be a strong cold air event sweeping East Asia (Tedesco et al. 2016). After the formation of the polar high, the polar cold air moves southward along its front and invades the middle latitude. Although the polar high is not a weather system that affects China's weather and climate directly, it can often affect China by affecting the blocking high in midlatitude in winter, resulting in extreme cold events. That is, the maintenance of the polar high will stabilize the long wave in the northern hemisphere, and the planetary wave interacts with the transient wave and modulates the transient wave activity, thus affecting the formation and maintenance of the blocking high. The connection between the polar high and the blocking the high pressure will make the polar cold air form a penetrating explosion from high latitude to low latitude (Liu et al. 2021), as well as an extreme low-temperature event and affecting most areas of China.

The direct performance of the Arctic amplification effect is the reduction of Arctic sea ice. When there is less Arctic sea ice in autumn and winter, the sea-level pressure in the Arctic is abnormally high. This local forcing related to Arctic amplification is one of the possible reasons for the generation and maintenance mechanism of polar high (Tedesco et al. 2016). Since the occurrence of Arctic amplification, there has been an increasing trend of the regional polar high variability in winter, so its impact on regional and long-range weather and climate is bound to change. For example, as a part of the polar high, the relevant research about Greenland blocking high can reflect the impact of the change of polar sea ice on the polar high to a certain extent.

The multidecadal peaks and troughs are evident in all four seasonal GBI (Greenland Blocking Index) series. The GBI has a peak in the late nineteenth century around 1880 (although the exact timing of this varies by season), with GBI having risen from a relatively low base in summer and autumn; a trough around 1910 or 1920; another peak or increasing trend around 1930–1940; a general decrease for autumn a somewhat flat-trend until the 1980s or very early 1990s; and an increase since then. The recent increase is most marked in summer and apparent for winter but displays a mixed signal for autumn and spring with high values present throughout the first decade of the twentieth century but combined with an apparent short-term negative trend during the past decade. Once the GBI data have been filtered, periods of around eight years become apparent in most months.

Trends in monthly and seasonal GBI for the full period (1851–2015) are all insignificant except for decreases in October, November, and autumn. The full period is split into sub-periods for more detailed analysis of monthly and seasonal GBI trends and variability. This shows several significant increases in GBI for 1851–1900 in May and June and for the year as a whole whereas there were no significant decreases. For 1901–1950, there are no significant changes. For 1951–2000, February, March, and winter show significant decreases. Between 1981–2010, all seasons, March, July, August, and October had significant annual GBI increases. For 1991–2015, these GBI increases were sustained for summer, July and August (the sustained summer increase over the last 25 years is clearly seen in Figure 2.3). There were no significant trends for the shorter recent 2001–2015 period except for an increase in July. Most monthly and all seasonal time GBI series do not show significant overall (1853–2013) trends in variability; the exceptions are May and December, both with significantly increasing variability.

2.3.7 Stratospheric Sudden Warming

Stratospheric Sudden Warming (SSW) is the sudden change of large-scale circulation structure of stratospheric atmosphere, which was first discovered by German scientist Scherhag (1952). When SSW occurs, the stratospheric temperature in the polar region increases suddenly, and the stratospheric circulation associated with the temperature change also adjusts sharply, the zonal-mean wind field reverses, and the polar vortex deviates from the polar region due to strong disturbance, or even completely collapses, thus establishing two polar symmetric flow patterns. This phenomenon occurs mainly in winter and spring, and in rare cases, SSW occurs twice in one winter, called a "double warming." SSW mostly occurs in the northern hemisphere, and the stratosphere in the southern hemisphere rarely shows explosive warming, according to the existing observation data, no strong SSW was observed in the southern hemisphere before 2000, and only one strong SSW occurred in September 2002.

According to the warming intensity, SSW can be divided into strong warming and weak warming. At 10 hPa or below, the poleward increase of zonal-mean temperature from 60°N and the related reversal of circulation, i.e. the collapse of polar vortex and the change of westerly circulation into easterly circulation, are called the strong warming process. Although the weak warming process can achieve a similar warming value, there is no circulation collapse. If the warming in late winter and early spring destroys the winter circulation and does not recover, but develops into the summer circulation, it is called the last warming. In addition, the warming caused by the splitting and eccentricity of the polar vortex due to the northward advance of the Aleutian anticyclone may also reverse the temperature gradient north of 60°N, and this warming is called Canadian warming.

The seasonal variation of stratospheric general circulation is obvious, with westerly winds prevailing in the northern hemisphere in winter and easterly winds in summer. The stratospheric polar region is dominated by a polar vortex in winter and a huge anticyclone in summer. Before the occurrence of SSW, the westerly winds prevailed in the polar region, and the strong cold polar vortex dominated the stratospheric circulation, occupying most of the middle and high latitudes in the northern hemisphere. The occurrence of SSW leads to abnormal stratospheric circulation, which leads to the change of the stratospheric weather system. SSW not only directly affects the changes of stratospheric general circulation, but also greatly affects the tropospheric atmosphere in winter, and even the whole middle atmosphere. The occurrence of SSW destroys the polar vortex, hindering the formation of polar stratospheric clouds, and cutting off the mechanism of ozone depletion. After the warming, the polar vortex gradually recovers, and sometimes serious ozone depletion events occur.

SSW greatly affects the structural characteristics and changes of the stratospheric atmosphere and even the whole middle atmosphere circulation in winter. Therefore, since this special warming phenomenon of the stratosphere was found in the analysis of radiosound data, meteorologists in various countries have carried out a lot of observation and research on it. The existing research on SSW mainly includes: the characteristics and mechanisms of SSW, the influence of SSW on atmospheric fluctuation and the interaction between upper and lower atmosphere, the influence of SSW on the distribution and transport of atmospheric chemical components, SSW and QBO, and El Niño-Southern Oscillation.

Stratospheric explosive warming is a phenomenon that stratospheric temperature increases sharply in winter, which can increase more than 40 °C in a short period of time. During an SSW event, there are many abrupt changes in stratospheric circulation. It is generally believed that stratospheric explosive warming has three main characteristics: (i) deformation and collapse of stratospheric polar vortex; (ii) abnormal warming of the air in the polar region causes the reverse of the meridional temperature gradient, and the temperature gradient eventually points from the polar region to the middle latitude; (iii) with the weakening or even disappearance of the westerly jet at polar night, the circumpolar easterly circulation appears, and easterly winds can extend from the polar regions to areas south of 60°N (Figure 2.8). The World Meteorological Organization stipulated in 1963 that explosive stratospheric warming in the northern hemisphere requires two conditions: (i) in the stratosphere of 10 hPa or below, the zonal-mean temperature gradient from 60°N to the polar region reverses, that is, from poleward to equatorial; (ii) the polar vortex weakened and moved southward, the polar regions were controlled by the warm high, and the westerly circulation in the polar region turned to the easterly circulation.

Early studies on SSW were mainly case studies, and many warming events have been studied separately. As early as 1959, Richard and Hering (1959) used radiosonde data to describe the explosive warming phenomenon that occurred in 1957. O'Neill and Taylor (1979) analyzed the strong SSW event of 1976/1977. Fairlie and O'Neill (1988) revealed the evolution and dynamic mechanism of the strong SSW in the winter of 1984/1985, and pointed out that an important aspect of the warming was that the flow was very asymmetrical when the stratospheric westerlies began to collapse rapidly. Zhang (1991) analyzed the circulation characteristics and energy changes during the SSW in February 1981. Manney and Sabutis (2000) studied a unique stratospheric warming in November 2000 and found that this warming was characterized by a one-wave intensification, no

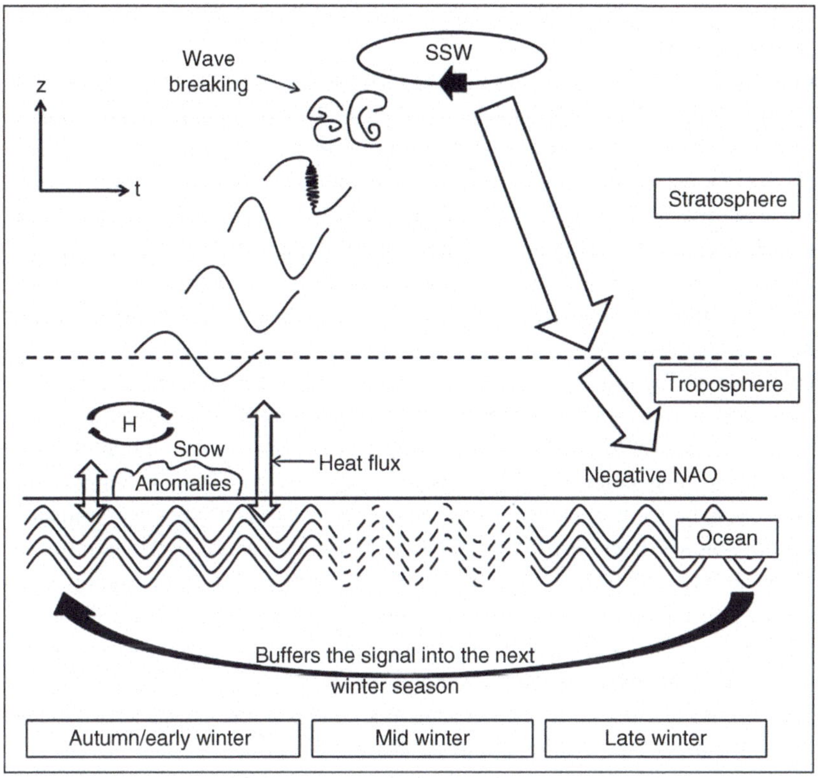

Figure 2.8 Mechanism for decadal variability of stratospheric sudden warming events (SSW) is generated. H denotes Meridional heat flux and NAO denotes North Atlantic Oscillation. *Source:* Schimanke et al. (2011) / John Wiley & Sons.

significant phase tilt with height, vortex deviation from the polar region, and moderate polar temperature increase, all of which are characteristics of early winter warming in Canada. Manney (2010) compared the SSW events of 2006 and 2009, and found that the wind direction in 2009 reversed easterly more rapidly and reversed westerly more slowly than in 2006. In 2009, the amount of mixed trace gases removed from the vortex during the polar vortex collapse was more than that before reaching the standard of strong warming condition. Persistent and clearly discernible polar vortex fragments and anticyclone occurrences in 2009 were more common. The SSW in 2009 affected the lower stratosphere more than any other observed SSW, and there was no significant polar vortex recovery in this region. The stratopause collapse and subsequent reconstruction at high latitudes, as well as the downward intensification into an upper stratospheric vortex, are similar in 2009 to 2006. There has also been a lot of analysis and discussion about the SSW in the southern hemisphere, especially the strong SSW in September 2002. For example, Ricaud (2005) found that the Antarctic polar vortex split in two under the influence of SSW in September 2002, and described the chemical and dynamic process of the polar vortex evolution in detail. Kruger et al. (2005) compared the strong SSW events in the southern hemisphere with the SSW events in the northern hemisphere in winter with similar wave disturbances, and pointed out that the eastward wave has an important influence on the forced interaction of quasi-steady planetary waves in the troposphere, which eventually leads to the collapse of the polar vortex. Siegmund and Velthoven

(2005) discuss the horizontal and vertical temperature advection during the strong southern hemisphere SSW in September 2002.

Through wavelet analysis of SSW time series, Schimanke (2011) found that the change period of SSW number was 52 years. In their study, the decadal variation of SSW is generated in the ocean-tropose-stratosphere system. The 100 hPa meridional heat flux (representing tropospheric fluctuations into the stratosphere) increased in proportion to the wave flux into the stratosphere due to anomalies of positive heat flux entering the atmosphere from the North Atlantic, increased Eurasian snow cover, and increased blocking events. Therefore, the polar vortex weakens and the number of SSW increases. In the ensuing weeks, the surface weather was affected by downward stratospheric anomalies. The anomalies generated in this way persist over a long period and lead to interdecadal variations in the SSW.

An important aspect of variability in the stratospheric circulation, and in particular the distortions just mentioned, is sudden stratospheric warmings. Scherhag (1960) first noted these in the 1950s and they were subsequently found to be a characteristic late winter phenomenon. Tropospheric eddies transport heat poleward. This acts to reduce the zonal-mean temperature gradient in the extratropic and hence the strength of the zonal-mean winds. Eddies also transport westerly zonal momentum poleward, tending to increase the zonal-mean winds in the extratropic. In the Transformed Eulerian Mean (TEM) framework (Limpasuvan et al. 2004), the net effect of these processes at a given latitude can be expressed in terms of the divergence of the Eliassen–Palm flux (EPF). The EPF has vertical and meridional components. The vertical component of the EPF is related to the strength of the meridional eddy temperature flux ($\overline{v'T'}$) while the meridional component is related to strength of the eddy momentum flux ($\overline{u'v'}$), where the primes indicate departures from the zonal mean and the overbars indicate the zonal mean.

For long-term mean winter conditions, through much of the depth of the atmosphere and north of about 30°N, the EPF divergence is negative (i.e. convergent). EPF convergence represents a westward force on the zonal wind, namely it induces slowing of the winds (Holton 2004). In a steady state condition, the zonal wind and temperature are constant. Slowing of the zonal-mean wind by the EPF convergence is balanced by the effects of a residual circulation in the meridional plane that is in a thermally direct sense (clockwise as viewed in a mean cross section with the North Pole to the right). This residual circulation (which is part of the total zonal-mean meridional circulation) is highly efficient in maintaining thermal wind balance in large-scale atmospheric motions, and keeping the ageostrophic flow small in comparison with the geostrophic flow. The Coriolis force on the residual poleward meridional wind speeds up the zonal winds aloft. The Coriolis force on the residual equatorward wind reduces the zonal winds near the surface where the EPF is divergent. Through continuity, there must be attendant vertical motions. A residual downward motion (adiabatic warming) balances radiative cooling in higher latitudes while upward motion (adiabatic cooling) in lower latitudes balances radiative warming. Departures from the steady state, seen as time changes in the zonal wind and temperature, arise from imbalances between the EPF divergence/convergence and associated residual circulations. The EPF divergence pattern is weaker in summer.

The mean circulation and thermal structure of the winter stratosphere examined earlier can be similarly understood from the presence of mean EPF convergence and meridional-plane circulations associated with the upward propagation of planetary waves. For example, in the extratropic, eddy forcing maintains the observed stratospheric temperature above its radiative equilibrium. Hence there is net radiative cooling, which is balanced by downward motion (adiabatic warming) associated with the residual circulation. In the tropics, the temperature is below radiative equilibrium, consistent with upward motion (adiabatic cooling). In turn, sudden stratospheric warmings (which as the name implies are transient events) can be understood from the anomalous upward propagation of planetary waves.

As outlined by Limpasuvan et al. (2004), when an anomalous vertically propagating planetary wave enters the stratosphere, it imparts an anomalous EPF convergence. The decelerated westerly flow is brought back toward balance by the Coriolis force acting on the poleward mean residual flow anomaly. This induces adiabatic temperature changes that also try and bring the flow back to balance. Through continuity, the poleward mean meridional flow across the axis of the EPF forcing requires sinking motion (adiabatic cooling) below and poleward of the forcing region. In turn, there is rising motion (adiabatic cooling) below and equatorward of the forcing region. These meridional-plane circulations quickly weaken the meridional temperature gradient, and give rise to rapid high-latitude stratospheric warming.

In extreme cases, stratospheric temperatures can rise 50 K and the circumpolar vortex can reverse to easterly flow over a few days. The presence of vertically propagating waves, while necessary for sudden stratospheric warmings, is not a sufficient condition. It appears that the stratospheric flow needs to be "preconditioned," such that wave activity is focused toward the polar vortex. The pre-conditioning occurs when the vortex is poleward of its climatological position (Limpasuvan et al. 2004). These are identified by Kodera and Chiba (1995) as representing, respectively, the prewarming, warming, and post-warming phases of the event. The prewarming stage shows a vortex center located well off the North Pole over the Norwegian Sea, with minimum temperatures of about −72 °C. During the warming phase, the vortex breaks down into four centers. The post-warming phase shows the further breakdown. The contrast in high-latitude (e.g. north of 70°N) temperatures between the prewarming and the post-warming phase are readily apparent. Between the warming and post-warming phase, the stratospheric zonal-mean zonal winds reversed at 70°N (Kodera and Chiba 1995).

There is mounting evidence from studies (Baldwin and Dunkerton 1999, 2001; Hartmann et al. 2000; Ambaum et al. 2002; Perlwitz and Harnik 2003) over the past several decades that sudden stratospheric warmings can have pronounced effects on the circulation of the troposphere and surface. While they are still being elucidated, the point to be emphasized is the apparent existence of two-way coupling while upward propagation of planetary waves influences the stratospheric circulation, changes in the stratospheric circulation can influence the troposphere and surface; for example, by changing the conditions for vertical propagation of waves. Baldwin and Dunkerton (2001) found that the surface signature of downward propagating anomalies strongly resembles the surface signature of the Northern Annular Mode (also referred to as the Arctic Oscillation), a mode of atmospheric variability that has pronounced impacts on Arctic climate and has emerged as an important issue in the climate change debate.

2.3.8 Weather Systems on the Tibetan Plateau

The TP, with an average elevation of more than 4 km above sea level, is known as the "third pole" of the Earth. It is the highest and most complex plateau in the world. It is the largest plateau cold region in midlatitude or the northern hemisphere as well. It also is a sensitive areas to climate change. Here the frost-free period is short, but it is a powerful heat source in summer. The coverage and strength various of permfrost, seasonal frozen ground and snow cover, making it is a dynamic, thermal and water vapor disturbance source high up in the middle troposphere, it has a great impact on the east Asian atmospheric circulation, catastrophic weather and climate events and even global climate change. The TP has the most complex atmospheric pressure system in the same latitude area of the northern hemisphere in summer, and the diversity of its weather system is comparable to that of the low-latitude tropical region. These weather systems mainly include Tibetan high, Tibetan low trough, Tibetan Plateau shear line (TPSL), Tibetan Plateau vortex (TPV), southwest vortex (SWV), northwest vortex, and plateau mesoscale convective system.

TPV is a unique meso-α scale weather system (sub-synoptic scale) in the main body of the TP. Figure 2.9 shows the vertical structures of a typical TPV. The cyclonic circulation of TPVs is confined primarily to the lower-middle troposphere, with maximum positive vorticity generally present at about 500 hPa. The typical horizontal and vertical scales of TPVs are about 400–500 km and 2–3 km above ground level, respectively. Most have a warm core, especially in their early genesis stage, along with a cloudless eye structure dozens of kilometers wide and spiral rainbands that are similar to tropical cyclones (tropical cyclone-like vortices). Above and below 400 hPa, there are obvious divergence and convergence within the TPV, associated with a strong updraft in the position of strong convections. A relative weaker downdraft exists in the cloudless eye region. Similar to tropical cyclones, the genesis and development of TPV are closely related to surface diabatic heating and atmospheric condensational latent heating effects, and the tropical cyclone-like vortex structures of TPV can be theoretically explained by singular solitary wave solution of nonlinear inertial gravity internal wave (Li and Lu, 1996; Li and Xu 2005). However, due to TP is a cold region with the relative colder and dryer atmospheric conditions, therefore, one prominent feature of TPV is that the convective rainbands associated with TPV are very different from convections in tropical regions. For instance, Wu et al. (2019) concluded that the convections over the TP are weak but deep. Convective available potential energy (CAPE) of a convection over the TP is lower than in most other continental regions (Fu and Liu, 2007), these characteristics contribute to convective clouds with short lifetimes (Chen et al., 2017). In addition, Luo et al. (2011) found that compared to the surrounding regions of the TP, the deep convection over the TP typically has a smaller horizontal scale and smaller CAPE values, and the deep convection is weaker and less well organized, leading to a weaker precipitation. Another prominent feature is that in TPVs, vortex Rossby waves dominate the inner part of the vortex, while inertial gravity waves dominate the outer part (Chen and Li 2014).

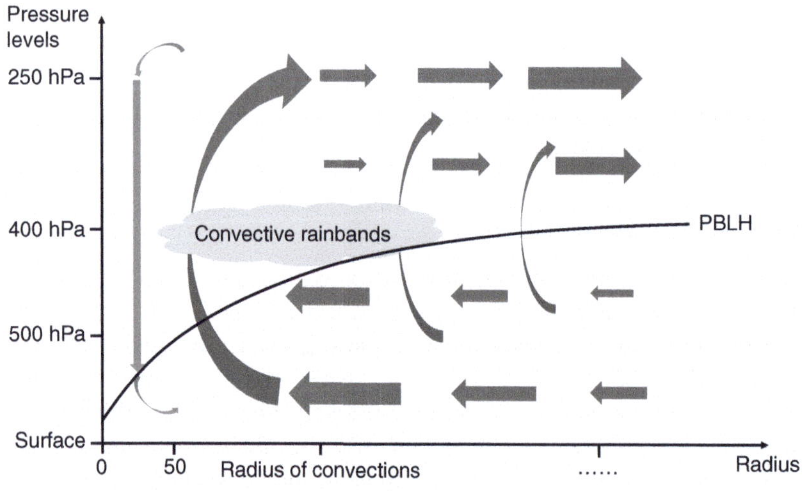

Figure 2.9 The vertical profile structure of tropical cyclone-like Tibetan Plateau vortex. PBLH denotes Planetary Boundary Layer Height.

The TPSL is the "hotbed" of the weather system over the TP (Figure 2.10). Unlike shear line in tropical regions, where East–West transverse shear line is prominent. The shear lines over the cold regions of TP are unique, according to the shape of the TP shear line, the TP shear line is commonly divided into quasi East-West transverse shear line and quasi north-south perpendicular shear line. The occurrence of the transverse shear line is twice that of the perpendicular shear line. The transverse shear line mostly appears between 30 °N and 35 °N and is most obvious over the TP. The perpendicular shear lines mostly appear in the steep slope area of 103 °E of the TP and the middle TP. Due to less movement of the transverse shear line, it lasts a long time over the TP, can cause a large amount of precipitation, but has marginal impact on the weather systems in the eastern TP, while the perpendicular shear line can move out of the TP, and has a significant impact on the weather in eastern China. The dynamic and thermal characteristics of the transverse shear line and perpendicular shear line are different (Yao et al. 2014). Under specific conditions, the shear line over the TP can induce the formation and eastward movement of the TPV, thus forming extreme weather events such as low vorticity rainstorms in the main body of the plateau and its downstream areas (Du et al. 2018).

SWV is also one of the most influential rainstorm weather systems over the TP. In the mature stage of the SWV, the temperature profile shows that there is an obvious temperature inversion layer, and the water vapor vertical distribution also presents the inverse humidity. However, the water vapor content in the mature stage of the SWV is obviously smaller than that in the early genesis stage, and the vertical flux of water vapor divergence can better explain the reason for the change of water vapor profile (Ni et al. 2017). The diagnosis of vorticity budget equation in SWV shows that the configuration of divergence term is basically opposite to that of advection term (Liu and Li 2014). The divergence term plays a leading role in the development and maintenance of the SWV, and torsion term also makes an important contribution to the formation of the SWV. Advection term and friction dissipation term are the main terms of vorticity consumption of the SWV, which is not conducive to the generation and development of the SWV. The diagnosis of the vortex energy budget equation indicates that the vortex energy maintained by the development of SWV mainly comes from the horizontal flux divergence term and the vortex energy production term, and the friction dissipation term and vertical flux divergence term are the main consumption terms. In both the genesis and mature stages of the SWV, the dynamic structure of convergence in the lower troposphere coupled with positive vorticity and high vortex center is maintained, and there is a strong upward movement. Meanwhile, there is a "warm center" structure with comparable potential temperature and a "wet center" structure with relative humidity. In the mature stage of SWV, the positive vorticity column can reach 300 hPa in the upper troposphere. In the vertical direction, the SWV first formed at 850 hPa in the lower layer, then developed upward and appeared in the closed vortex at 700 hPa and became mature. The dispersion of water vapor flux at low altitude is a good indicator of the

Figure 2.10 The (a) East–West transverse shear line and (b) north–south perpendicular shear line over the Tibetan Plateau.

intensity and movement of precipitation zone. The potential vortex budget diagnostic analysis shows that the vertical structure of the non-adiabatic action term is opposite to that of the vertical flux divergence term, and the non-adiabatic action term caused by latent heat release is beneficial to the growth of the potential vortex in the lower layer but inhibits the growth of the potential vortex in the upper layer, which plays an important role in the generation and development of the SWV.

In summer, there is a lower stratosphere semi-permanent anticyclonic circulation over the TP – the Tibetan high, which is strongest near 100 hPa. The high's center at 100 hPa is also frequently located over the Iranian Plateau. In winter, there is an anticyclonic circulation over the ocean, moving northwestward gradually with the change of season and forming a high center. By summer, the anticyclonic circulation reaches the TP and the Iranian Plateau, completing an annual cycle. Mason and Anderson (1958) suggested that, the South Asia High is the strongest, largest, and most stable anticyclonic circulation system at 100 hPa in the northern hemisphere. The South Asia High results from the heating effect of the TP (Flohn 1960). In the 1970s, meteorologists in China widely studied the climatic features, the formation mechanisms of South Asia High, and its relationship with precipitation in various regions of China. The activities of the South Asia High became an important basis for weather prediction in the rainy season (Sun and Song 1987; Luo et al. 1982; Krishnamurti et al. 1973; Zhang and Peng 1983; Zhou 1991).

2.4 Radiation

2.4.1 Surface Radiation Budget

Solar radiation is the basic energy for animal and plant life activities in ecosystems and the basic driving force for weather and climate formation and evolution. Due to its special dynamic and thermal effects, typical permafrost in cold regions is extremely sensitive to climate change. The characteristics of surface radiation composition, albedo changes and energy balance are key aspects of global climate change research and have a significant impact on the climate and environment of the entire cold region.

The surface radiation budget (SRB) is composed of surface shortwave (SW) and longwave (LW) radiation fluxes. Net shortwave (NSW) radiation is the difference between incoming SW fluxes and reflected SW radiation. Net longwave (NLW) radiation is the difference between downward and upward LW radiation. Among the four fluxes in the SRB, there are the downward shortwave (DSW) flux, the upward shortwave (USW) flux, the downward longwave (DLW) flux, and the upward longwave (ULW) flux. The SRB equation is calculated as below:

$$SR_n = SR_n^s + SR_n^l = (1-\alpha) \times SF_d^s + SR_n^l = \left(SF_d^s - SF_u^s\right) + \left(SF_d^l - SF_u^l\right) \tag{2.1}$$

where SR_n^s, SR_n^l, and SR_n, are the net SW, LW, and all radiation in the surface, respectively; α is the albedo parameter; and SF_d^s, SF_u^s, SF_d^l, and SF_u^l are the shortwave and longwave fluxes (DSW, USW, DLW, and ULW, respectively) in the surface.

The surface albedo, as the ratio between DSW and USW, is an integration of reflectivity over the shortwave spectrum (Liang et al. 2010), acting as an optical signature of the land surface. In general, snow has a high albedo while vegetation has a relatively low albedo. In addition, the albedo of vegetation is also affected by highly reflective backgrounds such as snow. The normalized difference vegetation index (NDVI) is an indicator of the present and health of living green vegetation and is based on the biophysical characteristics of vegetation that has low reflectivity in the red spectrum and high reflectivity in the near-infrared spectrum. The weak correlation between albedo and snow cover in spring may be related to the mosaic surface created by snow, vegetation, and the freezing thaw cycle of soil in spring. Changes in snow cover can alter albedo directly but can also affect albedo indirectly by altering moisture conditions to support plant growth. Interestingly, major peaks and drops of albedo anomalies at both the seasonal and annual scales are consistent with those for snow cover or NDVI. This indicates that albedo may be sensitive to extreme near-surface atmospheric conditions such as those detected from snow cover and NDVI extremes over the TP, a theory that has been suggested previously based on a case study of 2003 European heat waves at a regional scale (Teuling and Seneviratne 2008). In addition, vegetation growth as indicated from NDVI might be linked to the changes of the diffuse radiation components of DSW (Wild 2012), which can alter surface albedo substantially as reflectivity changes. In order to elucidate the effects of non-vegetated land cover, similar analyses were conducted in vegetated areas with the maximum annual NDVI > 0.1 (Piao et al. 2003). The result indicates a pattern of seasonal and annual variability that is similar to that of averaging over the entire TP.

The spatial distribution of albedo over the TP probably reflects the land-surface conditions over the TP. Annually, albedo is high and low in the northwestern and eastern TP, respectively. The annual standard deviation of albedo peaks in the Karakorum range, the western Himalaya range, and the southeastern ranges and valleys. Overall, albedo distribution is related to land-surface heterogeneity and land cover, especially for snow cover, which also influences the spatial distribution of standard deviation of albedo in highly complex terrain. Generally, decadal changes of albedo in most areas do not exhibit consistent directions, and the results capture only decreases of albedo that are located in the northwestern TP and the west Himalaya region in spring and in the center of the TP in winter. There is a potential seasonal dependency of the correlation among albedo, snow cover, and NDVI averaged over the TP. The correlation between albedo and snow cover is positive and statistically significant in autumn and winter, while NDVI is negatively correlated with albedo in summer and autumn.

The *in situ* measurements of surface broadband albedo are sparse, particularly over the Arctic. Satellite data therefore play an important role in assessing the climatology of this parameter. Surface reflectance can be specified as "inherent" albedo or "apparent" albedo. The inherent albedo is the "blackbody" albedo of the surface, which is independent of the changes in atmospheric conditions. The apparent albedo is measured by up- and down-looking radiometers and varies with the atmospheric conditions.

Unlike surface temperature, there are two peaks in the annual cycle of the surface albedo for the Arctic region north of 60°N: one in the early spring (about 60%) and the other in autumn (about 35%). The relatively low wintertime albedo corresponds to the low-latitude regions with less snow/ice coverage, because the dark, high-latitude areas are excluded from the statistics. The autumn peak is due to freeze-up, but more open water areas in autumn result in a lower albedo than in spring. Over the Arctic Ocean, the spring maximum surface albedo is about 60%, but the autumn second maximum albedo is only about 23%. For the Arctic landmasses north of 60°N, the autumn second maximum albedo is much higher than that for the Arctic seas and the entire Arctic region, with a value of approximately 45%. The Arctic annual mean surface albedo has a spatial distribution similar to surface temperature. The largest surface albedo is always in Greenland with a maximum value of 90% throughout all seasons.

The surface reflectance represents the reflectivity of the subsurface to solar radiation. The surface reflectance also shows a daily variation with the decrease of solar height due to the change of solar altitude angle, and surface reflectance is the smallest at noon and the variation is small. The surface albedo is closely related to the nature of the ground surface, which varies greatly in winter and summer, with the surface being bare sand and gravel in summer and covered by snow in winter. For the polar region, where snow-ice covers, observations in 2004 show that, the surface albedo of Zhongshan station is the smallest in January, which is only 0.24. However, the average surface albedo increases steeply in February, which is due to a strong snowfall process during February 5–6, and the surface is covered by new snow again, and the average albedo of new snow can reach about 0.9. In addition, there were more clouds in February, and it was shown that the increase in cloudiness also led to an increase in surface albedo. On the other hand, the fact that only observations from February 1 to February 15 are counted in February may also affect this result. In winter, the monthly average surface albedo was the largest in August, at 0.58. The monthly mean surface reflectance in July is the largest at 0.93, which reflects the reflectance of fresh snow after snowfall. From July to November, the surface reflectance continues to decrease, with the decrease in surface reflectance from July to October reflecting the effect of snow on surface reflectance, while the significant decrease in surface reflectance in November is mainly related to the snowmelt process, with a monthly average surface emissivity of 0.69 in November. Therefore, in the radiation balance, the biggest difference between the observations from sea-ice sites and land sites during the same period lies in the NSW radiation balance, where the albedo of the sea-ice surface is much larger than that of the land surface, making more solar radiation reflected by the surface, resulting in less NSW radiation on the sea-ice surface than that on the land surface. On the seasonal scale, the monthly mean daily variation of solar radiation varies considerably. Fifty-eight days of polar night are observed in the region in winter, including the whole month of June, when the solar radiation is $0\,\mathrm{W\,m^{-2}}$, while in summer, the solar radiation reaching the surface increases significantly. November, December, and January are the three strongest months of a year in terms of solar radiation, where the monthly mean daily maximum of solar radiation in December can reach up to 732 and $334\,\mathrm{W\,m^{-2}}$. The monthly mean daily variation of solar radiation varies significantly from month to month, while the monthly mean daily maximum of solar radiation in November can reach $659\,\mathrm{W\,m^{-2}}$ and the monthly mean solar radiation is $325\,\mathrm{W\,m^{-2}}$.

Seasonal DSW varies with latitude in the western TP and with elevation in the eastern TP, and there is a significant seasonal migration of high-DSW areas over the TP. The standard deviation of DSW exhibits the greatest spatial variation in autumn and the least differences in winter when the climate is cold and dry. Annually, DSW is high in the southwestern TP and low in valleys in the southeastern TP, while standard deviation is high in the northern TP and low in the southeastern TP. The changes of DSW are characterized by spring brightening in the western and southeastern valleys in the TP and summer dimming over most areas of the TP. Overall, DSW is related to a decrease of latitude from north to south, an increase of elevation from southeast to northwest, and the monsoon climate in the southeastern TP; conversely, the standard deviation for DSW reflects the contrasting pattern between the elevated plateau and ranges in the north and the valley ranges in the southeastern TP.

Comparison of the seasonal mean and annual mean DSW over the TP with cloud cover and temperature range anomalies indicated that, over the entire TP, DSW decreases by $-1.2\,\mathrm{W\,m^{-2}}$ $\mathrm{decad^{-1}}$, with a significant decrease of $-7.0\,\mathrm{W\,m^{-2}}$ $\mathrm{decade^{-1}}$ in summer and an increase of $2.0\,\mathrm{W\,m^{-2}}$ $\mathrm{decade^{-1}}$ in spring. The presence of cloud blocks the DSW from reaching the surface in daytime and shrinks the daily temperature range, thus, DSW is negatively correlated with cloud cover and is positively correlated with temperature range. It is noteworthy that cloud cover observations from ground stations often lack representative coverage at a regional scale owing to the high variability in cloud cover both temporally and spatially. In spring, the temporal variation in the DSW anomaly exhibits a negative relationship with cloud cover, except for 1997–1999, and 2004, although temperature range exhibits a positive relationship with DSW for those years. In summer, solar dimming is consistent with the decrease in temperature range ($-0.47\,°C\,decade^{-1}$), and cloud cover and temperature range exhibit the highest correlation with DSW. A significant negative (positive) correlation can also be found between DSW and cloud cover (temperature range) in autumn (winter). The decreased annual DSW is positively correlated with the decreased temperature range ($-0.17\,°C\,decade^{-1}$), while the DSW variation is negatively correlated with cloud cover. It could be argued that the previous paradox between the observed annual decrease in cloud cover and dimming could be explained by the marked summer dimming over most areas of the TP, which would dominate the annual dimming (Yang et al. 2012). Therefore, temporal variations in DSW and summer dimming over the TP are well correlated with cloud variability, including cloud cover, duration of cloud, and cloud type; this is supported by the strong correlations of summer DSW with temperature range and cloud cover anomalies.

Over the TP, both the direct impact of aerosol absorption and scattering and the indirect impact of aerosols on cloud formation have been suggested to be much lower than the observed dimming. The third factor contributing to the DSW is the absorption from various radiative active gases, which induces the summer dimming over the TP that is consistent with the observed moistening trend. However, when DSW and cloud cover anomalies covary (i.e. both exhibit large fluctuations in the same direction), temperature range is well correlated with cloud cover variation. This may be explained by changes of cloud types. The impacts of variations in deep convective clouds have been shown in a previous study to exhibit consistent negative correlations with DSW in 78 sites in the TP (Yang et al. 2012). Conversely, the covariability between DSW and temperature range implies the close connection between these factors by the solar (daytime) heating that affects the maximum temperature (Bristow and Campbell 1984). Unlike cloud cover, which is directly related to interannual DSW variability, the temperature range is considered to be an indirect confirmatory indicator of DSW and cloud cover variation (Liu et al. 2004; Ye et al. 2010; Wild 2012). This supports the importance of DSW variation in association with regional temperature changes over the TP.

Another feature is that the largest USW occurs in May rather than in June because relatively higher surface albedo in May reflects more DSW than that in June. The outgoing surface solar radiation is largely determined by the surface albedo and the amount of downward radiation. The spatial pattern of the annual mean DSW radiation with a minimum of 70–75 $W\,m^{-2}$ over the northern part of the Nordic seas is due to the synoptic transport of warm humid air and subsequent cloud formation in the area. The annual mean and seasonal cycle are determined by the length of the day, which gives zero direct-beam flux at the North Pole from the autumnal to spring equinoxes. The annual mean downward fluxes have a latitudinal gradient, which is modified by the occurrence of topography, clouds and their optical properties such as liquid water content, number of droplets and their size.

The seasonal characteristics of the monthly mean daily variation of USW are similar to those of the monthly mean daily variation of solar radiation. Although the solar radiation in December and January are the two weakest months of the year, the USW radiation in these two months is lower than that in November, reflecting the subsurface change. In December and January, the snow on the surface is almost melted and the subsurface is bare sand and gravel, when the surface albedo is lower, and the surface absorbs more solar radiation. Therefore, November is the month with the largest value of reflected shortwave radiation, the monthly average daily maximum value is $289\,W\,m^{-2}$, and the monthly average value is $128\,W\,m^{-2}$.

The seasonal characteristics of the monthly mean daily variation of USW radiation are similar to those of the monthly mean daily variation of solar radiation, where the monthly mean daily maximum of USW radiation in November reaches $449\,W\,m^{-2}$ and the monthly mean USW radiation is $199\,W\,m^{-2}$. This is related to the difference of surface properties, and the surface reflectance of sea-ice sites is much larger than that of land sites.

DLW displays a similar pattern for both seasonal and annual mean values: low in the central-western TP and high in the outer TP, with maximum values at the southeast boundary of the TP. The standard deviation of DLW is different for all four seasons, with high variation in autumn and low variation in winter and summer, respectively. The annual standard deviation has three hot spots in the southeastern, southwestern, and northeastern TP, while the lowest values are in the northwestern TP. Thus, DLW is related to elevation differences between the outer and inner TP and between the northwestern and southeastern TP; it is also related to the impact of the monsoon climate in the southeastern TP and the dry climate in the elevated plateau of the western TP. The annual changes of DLW are dominated by a large increase in most areas in summer, while increases of DLW are apparent over the northeastern TP in winter and autumn, and over the mid-south TP in spring.

The temporal variation in seasonal DLW average over the TP is associated with water vapor and cloud cover anomalies. Annual mean DLW increases by $2.5\,W\,m^{-2}$ decade^{-1}, with a maximum positive trend in summer ($5.3\,W\,m^{-2}$ decade^{-1}). Since water vapor is the most abundant greenhouse gas that emits LW radiation and cloud is an important radiator of LW radiation, DLW is positively related to water vapor and cloud cover. In spring, DLW varies with water vapor, with divergences from the overall covariation occurring in 1992, 1997, and 2004. Those divergences of DLW to water vapor and cloud cover in spring are possibly related to the variation in the hydrological cycle in the pre-monsoon phase. DLW is closely related to water vapor in summer, and both have exhibited positive anomalies since 1998. Cloud cover anomalies have the highest correlation by 0.57 in summer, with an exceptional peak and drop in 1988 and 2006, respectively. DLW is closely related to water vapor variation in autumn and winter, while cloud cover also regulates DLW. This relationship can explain most exceptional years in which DLW and water vapor did not covary, including autumn of 1988, 1992, and 2001 and the winter of 1986, 1999, and 2005. Although cloud cover decreases, DLW increases annually with increased water vapor ($0.21\,kg\,m^{-2}$ decade^{-1}). Temporal variation in annual DLW is closest to that of water vapor, with consistent drops in 1986, 1992, 1997, and 2007 and peaks in 1985, 1988, 1991, 1998, and 2003. Hence, the correlation between DLW and water vapor over the TP may be explained by the increased DLW caused by the warmer and wetter (especially in summer) climate in recent decades.

The increased DLW that is most pronounced in summer is related to the increased water vapor with positive anomalies since 1998. In most cases, water vapor has more power to explain the interannual variability of DLW, which is supported by the fact that water vapor is the dominant emitter of LW radiation (Wang and Liang 2009). In principle, DLW is determined by the atmospheric temperature and emissivity, which varies according to cloud and greenhouse gases that can absorb and emit radiation in the thermal infrared spectrum. Cloud might be another important factor influencing DLW interannual variability over the TP. This relationship between DLW and cloud cover is more obvious when a divergence or significant drops and peaks occur in the interannual variation in DLW and water vapor. Similar to the method by which all-sky DLW can be estimated based on the clear-sky emissivity from water vapor pressure and the adjustment of cloud effects using the cloud fraction (CF), the correlation between cloud cover and DLW may be explained by the positive relationship between the longwave radiation emitted from cloud, which contributes to cloudy-sky DLW; conversely, the clear-sky DLW is determined primarily by water vapor. A jump of DLW is clear for the summer of 1997–1998; this is consistent with specific humidity patterns from ground observations averaged from China Meteorological Administration stations (Yang et al. 2012). The Clausius–Clapeyron equation, which governs the warming and moistening over the TP in summer, could be applied to this phenomenon. However, the cloud cover varied primarily in the same direction as the water vapor, amplifying the interannual variability of DLW consistently.

Unlike those of DLW, which display a uniform and smooth pattern for all four seasons, the spatial characteristics of seasonal ULW are impacted by surface temperature, land cover, and elevation. In winter, ULW decreases from the southeastern TP, reaching low values in the west and north TP. In all four seasons, high values of ULW are displayed in the mid-west plateau and at the low elevation boundaries, with low values in the northwestern TP. Annually, ULW increases from the north to the south of the TP, with the lowest values in the northwest and the highest in the eastern TP. Standard deviation of ULW is extremely low in winter and summer, and the annual standard deviation distribution exhibits a decreasing trend along the elevation gradient, with peaks in the western plateau and the northeast basin and ranges. Corresponding to the warming of the TP, ULW has exhibited a gradual increase in the last two decades; this increase is distributed over most areas in winter, in the western TP in spring, and in the middle and southern TP in summer and autumn. Therefore, ULW is high in places with higher temperatures than the surrounding areas. In addition, spatial patterns in standard deviation of ULW correspond with surface landforms and land cover, with high values in the western area, which is sparsely vegetated, with low values in the moist and warm areas (e.g. the southeastern area), which are covered by forest.

ULW covaries with temperature and NDVI over the TP at both the seasonal and annual scales. The TP experienced annual increases in ULW (2.6 $\mathrm{W\,m^{-2}}$ $\mathrm{decade^{-1}}$) from 1984 to 2007, and ULW exhibits significant increases for all seasons, ranging from 2.1 $\mathrm{W\,m^{-2}}$ $\mathrm{decade^{-1}}$ in summer to 3.2 $\mathrm{W\,m^{-2}}$ $\mathrm{decade^{-1}}$ in winter. In spring, ULW increases with increasing temperature with Cor as high as 0.90. In summer, ULW varies with fluctuations of temperature and NDVI, which agrees with peaks of ULW in 1988, 1994, 2002, and 2006. In autumn, ULW exhibits the largest jump from 1997 to 1998; Temperature also switches to higher values at this time, and NDVI covaries with similar dramatic changes. In winter, ULW is highly correlated with temperature with maximum surface warming by 0.7 °C $\mathrm{decade^{-1}}$; moreover, high values of ULW in 1988, 1994, 1999, and 2006 agree with peaks in temperature. Annually, ULW increases with increasing temperature and NDVI. Overall, the temporal variation in ULW is positively related to annual or biannual variations in temperature and to the variability of land cover as indicated by NDVI, which exhibits higher values for the more recent decade.

ULW is the only SRB component that has exhibited consistent changes over most areas in TP for all four seasons. ULW is governed by emissivity and the Stefan–Boltzmann law, which states that the ULW is determined by the surface emissivity and surface skin temperature. A recent study has shown that the skin temperature over the TP exhibits very similar annual and monthly variations to those of the 2 m air temperature measured in the morning (Salama et al. 2012). There is no observation system in existence that collects frequent and continuous observations of high-quality skin temperature that cover the study period; therefore, the air temperature is used as a proxy for the skin temperature. Spatially, increases of ULW are larger for the high-elevation area, a trend that has also been noted in recent studies to quantifying the elevation dependency of warming (Qin et al. 2009). Based on ground observations, the magnitude of the warming over the TP is found to be larger in the north than the south (Yang et al. 2011; Guo and Wang 2012); whereas

ULW from the fused data set illustrates that the increase of ULW only dominates a fraction of areas in the northern ranges (Shi and Liang 2013). This phenomenon could be explained by the sampling of differences between the data sources: the largest warming from ground observation partly contributed from stations below 3000 m in the northeastern TP, and the uncertainty of the fused ULW originated from a lack of ground observations over the north ranges of the TP. Temporally, the decadal increase of ULW is largest in winter, when the observed 2 m air temperature also exhibits its increase. The presence and density of vegetation cover can alter the surface emissivity, making it different from that of bare soil; moreover, warmer temperature favors vegetation growth, especially promoting the growth of vegetation under cold stress conditions (Bounoua et al. 2000). Those processes are supported by patterns for which major peaks and drops in ULW correlate well with changes in NDVI. Such a relationship between NDVI and ULW is more dominant at the beginning and end of the growing season (i.e. in spring and autumn), usually amplifying or counteracting changes of temperature that can be related to changes of DLW.

The main factor that determines the annual mean and seasonal cycle of the upwelling LW radiation terms is the surface temperature. The primary determinants of the downwelling surface LW radiations are the boundary layer humidity and temperature; its stratification; and the amount and optical properties of clouds. LW radiation transfer in high latitudes is somewhat different from the lower latitudes. Due to the small amount of water vapor the opacity of the water vapor rotation band is smaller; also, the lower temperatures shift the maximum blackbody intensity to lower frequencies and therefore toward the low-frequency rotational band of water vapor (Stamnes et al. 1999). For a clear sky the downwelling LW radiation reaching the surface comes from a very shallow layer of the atmosphere (90% of the accumulated contribution comes from the lowest 500–1000 m of the atmosphere). Thus, high vertical resolution in the boundary layer may be required in order to capture both the annual mean and especially the seasonal cycle of this element, making it a challenging task for climate models.

As the LW radiation dominates the SRB during much of the year, the quality of the simulation of this element is crucial for an accurate representation of the Arctic mean climate and its seasonal cycle. The main observed features are well represented, with the North Atlantic currents and the high storm track density of the Nordic seas contributing to maximum over the Northern Nordic seas of 250–280 and 300–330 $W m^{-2}$ for the downward and upward fluxes, respectively, with a gradual reduction to 215–225 and 230–250 $W m^{-2}$ over the central Arctic.

The observational mean DLW is 220.4 $W m^{-2}$. There is no evident relationship between the annual cloud cover/sea-ice fraction and annual DLW, but a relationship between estimated cloud cover and summertime DLW is evident, with a large cloud cover having more surface LW radiation. A large Barents Sea (15–65°E and 70–85°N) annual sea-ice fraction emits less downward radiation in the Barents Sea area. A positive bias is seen over Greenland and the North American Arctic (5–10 $W m^{-2}$). Averaged over all five observational estimations, the annual mean upward LW flux averaged over 70–90°N is 258.4 $W m^{-2}$. A large sea-ice fraction tends to have smaller upward LW radiation.

It should be noted that the relations between the downward radiation components and cloud cover should not be taken as the direct influence of the cloud cover as the correlations do not give any causal relationships. Cloud cover changes are related to changes in both heat and moisture transport which, in addition to changing the cloud cover, may change the atmospheric temperatures and water vapor content, Consequently, it is difficult to distinguish between the direct influence of CF and the influence of atmospheric water vapor and temperature, which may covary with the cloud cover fraction and therefore lead to too strong statistical cloud–radiation relations.

The monthly mean daily variation of atmospheric longwave radiation throughout the year, and atmospheric longwave radiation mainly depends on the temperature of the whole atmosphere. There is no significant daily variation in either winter or summer. In addition to the air temperature, atmospheric longwave radiation is also affected by cloudiness. January is almost sunny for the whole month (total monthly precipitation is only 0.4 mm), so the atmospheric longwave radiation is significantly smaller in January compared to December and February. The monthly average daily variation of surface longwave radiation throughout the year depends mainly on the surface temperature. The seasonal variation of surface longwave radiation follows the seasonal variation of surface temperature, and the surface longwave radiation is significantly stronger in summer than in winter, showing the same daily variation with the surface temperature.

The monthly mean daily variation of atmospheric longwave radiation, shows that there is no significant daily variation of atmospheric longwave radiation during the observation period, with the strongest monthly average of 227 $W m^{-2}$ in November. The average temperature in July is lower than that in August, but the atmospheric longwave radiation is significantly stronger than that in August, due to the strong snowfall process in July and the increase of cloudiness in the sky, resulting in the enhancement of atmospheric longwave radiation. The monthly average daily variation of surface longwave radiation, unlike atmospheric longwave radiation, which shows a more regular daily variation in September, October, and November, with the strongest monthly average surface longwave radiation of 296 $W m^{-2}$ in November and the smallest monthly average surface longwave radiation of 222 $W m^{-2}$ in September throughout the observation period due to the influence of cold air transit.

NR displays seasonal cycles over the TP, with the lowest and highest values in winter and summer, respectively, and the second-lowest values in autumn. Each season, the northern TP exhibits the lowest values, while the eastern and southeastern Himalayas, and valleys exhibit the highest NR. Annually, NR decreases from southeast to northwest in the eastern TP, while an anomaly exists between the south and north in the western TP. The southern and eastern outer TP display higher NR than the inner plateau. The standard deviation of NR is low in winter and summer and displays a similar pattern to spring in autumn, with a high value in the inner TP and a lower value at the boundaries. Annually, the standard deviation of NR is low in the southwest and high in the middle and northern parts of the eastern TP. Changes in NR can be seen along with the ranges at the outer edge of the TP for all seasons, with increases in the western region, except in summer, which is dominated by decadal decreases in most areas. In contrast to spring, when increases and decreases of NR are apparent in the western TP and along the middle of the Himalaya range, NR increases and decreases in the northeastern and southern boundary ranges of the TP, respectively, in autumn and winter. Hence, there are multiple potential factors associated with the spatial pattern of NR including the decrease of elevation from the northwest plateau to the southeast valley, the impact of the monsoon on NR in the eastern TP (i.e. higher NR in summer and higher STD (standard deviation) in spring and autumn), the impact of latitude on the overall north–south pattern of mean NR, and the distribution of different land cover types (i.e. higher NR in the forest area in the southeastern TP and lower NR in the sparse grassland area in the western plateau).

Annual NR over the TP decreases by $-1.0 W m^{-2} decade^{-1}$, with a larger decrease in summer ($-3.6 W m^{-2} decade^{-1}$) and a smaller decrease in autumn ($-0.8 W m^{-2} decade^{-1}$). In contrast to each SRB component (i.e. SW and LW fluxes), NR is regulated by multiple factors in association with both surface and atmospheric conditions. NR represents the amount of available radiative energy that can be partitioned into sensible heat, latent heat, and ground heat fluxes according to the surface energy balance. Therefore, NR has a smaller absolute value than the SW or LW components, which are related to fewer variables. Annually, NR is more loosely and inversely related to warming but correlates most strongly with NDVI, a possible association between the average vegetation conditions indicated by NDVI and the available radiation energy. In spring, a significant positive correlation between water vapor and NR based on the emitted infrared radiation from water vapor was observed; temperature changes are also associated with NR. In summer, the covariation of NR and temperature range is strong for the period before 1998, and NR is also negatively correlated with water vapor through absorption of DSW. In autumn, NDVI is closely related to NR (with exceptional years in 1987 and 1991) and water vapor is positively related to NR anomalies through regulation of DLW. In winter, NR is related primarily to NDVI and snow cover, with a major fluctuation between 1995 and 1999.

In contrast to the positive correlation between NR and water vapor in spring and autumn, the negative correlation between NR and water vapor in summer indicates the shift of the relative relationship between water vapor and interannual NR variability from DLW to DSW. The water vapor absorption of solar radiation and its association with cloud formations regulate the summer variability of NR over the TP substantially. The correlation between WV and NR is not significant in winter, possibly owing to the dry climate that prevails over the TP, when water vapor has less power to explain the variation in NR than other related conditions. Winter snow cover is inversely related to NR by the dominant impact from the positive relationship of snow cover to albedo and the negative relationship of albedo to NR, while DLW counteracts ULW with a much smaller impact on NR interannual variability. Interestingly, NDVI has a larger positive correlation with NR in autumn and winter, which is supported primarily by the inverse relationship between surface albedo and NDVI. This process would generate positive feedback in a warming situation, where more vegetation results in lower albedo and more NR, accelerating warming. Alternatively, increased vegetation cover can also change ULW and could result in a negative correlation with NR. Such feedback is less pronounced than the albedo effect in autumn, but it can offer a possible explanation for the lack of significant correlation in summer and spring. Furthermore, it is worth noting that other links exist between NR and NDVI, such as the photosynthetic active radiation that is an important factor in the photosynthetic activity of vegetation (Zhou et al. 2007), and the capability of vegetation to regulate the partition of NR into sensible heat flux and latent heat flux that is related to vegetation growth. Therefore, the correlation between NDVI and NR indeed reflects the fact that vegetation interacts with the energy cycles (Wang et al. 2007).

Regarding the NSW, NLW, and net all-wave radiative fluxes at the surface, one feature is that the largest loss of the longwave radiative energy from the surface occurs in April, corresponding to the lowest CF and hence the lowest downwelling longwave flux. The net all-wave radiation follows the net shortwave radiation, though lower in magnitude. For the landmasses, the greatest loss of longwave radiation is in July due to the high surface temperature.

2.4.2 Aerosol–Cloud Radiative Effect

The sun is the major source of energy that drives Earth's climate. Solar energy that begins in the top of the atmosphere traverses through the atmosphere and reaches the Earth's surface. Nevertheless, not all of the energy that originates at the top of the atmosphere finds its way to the surface, because a large fraction of solar energy can be reflected back to space or absorbed by aerosols. The process throught which energy is reflected back to space is called scattering and that absorbed by aerosols is called absorption. These processes have a direct effect on Earth's radiation balance via so-called "aerosol–radiation interactions" (Figure 2.11). The aerosol–radiation interactions lead to climate forcing by changing the planetary albedo. The magnitude of scattering and absorption of solar radiation depends on the physical–chemical properties of atmospheric aerosol.

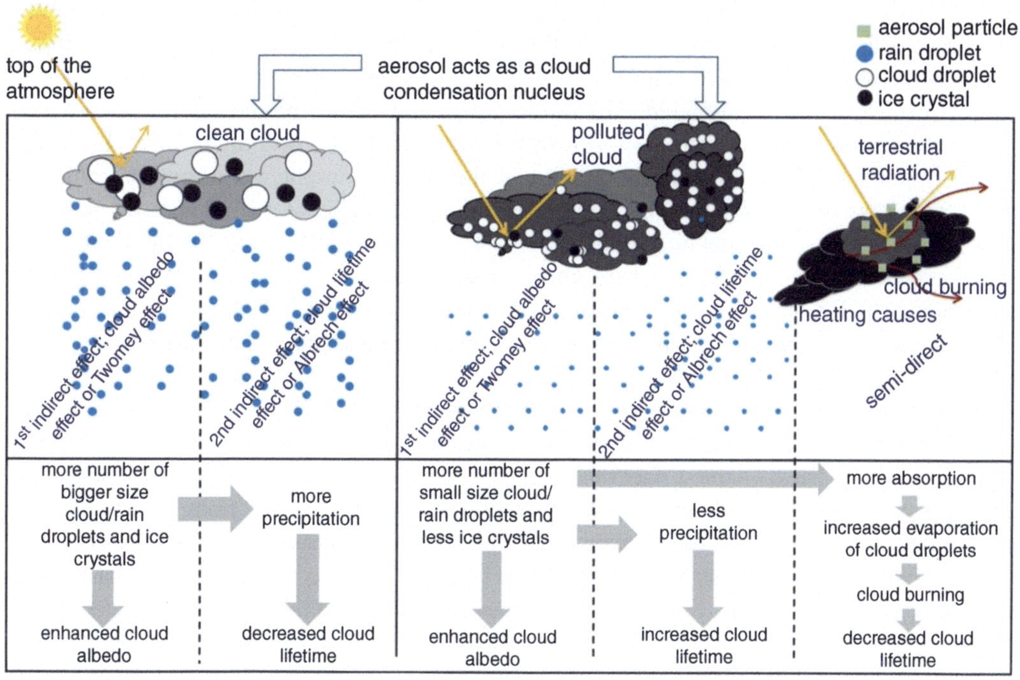

Figure 2.11 The various radiative mechanisms associated with aerosol–cloud interactions.

Aerosol absorbs a large fraction of solar radiation and is considered to be an important component of climate forcing. The sulfate aerosol scatters a large fraction of solar radiation and is also an important part of climate research. Aerosol produces a cooling effect through scattering solar radiation, while producing a net warming effect via absorbing solar radiation. In general, the mixture of aerosol particles in the atmosphere is dominant. Therefore, the net effect of mixing aerosol particles on the Earth's energy budget is determined by surface and cloud characteristics. Scattering aerosols above a dark surface are relatively more efficient, whereas absorbing aerosols are less efficient. A dark surface absorbs a large fraction of solar radiation, and thus, absorbing aerosols have a marginal effect. On the other hand, absorbing aerosols above a bright surface are most efficient, particularly when positioned above the clouds, whereas scattering aerosols are less efficient. A bright surface amplifies the total reflectance of solar radiation. Thus, scattering aerosols produces a marginal effect.

The black carbon (BC) aerosols modify the microphysical properties of clouds, their number and lifetime. This effect of aerosols on solar radiation is called "aerosol–cloud interaction." Aerosol particles act as cloud condensation nuclei and are essential for determining the aerosol–cloud interactions. The clean clouds contain larger cloud drops. The large concentrations of natural and anthropogenic aerosols increase cloud droplet number concentration and hence form polluted clouds. The polluted clouds contain a greater number of smaller cloud droplets. This microphysical induced effect of aerosol with constant liquid water content (LWC) is called "first indirect effect," "cloud albedo effect," or "Twomey effect." The microphysical induced effect of aerosol increases liquid water content, cloud height, and lifespan of clouds. This microphysical induced effect of aerosol is called as "second indirect effect," the "cloud lifetime effect," or the "Albrecht effect." It is to be noted that terminologies such as "aerosol direct effects," the "cloud albedo effect" and the "cloud lifetime effect" were frequently used in third and fourth Intergovernmental Panel on Climate Change's (IPCC) reports. In the fifth IPCC climate assessment report, these radiative effects of aerosol particles are considered as an "interaction" rather than an "effect." "Direct effects" are referred as "aerosol–radiation interactions," and cloud "indirect effects" as "aerosol–cloud interactions." The absorbing aerosols lead to heating of the lower atmosphere. In turn, it changes the relative humidity and the stability of the lower atmosphere and thereby influences cloud formation and its lifespan. This microphysical induced effect of aerosol is called as "semi-direct effect." However, the net interaction of aerosol is uncertain and largely depends on the horizontal as well as the vertical profile of absorbing aerosols such as BC. When BC is deposited on the snow surface, it darkens it. This mechanism reduces the snow albedo and increases the absorption efficiency of solar radiation, which in turn warms the surface and therefore contributes to glacier retreat.

The TP is one of the largest sources of BC soot particles in the world. The BC deposit on the surface of the Tibetan glaciers has been a significant contributing factor to accelerating the melting rate of glaciers. BC deposited on the surface of glaciers in China has been found to darken them by up to 5% which accelerates their melting rate. Warming of the atmospheric layers from 3 to 6 K observed at several locations in India is due to mixing state of BC aerosols. In cases where BC exists in a mixing state, its radiative impact significantly becomes larger. In such a condition, the aerosol-induced warming could be much more than our expectations. Recently, the aerosol-induced elevated warming was observed below 2 km over the northern Indian Ocean, whereas over central India it was observed at around 4 km. This exhibits strong meridional gradients ~4 K at atmospheric levels above 2 km, which is considered to be a sufficient cause to alter the hydrological cycle and hence the Indian monsoon via the "elevated heat pump mechanism." The significant concentration of the mixing state of BC aerosols above the bright clouds is observed along with the elevated sites of the Indian Himalayas. This has enhanced the absorption efficiency of solar radiation and plays an important role in shrinking the Himalayan glaciers. The glacier area retreated by 21% from 1962 to 2001. However, the fractional contribution of BC aerosols in total composite aerosol optical depth (AOD) is small, but its continuous deposition on snow or ice surfaces may accelerate the melting rate of polar ice or the Himalayan glaciers and in the near future, may raise several questions such as what are the climate implications of extensive warming by elevated absorbing aerosols in most the regions of the world in particular the Himalayan regions? However, BC-induced reduction of snow albedo and its contribution to glacier retreat has only started to receive attention.

Over the Indian region, the BC contribution in AOD is significant (i.e. 10–20%) with notable lower atmospheric heating. The significant absorbing aerosol load over the northwestern part of the Indian Himalaya results in notable surface cooling associated with strong solar heating (0.41–0.73 K day^{-1}), which may be capable of affecting the atmospheric dynamics. In most of the studies, the effect of surface reflectance in atmospheric warming is overlooked. Earlier estimates have shown that the BC mass concentration over India is of the order of 10–20 $\mu g\, m^{-3}$, which is quite large.

The mean aerosol concentration over the TP is significantly low and overall comparable to other remote areas, and the anthropogenic aerosol content is extremely low. Dust aerosols are the most dominant aerosol type over the TP, and other types of aerosols have less impact on the atmospheric environment over TP. Since 2006, surface dust events have been frequently observed on the TP, and the dust over TP mainly originates from the rise of sand and transport on the surface of the Taklamakan Desert, and accumulates on the northern slopes of the TP. AOD increases gradually from the southern to the northern slopes of the TP, with the smallest AOD (0.02) in the southern TP and the largest AOD (0.62) at the northern edge of the TP. The higher AOD values on the northern TP correspond to dust aerosols, while the aerosol types on the southern and eastern TP mainly correspond to anthropogenic aerosols. Daytime and nighttime AOD distributions on the TP are similar, but daytime AOD values are higher than those at nighttime.

Aerosols perturb the radiative balance of the atmosphere through scattering and absorption of solar radiation, reduction of snow albedo, and alteration of cloud properties. The radiative impact of Arctic aerosols is modulated by the large seasonal amplitude in available radiation, which peaks in summer (Quinn et al. 2008), and in cloud coverage, which peaks in summer to autumn (Shupe et al. 2011). Aerosol abundance at remote Arctic stations also shows a distinct seasonal cycle, with a maximum in late winter and early spring knew as "Arctic haze" and a minimum in summer. Aircraft measurements revealed that the aerosol seasonal cycle extends into the Arctic middle to upper troposphere (Fisher et al. 2011; Wang et al. 2011). The spring peak in aerosol loading arises from efficient pollution transport along isentropic surfaces from the midlatitudes to the Arctic.

Observations of sulfate and BC at remote Arctic stations have shown rapid decreases of 1–3% yr^{-1} since the 1980s. Consistent with these trends, ice cores in Greenland and Svalbard Island indicate that sulfate deposition peaked in the 1970s, followed by a decline. At Barrow, Alaska, observed background AOD decreased from 1977 to 2002 but increased slightly between 2002 and 2010. Many studies have attributed the observed decreases in sulfate and BC in the Arctic to reductions in anthropogenic emissions of primary aerosol and aerosol precursors in Russia and West Eurasia. Most Arctic surface and ice core observations show decreases in sulfate and BC concentrations of 2–3% yr^{-1}, consistent with estimated emissions reductions in Eurasia and North America. Observed springtime (April–June) Arctic AOD at Resolute Bay, Canada, decreased over 1975–2010, but the trend is relatively weak (−0.8% yr^{-1}). The weakness of this trend can be traced to the competing influences of decreasing aerosol load in the lower troposphere, driven by emissions trends at high northern latitudes, and increasing aerosol in the middle to upper troposphere, driven by trends in China and other developing countries at lower latitudes. Anthropogenic aerosol yields a negative forcing over the Arctic. The 1980–2010 emission reductions in anthropogenic aerosols in the developed world may have contributed $+0.27 \pm 0.04$ K warming to present-day Arctic temperatures at the surface or approximately 25% of the observed Arctic warming. About two-thirds of the warming can be attributed to forcing from anthropogenic aerosols over the Arctic, with the rest to forcing from midlatitudes.

Cloud cover plays an important role in climatic systems, having a significant effect on the radiation budget and corresponding water cycles. The main characteristics and physical processes of clouds, such as various macrophysical effects (e.g. CF, cloud thickness) and microphysical properties (e.g. cloud droplet number concentration, particle size, phase), as well as the complicated dynamical and microphysical processes, significantly impact on the cloud radiation effect. The key parameters of the cloud are as follows:

Cloud cover (also known as CF or cloud amount) refers to the fraction of the sky obscured by clouds on average when observed from a particular location. The cloud cover is correlated to the sunshine duration as the least cloudy locales are the sunniest ones while the cloudiest areas are the least sunny places, as clouds can block sunlight, especially on sunrise and sunset when sunlight is already limited. Clouds play multiple critical roles in the climate system and diurnal cycle. In particular, being bright objects in the visible part of the solar spectrum, they efficiently reflect light to space and thus contribute to the cooling of the planet, as well as trapping remaining heat at night. Cloud cover thus plays an important role in the energetic balance of the atmosphere and a variation of it is a factor and consequence of and to the climate change expected by recent studies. A CF of one means the pixel is completely covered with clouds, while a CF of zero represents a totally cloud-free pixel. CF is important for the modeling of downward radiation.

A worthy note phenomenon in the TP is that it is often observed at the surface that the global solar radiation reaching the surface exceeds the solar constant(1353 ± 21 W/m^2). This phenomenon is usually explained by the component of the radiation that has been reflected and refracted many times into the clouds, which is appended to the downward shortwave radiation. With the increase of observational methods and data, the explanation of this phenomenon will become more detailed and in-depth.

The LWC is the measure of the mass of the water in a cloud in a specified amount of dry air. The classification of the cloud is highly related to the liquid water content as well as the origin of the cloud. Different cloud types have different effects on radiation.

$$LWC = \frac{4\pi}{3}\rho_w \int a^3 n(a)\,da \tag{2.2}$$

where ρ_w is the density of water, a is water drop radius.

Liquid water path (LWP) is an important quantity in understanding radiative transfer in the atmosphere. It is defined as the integral of liquid water content between two points in the atmosphere. The LWP also contributes to important cloud properties. As the value of the LWP increases, so does the albedo of the cloud. This increase in albedo is seen most quickly at the lower end of the LWP spectrum, that is, the smaller the amount of total water, the quicker the albedo increases. The radiative absorption of clouds is also dependent on the LWP. An increase of LWP leads to an increase in absorption. Again, the largest increase is seen with lower levels of LWP. These connections are due to the proportionality between the LWP and the optical depth of the cloud.

$$LWP = \frac{4\pi}{3}\rho_w \int_0^z \int a^3 n(a)\,da\,dz \tag{2.3}$$

The ice water content (IWC) and ice water path (IWP) are same as LWC and LWP, but the measure of the mass of the ice in a cloud.

$$IWC = \rho_i \int V n(L)\,dL \tag{2.4}$$

$$IWP = \rho_i \int_0^z \int V n(L)\,dL\,dz \tag{2.5}$$

where ρ_i is density of water, L is the maximum size of ice.

The cloud drops' effective radius (alternatively effective cloud radius or simply effective radius when in context) is a weighted mean of the size distribution of cloud droplets. This is redefined as the ratio of the third to the second moment of a droplet size distribution to aid in the inversion of remotely sensed data. Physically, it is an area-weighted radius of the cloud drop particles. Mathematically, this can be expressed as:

$$\mathrm{Re} = \frac{\int \pi r^3 (nr)\,dr}{\int \pi r^2 (nr)\,dr} \tag{2.6}$$

The cloud optical depth (thickness) is a key parameter in the study of cloud optical properties, describing the extinction effect of clouds, defined as:

$$COD = \Delta Z \int Q_e \pi a^2 n(a)\,da \tag{2.7}$$

where Q_e is the extinction efficiency factor, which is a function of cloud droplet radius, wavelength, and refractive index. ΔZ is cloud thickness. $n(a)$ is concentration of the number of particles with radius a in a unit volume.

Cloud cover plays an important role in climatic systems, having a significant effect on the radiation budget. Clouds are a very important component of the cold regions' climate system. Clouds strongly modulate the Earth's radiation balance. Clouds trap outgoing long-wave radiation, inducing a warming effect, and also reflect incoming solar radiation back to space, leading to a cooling effect. Meanwhile, clouds can affect atmospheric thermal and dynamic states through radiative heating and latent heat release. To effectively measure the bulk radiative role of clouds in the atmosphere–surface system, the concept of cloud radiative effect is widely in cloud-radiation feedback, climate model evaluation and climatic change research (Ramanathan 1987; Stephens 2005; Trenberth et al. 2009).

The distribution characteristics of cloud properties over the TP include six cloud characteristic parameters: CF, cloud optical thickness, cloud height, cloud particle radius, IWP, and LWP. The high value of CF is mainly distributed in the southeastern part of the TP, and the low value is located in the Qaidam Basin region and a small part of the central and western part of the plateau. Compared with the surrounding areas, the plateau CF is a distinctly high-value area. The maximum value of the CF can reach 100%, and the minimum value is as low as 30%. Except for a portion of the high-value area on the eastern and southern edges of the plateau, the cloud optical thickness is more uniformly distributed over the plateau, remaining between 2 and 5. Similar to the CF, the cloud height on the TP is also an obvious high-value area compared with the surrounding areas, and the cloud height gradually increases from the edge of the plateau to the center, up to 10 km. In addition, the cloud particle radius is distributed more uniformly on the plateau, and the cloud particle radius of the main plateau is mainly concentrated around 39–42 μm, and the high-value area is concentrated in the northeastern part of the plateau, with the maximum value reaching 45 μm. The distribution of LWP and IWP is similar, with the high-value areas concentrated in the eastern part of the plateau and a clear high-value boundary separating the main part of the plateau from the surrounding areas at the southern edge of the plateau, with the maximum values of LWP and IWP being 90 and 280 $g\,m^{-2}$, respectively. The cloud amount over the TP has a large value between 7 and 11 km from January to April, and the largest value (more than 44%, 9–10 km) occurs in the last pentad of March. During this period, the cloud amount has only one peak vertically (Wang et al. 2011). Since mid-May, the large value signal moves to lower altitudes (5–8 km). Meanwhile, the cloud amount (8–20%) among 11–17 km increases, mainly due to the development of deep convections in eastern TP and leading to vertically double peaks in summer. After September, the cloud amount reverts to a single peak (8–28%, 5–9 km).

The ground to 700 hPa height corresponds to low clouds, 700–500 hPa height to low medium clouds, 500–300 hPa height to high medium clouds, and above 300 hPa to high clouds. The annual mean low-CF is minimal in both warm and cold seasons due to the influence of topography height of TP. In the warm season during 2000–2015, excluding a small area in the central and western part of the TP, the medium-CF shows high values over the main part of the TP. The distribution pattern of medium-CF in the cold season is similar to that of the warm season. Medium clouds are mainly distributed on the edge and the central-eastern part of the TP, with maximum values up to 90%. The distribution of high clouds in the warm and cold seasons differs greatly on the TP. The center of high CF in the warm season is located above the southern TP, while in the cold season it moves to the central-eastern TP. In the warm season, high clouds cover most of the main part of the TP, except for a small part of the western TP. The annual mean CF gradually increases from north to south, reaching a maximum of 44%. The high CF in the cool season is much lower than that in the warm season, with a large value zone in the central-eastern TP, with a maximum value of 20%. In general, medium clouds dominate on the TP in both warm and cool seasons.

Compared to the water cloud, ice-CF is larger on the TP. The ice-CF high value area is mainly distributed in the central TP. The distribution of the two is in opposite form: the water-CF decreases gradually from the edge of the TP to the center, while the ice-CF increases gradually from the edge of the TP to the center of the TP. Compared with the peripheral regions of the TP, the water-CF is the obvious low-value center, while the ice-CF is the obvious high-value center. Similar to aerosols, diurnal variation exists for clouds over the TP. For water clouds, the CF of water clouds on the TP is significantly higher during night than during the day. Compared with the surrounding areas, the CF of water clouds during daytime and nighttime is a low-value center, and the high-value area is located at the edge of the TP, with a maximum value of 12%. For ice clouds, the high-value center of the daytime ice-CF is located in the central TP, and the high-value center of the nighttime ice-CF moves eastward to the eastern TP. Meanwhile, the ice-CF is higher during the night than during the day in the western TP. In contrast to the distribution of water clouds, the daytime and nighttime ice-CFs are a clear center of high values relative to the surrounding areas of the TP.

Both ice clouds and water clouds occur mainly at the edge of the TP, and their frequency decreases from the edge of the TP to the center of the TP, with the highest frequency of cloud occurrence reaching 100%. Compared with water clouds, ice clouds occur more frequently on the TP, especially on the northern and western slopes of the TP, and the high value of IWP is located in the northern and western TP, where ice clouds dominate. Dust aerosols are mainly distributed on the northern slopes of the TP, while anthropogenic aerosols are mainly distributed on the southern slopes of the TP. Sand and dust aerosols, which are the main ice nuclei, can promote the generation of ice clouds on the northern slopes of the TP. When the dynamical conditions are sufficient, the upward motion of condensation drives the warm humid air to higher altitudes, which leads to an increase in upper-level clouds. Therefore, the northern slopes of the TP are more favorable for ice-cloud formation than the southern slopes. Ice clouds occur much more frequently during the day than during the night. During the day, the ground temperature and sensible heat increase with the increase of solar radiation from the ground. As a result of this heating effect, upward motion is stimulated, which drives moist, warm air to higher altitudes. It is known that the air temperature decreases at a rate of $0.65\,\text{K}\,\text{km}^{-1}$ with increasing altitude. Therefore, the higher the warm and humid airlifted from the ground, the more ice clouds are formed accordingly. On the contrary, the absence of solar radiation and the presence of upward thermal radiation at night both cool the surface and weaken the lifting motion compared to the day, which is not conducive to the formation of ice clouds at night. The diurnal difference in radiative heating is one of the main reasons for the large difference in the frequency of daytime and nighttime ice clouds on the TP. The monthly variation of ice clouds during the day and the night is similar, with the highest frequency in winter and the lowest frequency in summer. During the day, the frequency of ice clouds is highest in January (88%) and lowest in July (79%). Similarly, the frequency of ice clouds at night is highest in January (72%) and lowest in July (62%). The center of the low frequency of ice clouds is located in the central region of the TP, while the center of the high frequency is located at the edge of the TP, both during the day and during the night. Similar to the diurnal variation of ice clouds, there is also a diurnal variation in the frequency of water clouds on the TP. Water clouds occur more frequently at night than during the day. Similar to ice clouds, the diurnal variation of water clouds is mainly due to the diurnal variation of radiative heating. During the day, due to radiative heating, the lifting motion is strong and the warm humid air is lifted to a higher altitude, forming more ice clouds than water clouds. On the contrary, at night, weakened radiative heating leads to weaker lifting motion, and warm humid air is lifted to lower heights, so that water clouds with lower heights are more likely to form. The monthly variation of daytime and nighttime water clouds is similar, with the highest frequency occurring in summer and the lowest frequency in winter. During the day, the frequency of water clouds is highest in July (90%) and lowest in February (72%). At night, the frequency of water clouds was highest in August (92%) and lowest in February (78%). The frequency of water clouds is higher at night than during the day. The frequency of water clouds is low in the central region of the TP, while it is highest at the edge of the TP.

Liquid-containing clouds are present over the Arctic Ocean and North Atlantic in all seasons. These liquid-containing Arctic clouds exist in an environment with temperatures that are frequently well below 0 °C during fall (September, October, November), Winter (December, January, February), and spring (March, April, May). The largest liquid-containing CFs occur during SON. Even during DJF, the season with the least supercooled liquid-containing Arctic clouds, liquid-containing Arctic CFs exceed 0.2 over the entire ocean-covered Arctic domain. DJF and MAM have relatively large liquid-containing CFs (>0.5) associated with the semi-permanent Aleutian and Icelandic lows, and relatively small liquid-containing CFs over the central Arctic Ocean (0.2–0.4). The contrast between the semi-permanent lows and the central Arctic Ocean is reduced during summer (JJA) and fall (SON) when large liquid-containing CFs (>0.5) are present over all high northern latitude ocean areas. Liquid is only present above the homogeneous freezing temperature (~−42 °C) and ice is only present below the melting temperature (0 °C).

Over the high northern latitude oceans, most liquid-containing clouds occur below 3 km and at temperatures between −25 and 0 °C (de Boer et al. 2011; Shupe 2011). Temperature is only one of many factors controlling the presence of supercooled liquid in Arctic clouds, and prescribing temperature-dependent cloud phase in numerical models is not appropriate. Liquid-containing cloud occurs over the largest height range during summer (June, July, August), from the surface to 7 km. During the transitional seasons (spring and summer), the most common liquid-containing clouds occur at a higher height and a lower temperature than during JJA. Over the high northern latitude oceans, ice clouds occur above 4 km and at temperatures between −30 and −60 °C. Many Arctic clouds are liquid-containing clouds that precipitate snow.

The annual average CFs derived from radar-lidar and ceilometer measurements are 0.78 and 0.75, respectively. In general, CF increases significantly from March to May (0.57 → 0.84), remains relatively high (~0.80–0.9) from May to October, and then decreases from November to the following March (0.8 → 0.57). There are two distinguishable periods with low CF (~0.6) from January to April and high CF (~0.9) from August to October, respectively. The CFs in other months basically fall within these two periods with a range of CF from 0.7 to 0.8.

The CRF (cloud radiative forcing) can be thought of as the radiative effect of clouds on the Earth's radiation budget and quantified by comparing the surface radiative fluxes under all skies with clear skies. The LW, SW, and all-wave CRF parameters are given as:

$$LW\,CRF = LW\,net(AS) - LW\,net(CS) \tag{2.8}$$

$$SW\,CRF = SW\,net(AS) - SW\,net(CS) \tag{2.9}$$

$$TOT\,CRF = LW\,CRF + LW\,CRF \tag{2.10}$$

where AS and CS (clear sky) represent all and clear skies, respectively. The LW CRF is typically positive because clouds absorb and emit LW radiation efficiently in the clear atmosphere, and the SW CRF is typically negative as clouds attenuate downwelling SW radiation via scattering and absorption. A positive all-wave CRF indicates that the NR at the surface is enhanced with the presence of clouds and vice versa.

The total CRF on the ground in the western part of the TP is positive with a maximum value of 67 Wm^{-2}, but the total CRF on the ground in the eastern part of the plateau is negative with a minimum value of −95 Wm^{-2}. Positive and negative values of CRF indicate heating and cooling effects, respectively. Therefore, clouds are a warming effect in the western part of the TP, but a cooling effect in the eastern part of the TP. In other words, the cloud radiation effect that dominates the western part of the TP is a long-wave greenhouse effect, while the eastern part of the TP is dominated by a shortwave cooling effect. Total CRF is negative (minimum value of −124 Wm^{-2}) over most of the TP during the warm season of 2000–2015, except for some areas of the western part of the TP where CRF is positive (maximum value of 50 Wm^{-2}). The total CRF is positive in almost all regions of the main part of the TP during the cold season, with a maximum value of 126 Wm^{-2}. Clearly, clouds exhibit a significant heating effect during the cold season, but a cooling effect during the warm season. The regional average total CRF, shortwave CRF, and long-wave CRF in the cool season are 24.28, −25.64, and 49.87 Wm^{-2}, respectively. The shortwave cooling effect is dominated by low-level clouds, while the long-wave warming effect is dominated by high-level clouds. The significant warming over the TP due to the increase of high clouds and decrease of medium clouds may be attributed to the weakening of albedo effect due to the decrease of medium clouds and the enhancement of cloud greenhouse effect due to the increase of high clouds. With global warming, cloud changes may play an essential role in accelerating the warming of the TP. High cloud variability trends are altitude-dependent, that is they correspond to higher rates of change at higher altitudes. With increasing altitude, both total CRF and long-wave CRF show amplified growth rates, namely the rate of increase is greater at higher altitudes. The radiation effect of clouds gradually increases with increasing altitude.

The LW CRFs are primarily determined by the differences in LW-down fluxes between all- and clear-sky conditions because the differences between all- and clear-sky LW-up fluxes are small relative to their downward counterparts. LW CRFs with an annual average of 30 Wm^{-2}, having a minimum value of ~6 Wm^{-2} in March, increasing monotonically to a maximum of ~63 Wm^{-2} during August. LW CRF decreases through the winter and early spring when clouds tend to be optically thin and CF is at a minimum.

Variations in SW CRFs are approximately the inverse of their LW counterparts and have the greatest negative impact during the snow-free months of July and August. Because there is essentially no incoming SW radiation during November–January, values of SW CRF are zero during these months and are always negative during sunlit months. Values of SW CRF slowly decrease from winter to spring, and reach their maximum negative (~−90 Wm^{-2}) during July–August when the cooling effect of clouds is greatest, and the surface albedo is the lowest. The only difference in calculating global and total SW CRFs is the global and total SW-down fluxes. The more negative SW CRF derived from the US NOAA (National Oceanic and Atmospheric Administration) total flux measurement during June is due to lower total SW-down flux (~7 Wm^{-2}) than NOAA global SW-down flux. The annual average SW CRF is approximately −27 Wm^{-2} on the basis of the combined atmospheric radiation measurement and Barrow Observatory analysis.

The monthly variations of net CRFs, the sum of SW and LW CRFs, generally follow the cycle of SW CRFs but have values modified by LW CRFs. On annual average, the negative SW CRFs and positive LW CRFs nearly cancel. The annual average net CRF is about 3.5 Wm^{-2}.

LW CRFs increase and SW CRFs decrease with increasing CF. CRF represents the bulk effects of the seasonal variations of cloud and surface parameters, that is, the surface CRF does not solely represent the radiative effect of clouds, but rather the mixed effects of changes of cloud and surface (atmosphere) properties.

The radiative effect of changes in cloud properties because of the surface (atmosphere) is generally less critical in environments where the variations of the clear-sky fluxes are relatively small compared to the cloud-induced changes. However, the radiative effect due to the seasonal variations of surface parameters, such as surface albedo, in the Arctic are nearly as large as the cloud-induced radiative effect. In other words, the annual CRF would vary significantly even if the cloud properties were constant over the annual cycle in the Arctic. LW CRF increases with increasing CF. CF plays an important role in determining LW CRF at both middle latitudes and Arctic regions. LW CRF has a strong positive correlation (0.90) with cloud radiating temperature and a moderate negative correlation (−0.68) with cloud-base height. These relationships make physical sense because the all-sky LW-down flux is proportional to the cloud radiating temperature, while the cloud radiating temperature decreases with increasing cloud-base height. LW CRF linearly increases with increasing cloud LWP with a strong positive correlation of 0.83. LWP should have no further direct impact on LW radiation for LWP greater than 30–50 g^{-2} because clouds essentially emit as blackbodies. Higher LWPs are due to related increases in the cloud radiating temperature as the LWP increases in the summer. Note that CF increases from 0.52 to 0.89, while LW CRF increases from 6 to 63 Wm^{-2} from March to August. LW CRF represents the bulk integrated effects of the seasonal variations of cloud, surface, and atmospheric parameters, that is, the LW CRF at the surface does not only change with CF only but also other parameters as well. LW CRF increases exponentially with increased CF and cloud radiating temperature and linearly with cloud LWP. In addition to the seasonal variation in cloud properties themselves, there are also significant differences in aerosol loading, atmospheric profiles of temperature and water vapor, and surface background between spring and summer. The sensitivities of SW CRFs to cloud and surface properties are complicated. A strong positive correlation (0.87) between SW CRF and surface albedo exists because SW CRF values are slightly negative for high surface albedo and largely negative for low high surface albedo.

2.5 Precipitation and Snow

In terms of climatological mean, distinguished from rainfall in the tropical regions, which are dominated by the monsoons, precipitation in high latitudes of cold region migrates northward and southward with the seasons (Figure 2.12), following the solar radiation, compared with the most equatorial regions where rain falls all year round. The belts that lie 10–20° away from the equator

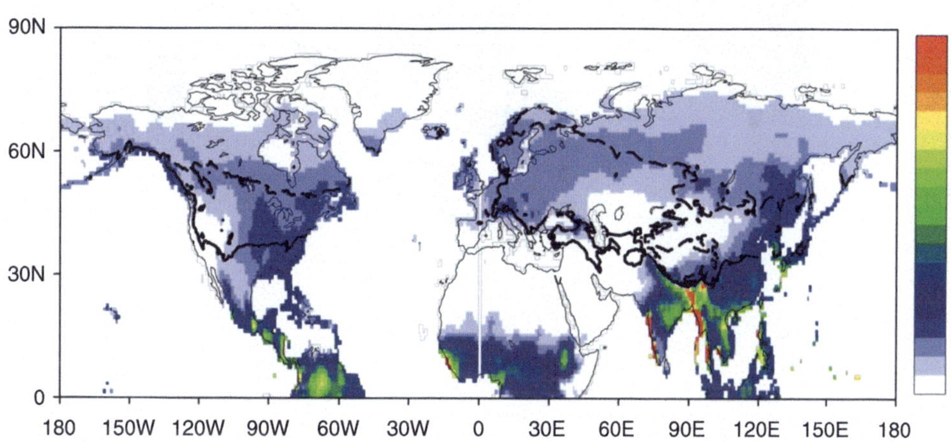

Figure 2.12 Distribution of summer (June, July, August) precipitation (mm day^{-1}) over the northern hemisphere.

experience pronounced dry seasons that correspond to the time of year when the sun is overhead in the opposing hemisphere, this decides the seasonal variations of precipitation in cold regions. The rainy season in India and Southeast Asia coincides with the time of year in which the surface winds over the northern Indian Ocean blow from the west. Analogous relationships exist between wind and rainfall in Africa and the Americas. The onset of the rainy season, a cause for celebration in many agricultural regions of the subtropics, is remarkably regular from one year to the next and it is often quite dramatic. Compared to the precipitation in the tropics and subtropics, precipitation in cold regions has its unique characteristics. The flow of warm humid air around the western flanks of the subtropical anticyclones brings copious summer rainfall to eastern China and Japan and the eastern United States. In contrast, Europe and western North America and temperate regions of the southern hemisphere experience dry summers. These regions derive most of their annual precipitation from winter extratropical cyclones that form within the belts of westerly surface winds over the oceans and propagate eastward over land.

Pure snow cover is the whitest natural surface on Earth, and due to its light color, it reflects about 90% of sunlight that reaches it. A snow-free ground surface absorbs four to six times the amount of solar radiation than a snow-covered ground surface. Beyond its cooling impact on global climate, snow also affects soil moisture and surface runoff – including critical drinking and irrigation water in dry areas.

Nearly all of Earth's seasonal snow cover is confined to the northern hemisphere (Figure 2.13). Distribution of winter (December, January, February) snow depth over the northern hemisphere

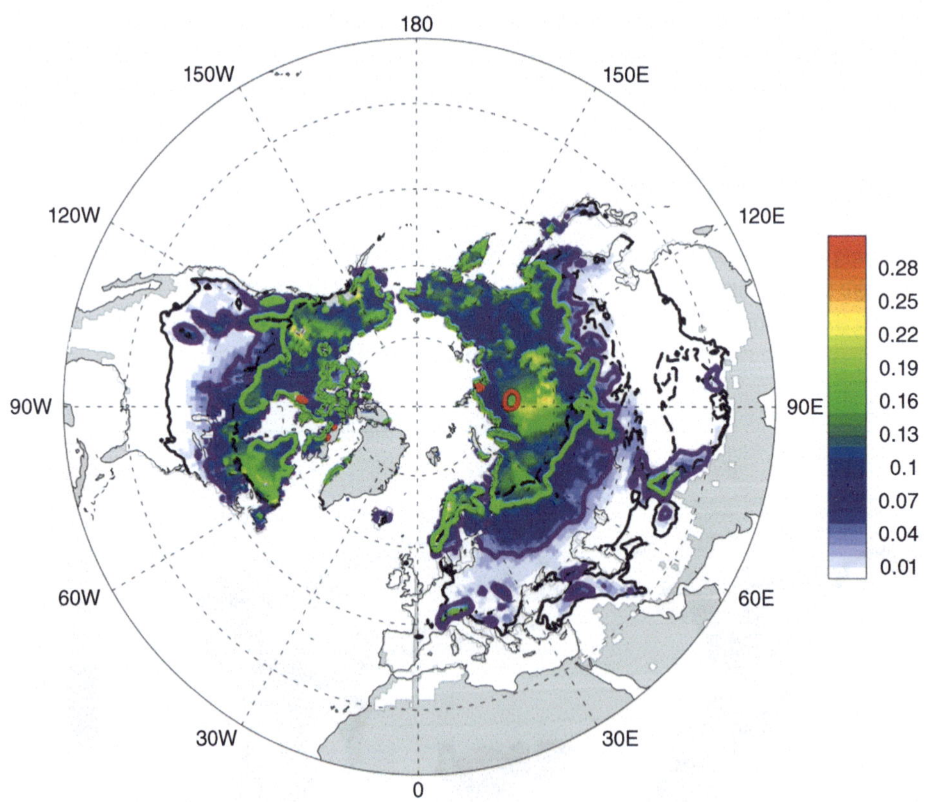

Figure 2.13 Distribution of winter (DJF) snow depth (unit: m) over northern hemisphere, the purple, blue, and green line represent the snow boundary line (snow depth >0.01 m) in spring (MAM), summer (JJA), and autumn (SON), respectively.

indicate that the snow cover on land (some snow cover occurs over sea ice) averaged roughly 24.9 million km^2 (1967–2012 average) in the northern hemisphere. Unless snow cover is thin, its overall effect is to insulate soil and keep soil water from freezing. Over the past 20 years, the snow area extent has decreased in the northern hemisphere, leaving more ground exposed to the cooling effects of air temperatures.

2.6 Frozen Ground

Fifty-five to sixty percent of the exposed land surface in the northern hemisphere freezes and thaws seasonally, with a tremendous impact on ecosystem diversity and productivity. Freezing and thawing cause changes in soil structure, affect the cycle of water on and below the land surface, and regulate the availability of nutrients in the soil for plants and biota that depend upon them.

Permafrost is defined as ground (soil or rock and including ice or organic material) that remains at or below 0 °C for at least two consecutive years. Seasonally frozen ground refers to those areas where soil is frozen for 15 days or more per year, while intermittently frozen ground refers to areas where the soil is frozen for fewer than 15 days per year. The distribution of permafrost and the average maximum extent of seasonally and intermittently frozen ground in the northern hemisphere indicate that, permafrost occupies about 22.79 million km^2 or 23.9% of the exposed land surface. On average, the maximum extent of seasonally frozen ground is about 55 million km^2 or 55% of the total land area, while the maximum snow cover extent is about 47 million km^2 or about 47% of the total land surface in the northern hemisphere (Figure 2.14).

Even a thin frozen layer can essentially halt the moisture exchange between the atmosphere and soils. Frozen soil possesses a large apparent heat capacity, consuming or releasing about 80 times the amount of energy in freezing or thawing than is consumed or released by water during heating and cooling. Additionally, the processes of freezing and thawing soil can affect the decomposition of organic substances in the soil as well as impact the migratory patterns and physiology of biota

Figure 2.14 Distribution of permafrost and extent of permafrost and seasonally frozen ground from observation (Zhang et al. 1999, 2003; which is obtained from NASA Earth Observatory, https://earthobservatory.nasa.gov/Features/FrozenSoils).

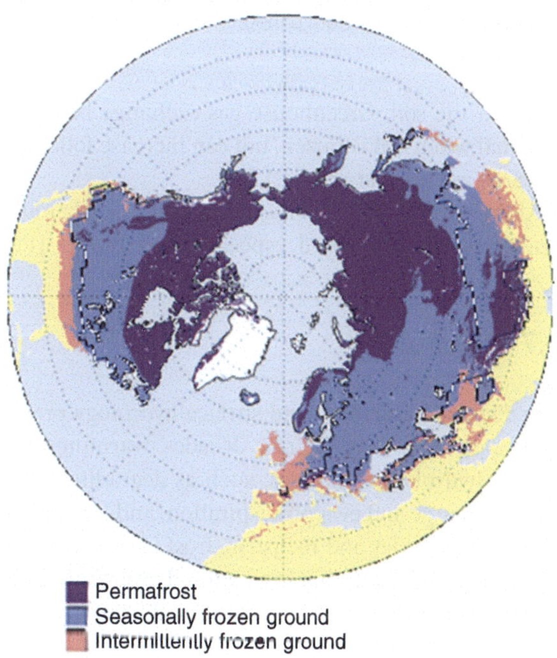

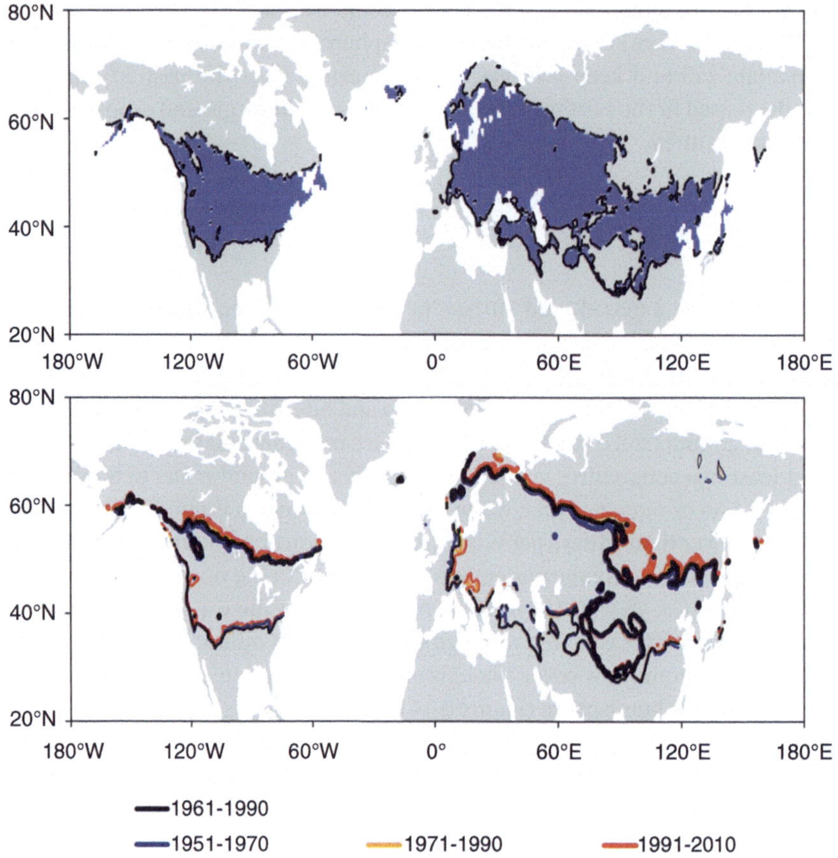

Figure 2.15 Distribution of climatological (1961–1990) seasonally frozen ground (SFG) (top panel). Variations of SFG boundary, the position of annual average zero isotherms (thick line) and January zero isotherm (thin line) (bottom panel).

living in the soil. Greenhouse gas exchange between the atmosphere and land surface may be minimal when soil is frozen, but can increase following spring thaw in a dramatic release noted in research as "respiratory burst."

Along with global warming, the positions of annual average zero isotherms and January zero isotherms shift northward, especially over Europe (Figure 2.15).

2.7 Hydrology

The distinct features of land-surface hydrology processes in clod region are the soil freeze–thaw cycle and snow. Over the past decades, warming in cold regions has affected changes in the terrestrial hydrologic processes, such as degrading permafrost, decreasing snow extent, increasing river discharge and evapotranspiration, and increasing streamflow temperature. The active layer's freeze–thaw cycles and the permafrost layer's aquiclude effect control the land surface hydrology in cold regions by altering seasonal soil moisture, water storage, evaporation, and water movement through soil and vegetation (Zhang et al. 2003).

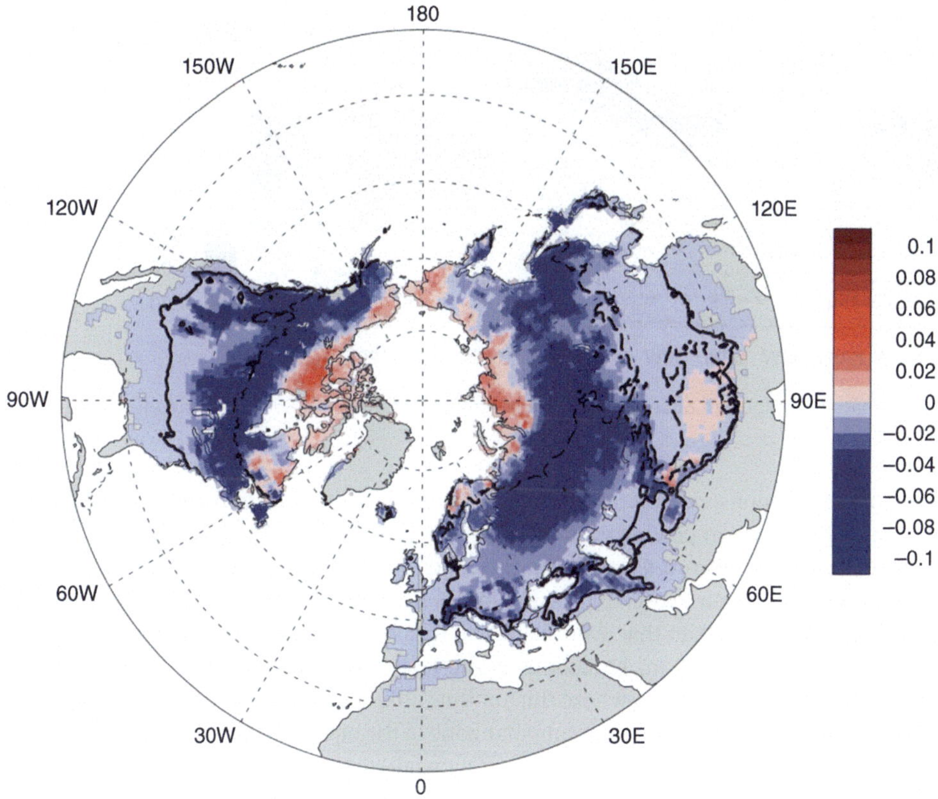

Figure 2.16 Distribution of snow melting in spring (MAM snow depth minus DJF snow depth; unit: m), the negative values mean the snow-melting and positive values mean snowfall.

The distribution of snow melting in spring (winter snow depth minus spring snow depth; Figure 2.16) indicates that, except for the circum-Arctic region, there is obvious snow melting in the cold region in the northern hemisphere, especially in North America, Western Europe and Northern Asia. A portion of the melting snow forms the surface runoff, and the other part penetrates into the soil, increasing the soil moisture. The snow melting mainly occurs over the circum-Arctic region (i.e. Canada, Siberia) from the winter to spring, while the surface snow increases in polar regions in spring.

The freeze–thaw process and snow in cold regions change the soil-moisture content, affecting the surface evaporation, the partitioning of the incoming energy in the latent and sensible heat fluxes and other surface hydrology processes.

2.8 Winter Monsoon

Along with the alteration of sea-land thermal difference, the atmospheric circulation changes from the summer to the winter pattern. All winds blow as a result of pressure imbalances between two locations. In the case of monsoons, this pressure imbalance is created when temperatures across vast landmasses such as India and Asia, are significantly warmer or cooler than those over

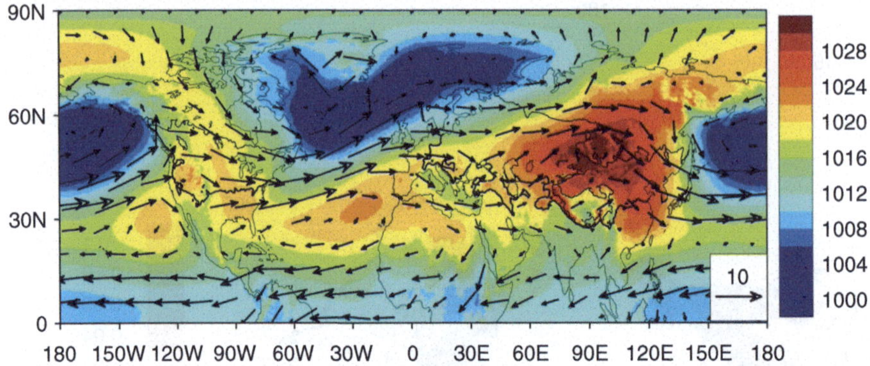

Figure 2.17 Spatial distribution of sea-level pressure color; unit: hPa and 850 hPa horizontal wind (vector; unit: $m\,s^{-1}$).

neighboring oceans. A monsoon is actually a seasonal shift in wind direction and pressure distribution that causes a change in precipitation.

In the colder months, winds reverse and blow in a land-to-ocean circulation. As the land masses cool faster than the oceans, an excess in pressure builds over the continents, causing the air over land to have higher pressure than that over the ocean. As a result, air over the land flows into the ocean. One of the most significant monsoons is the East Asian winter monsoon (EAWM), a distinct meteorological phenomenon whose fluctuations influence the weather and climate in the Eurasian and Asia-Pacific region from the subpolar zone to the equator (Figure 2.17).

The EAWM can significantly affect precipitation (snowfall) and near-surface air temperature in winter. On the interannual time scale, the EAWM variability features two major modes: the strength and pathway of the EAWM, which are linked to the changes in the amplitude and phase of atmospheric planetary waves respectively. The blocking activity in the Ural and Pacific regions significantly impacts the strength of the EAWM in the form of Eurasian and Western Pacific patterns. The former also contributes mainly to the recent decadal amplification of the EAWM. ENSO and Arctic sea ice are a crucial external force for the EAWM variability, while the impacts of ENSO on the EAWM vary with decades. On the intra-seasonal time scale, the SH shows clear out-of-phase relationship between early and peak winter for the past 30 years. The stratospheric variability and the Madden-Julian Oscillation could influence the characteristics of cold surges over East Asia. All these provide intra-seasonal prediction potentials for the EAWM.

The interannual variability of the EAWM can be described by the major modes of the circulation or the near-surface temperature. The spatial distribution of sea-level pressure and 850 hPa horizontal wind indicate that, near the surface, it is characterized by an intense Siberian High with the Aleutian Low to its east and the Maritime Continent low pressure to its south. The resultant strong northwesterly and northeasterly winds are observed to the north and south of about 30°N, respectively.

The EAWM greatly influences the climate (i.e. near-surface temperature and precipitation) of cold regions over the northern hemisphere. The changes of temperature modulate the soil hydro-thermal regime. The changes in rainfall or snowfall affect the soil water content and snow distribution.

3

Land-Surface Process in Cold Regions

In cold regions, a basic and primary characteristic of land surface is the abundance of frozen ground and snow; with the changes in the reception of solar radiation and longwave radiation at land surface, the freeze–thaw cycle of soil is prominent. Most areas in cold regions are covered by snow. Changes in snow cover or snow depth will influence the land-surface process in cold regions through albedo effects and hydrological effects of snow.

This chapter will discuss the features of land-surface process in cold regions, including soil freeze–thaw process, snow-melting process and their impacts on hydrology, surface diabatic heating at different spatial–temporal scales.

3.1 Land-Surface Characteristics in Cold Regions

3.1.1 Overview

The climate system in cold regions is characterized by its low thermal energy state and intimate couplings between atmosphere and land surface. This makes the cold region a challenging and fascinating region to study. Given the complex nature of the cold regions climate system, where does one start describing it? A basic foundation is needed which is provided in the present chapter through an overview of some basic land-surface characteristics in cold regions, the most typical and representative region is the Arctic region. Below, we will introduce the full work of Serreze and Barry (2005).

The most formal definition of the Arctic is the region north of the Arctic Circle (66.5° N). When defined in this way, the most fundamental characteristic of the Arctic is 24-hour daylight in summer and 24-hour darkness in winter, the number of days with these two extremes increasing with latitude. For the extreme case of the North Pole, the sun is above the horizon for six months between the spring and autumn equinoxes and below the horizon for the remaining six months. In a more usual sense, the Arctic does not have strict geographical boundaries. Arctic conditions (e.g. extreme winter cold or the presence of permafrost) may be found well south of the Arctic Circle. The terms Arctic and North Polar Region are often used interchangeably. The geography of the North Polar Region is in striking contrast to its southern counterpart. The Arctic Ocean occupies most of the area north of latitude 70° N. Apart from the sector between 20° E and 20° W, it is almost entirely surrounded by the landmasses of Eurasia and North America.

The Arctic climate is characterized by long, cold winters and short, cool summers. The Arctic region has diverse climate changes, but experiences extreme changes in insolation during both winter and summer. Some areas are covered with ice (sea ice, glaciers, or snow) year-round, and

Climatology in Cold Regions, First Edition. Chenghai Wang.

the vast majority of areas are covered by snow and ice. Average air temperatures in January range from −40 to 0 °C in Arctic regions, with lower air temperature less than −50 °C in most areas during the winter. The average air temperature in July is between −10 °C and 10 °C, with summer highs in some areas reaching over 30 °C.

The central part of the Arctic region is the Arctic Ocean, surrounded by the continents of Europe, Asia, and America. As a result, the climate of the Arctic region is milder due to the ocean effects. The temperature of the sea water in the Arctic region is basically maintained at −2 °C or higher. Because of this relatively warm water, the Arctic is not the coldest region in the northern hemisphere in winter, and is significantly "warmer" than the Antarctic, despite the huge ice caps floating on the Arctic water. In summer, the cooler waters also keep the Arctic Ocean coastal areas cooler. The Arctic winter, which is mainly controlled by a strong and persistent surface inversion, has surface temperatures below −30 °C, however, at 500 m above the surface, the temperature rises to about −25 °C. At 1500 m, the temperature is about −20 °C, between 500 and 1500 m, there is usually a deep isothermal layer. In summer, the snow on the ground and on the ice surface starts to melt. The ground becomes wet, water evaporates, disturbs the mixing and establishes a ground mixed layer. Stratospheric inversions remain in the mesosphere for 40% of the summer, but they are weak and the frequency of surface inversions over the Arctic basin is 15–20%.

Throughout the year, much of the Arctic Ocean is covered by floating sea ice (Figures 3.1 and 3.2). Both the sea-ice surface and land surface in the Arctic are covered by snow for most of the year. Permanent land ice does not occupy a large area outside of Greenland. Subpolar ice caps and

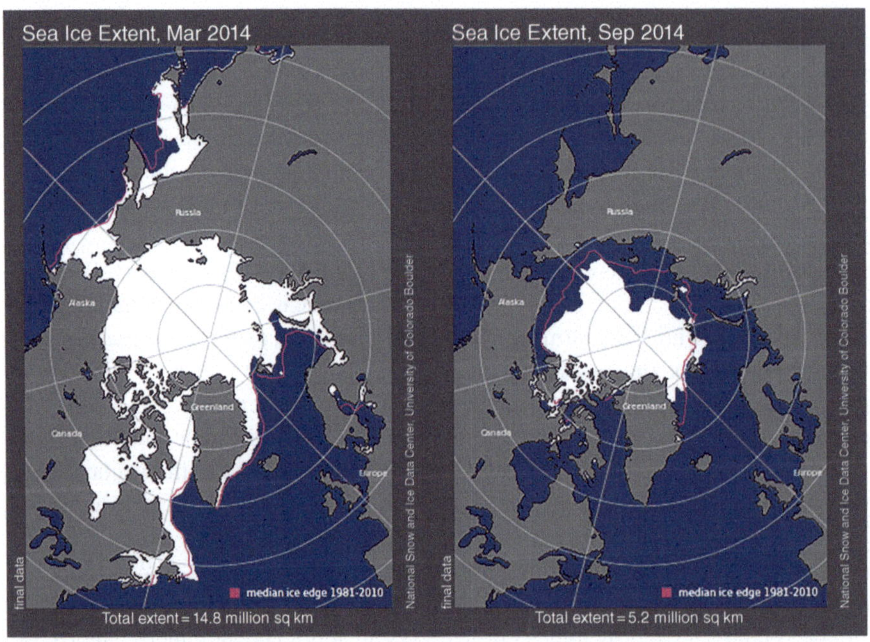

Figure 3.1 Mean sea-ice extent for March (all shaded areas) and September (dark shading), based on data from the Special Sensor Microwave/Imager (SSM/I) over the period 1979–2001. SSM/I has a nominal resolution of 25 km. *Source:* Courtesy of K. Knowles, NSIDC, Boulder, CO.

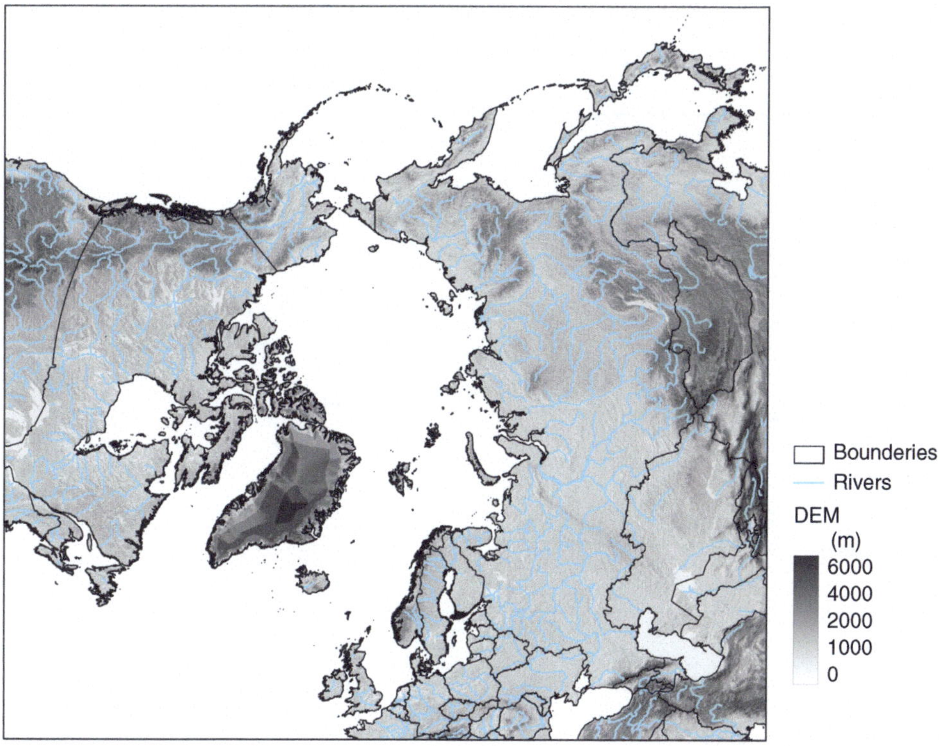

Figure 3.2 Physiography of the Arctic lands, showing topography and major river systems.

glaciers are found mainly in the mountains of the Siberian and Canadian archipelagos, and in Svalbard and Iceland (Table 3.1). However, most of the lands are underlain by permafrost (perennially frozen ground), some of which is rich in ground ice. Based on broad latitudinal, climatic, and biological characteristics, the Arctic is sometimes divided into High Arctic and Low Arctic. The land surface in the High Arctic is often characterized as polar desert (Figure 3.3), an extreme form of tundra. In the Low Arctic, the tundra commonly includes shrub vegetation of birch and willow, with abundant lakes and wetlands.

Most of the Arctic land and sea-ice surface is covered by snow for at least six to eight months of a year (Figure 3.4). Snow cover is a key climatic variable in the Arctic. As we mentioned, snow has high albedo (reflectivity in solar wavelengths), typically 0.80–0.90 for new snow; the insulating effect of snow on the underlying tundra or sea ice is not prone to land–air interaction; snow plays a role in storing precipitation over terrestrial regions, and is released as river discharge during spring and summer. These features strongly impact on the vertical structure of the Arctic Ocean and hence the sea -ce cover.

The impact of latitude is obvious in northern Canada, Alaska, and Asia, where snow is typically present 30–40 weeks per year. Terrestrial snow cover categories have been developed according to climatic and vegetation zones, as well as on the basis of depth, duration, and density of snow. A classification for land that takes account of the stratigraphic and textural attributes of a snowpack is proposed (Sturm et al. 1995). The classification has a number of categories of seasonal snow cover: tundra, taiga, maritime, alpine, prairie, and ephemeral. There is also a special classification for snow cover in mountainous regions. While several of the names follow earlier schemes based

Table 3.1 Major Arctic ice caps (Field 1975; Govorukha 1988; Williams and Ferrringo 2002).

Name/Region	Surface area (10^3 km^3)
Devon Island	16.6
Ellesmere Island	77.2
Axel Heiberg Island	12.6
Barnes (Baffin Island)	6.0
Penny (Baffin Island)	6.0
Iceland	11.3
West Spitsbergen	36.6
Northeast Land	14
Franz Josef Land	13.7
Novaya Zemlya	23.6
Severnaya Zemlya	18.3
Greenland Ice Sheet	1710
Greenland glaciers/ice caps	48.6

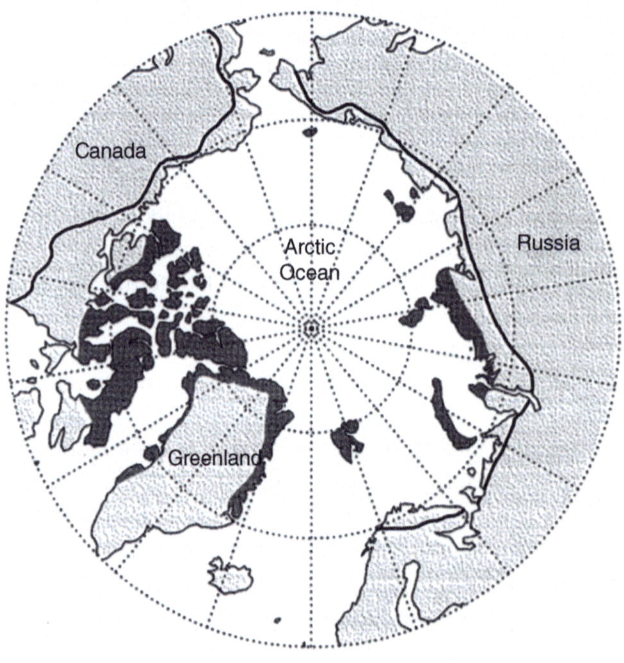

Figure 3.3 Distribution of Arctic polar desert (dark shading) and approximate southern limit of tundra (bold line). *Source:* Courtesy of N. Saliman, NSIDC, Boulder, CO.

on climatic/vegetation zones, these designations do not require such geographical associations; only the relevant physical characteristics are considered (Sturm et al. 1995). The classes were identified on the basis of a statistical analysis of field measurements and climatic data for Alaska. Five of the six classes are found in Alaska. Tundra and taiga snow dominate most of the Arctic. Tundra snow is a thin, wind-blown snow, with maximum depths of about 750 mm, and a bulk density of typically 380 kg m^{-3}. It consists of a basal layer of depth hoar, overlain by multiple wind slabs.

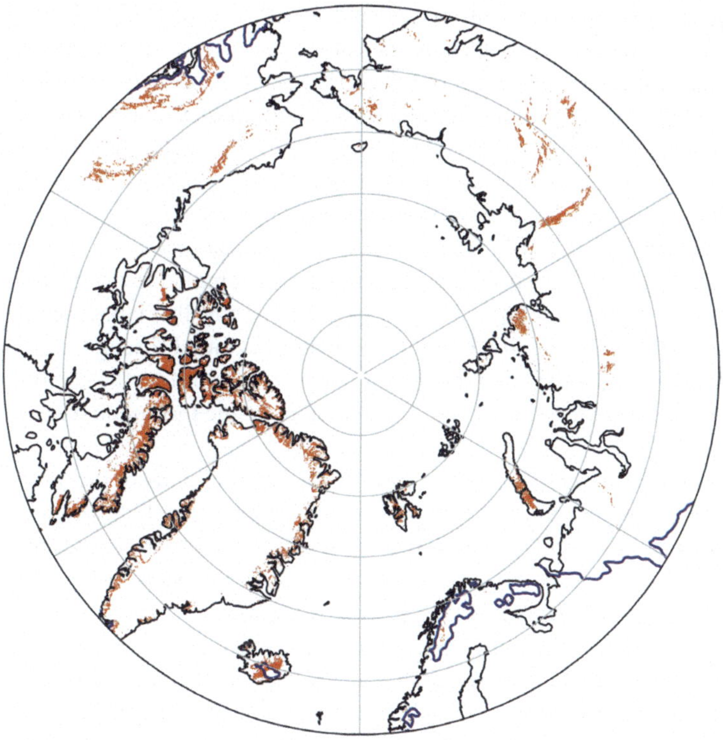

Figure 3.4 Average number of weeks of snow cover in the northern hemisphere, based on the NSIDC blended weekly product for 1972–2001.

Surface sastrugi are common. Sastrugi refers to drifts of wind-packed snow aligned along the prevailing wind direction. Depth hoar represents a low-density (low cohesion or poorly bonded) layer formed by sublimation associated with a strong temperature gradient in the snow. Taiga snow is thin to moderately deep (300–1200 mm) with a lower density than tundra snow (260 kg m^{-3}). In forest climates, wind, initial snow density and air temperature are all low. There are a few terrestrial snow depth datasets for land areas though they have limited spatial coverage and resolution. High Arctic coverage is poor, especially over the Canadian sector. The maximum annual mean snow depth in Eurasia was obtained. Depending on the region, maximum snow depth tends to occur between February and early April. The maximum snow depths range from 300 to 800 mm north of 60° N. Assuming an average snow density is about 300 kg m^{-3} across Eurasia, then it will yield maximum snow water equivalents of about 100–270 mm. At local scale, the snowpack thickness varies considerably in relation to terrain. A seven-year survey in a small basin near Resolute Bay, showed that snow in gullies and valleys has the greatest accumulation while in the least in exposed hilltops, in addition, this basic pattern tends to hold from year to year (Woo et al. 2008). Snow amounts on slopes vary according to exposure to the prevailing wind of locally concave/convex surfaces. The snow melt season in the tundra is typically short, lasting only about two weeks (Weller et al. 1972). When bare ground begins to appear, local heat advection from the warmer exposed surface accelerates the melting of the shrinking snow patches.

The records provide average snow thickness on the predominant sea-ice type in the landing area, in sastrugi behind pressure ridges, and an average for the windward and leeward side of hummocks (Colony et al. 1998). The standard deviation of snow depth is fairly large and naturally increases with mean snow depth. This represents both spatial and year-to-year variability. During April, snow depths range from near zero to over 1000 mm, indicating an existence of drifting and wind scour effects. Since the snow lines are located over flat ice, the standard deviation of snow depth is lower than regional averages. Taking the mean snow density and depth, the average water equivalent during April and May is about 100–110 mm. Locally, the snow depth also depends on the age of the ice; when a lead forms in the ice in winter, new ice growth will take place. The snow then accumulates on the surface of the new ice from snowfall and drifting.

The general definition of the term "snow line" is the boundary which separates areas covered with snow or not. However, the term can have different definitions depending on the context in which it is used, therefore, there exist different types of snow lines. Some snow line types include the permanent snow line (used to describe the line beyond which snow is found throughout the year), the annual snow line (used to define the boundary found in glaciers separating the ablation zone from the accumulation zone), and the regional snow line (used to define the boundary over large areas).

Latitude is one factor that dictates the position of the snow line in different parts of the globe. The position of snow line varies as one moves away from the equator and reaches 18 700 ft above sea level in the Himalayas which lie near the Tropic of Cancer. The effect of latitude eclipses that of altitude in the position of the snow line in the earth's Polar regions. In the polar regions, the snow line is found at the sea level which is evident in the presence of ice caps.

Another factor that affects the position of the snow line is the distance of a region from the coast. Regions that lie near the coast usually receive more snowfall than areas located inland. As a result, the snow line of coastal regions is usually at a lower altitude than that of regions far away from the coast. Ocean currents also affect the altitude at which the snow line is found. Large ocean currents have a warming effect on the surrounding land, which drives the snow line to higher altitudes. The orientation of a slope also affects the position of the snow line, that is, north-facing slopes in the northern hemisphere receive less solar energy, and therefore, the snowline is at a lower altitude than southern-facing slopes.

The snow line can be measured remotely by modern technology such as satellite retrieval, radar detection, and so on. It can be measured in different ways including the use of satellite images, aerial photography, and automatic cameras. These high-tech measurement methods are highly accurate and make it possible to understand the history of snow line under the climate change. The snow line is not fixed and responds to changes in global temperatures. The ongoing climatic change has affected the position of the snow line around the world. Some snow lines have disappeared in the past few decades, which is attributed to increasing global temperatures caused by human activities. Studies indicated that the current snow line is much higher than it was thousands of years ago. It was between 600–1200 m (2000–4000 feet) lower in the last glacial period than it is today.

The average climatic snowline is the elevation above which snow will survive the summer melt season. It is an indicative of both amount of snow and temperature, the latter is especially important in summer. It is around 1200–1400 m in Iceland and 700–900 m in Jan Mayen Island, decreasing northeastward. In Svalbard, it is only 200 m in western coastal regions, rising inland to 800 m (Koryakin 1990).

Perennially frozen ground, also known as permafrost, underlies nearly all of the Arctic land area. Permafrost is generally present whenever ground temperatures are below freezing through two summer seasons. Ground ice need not be present, although in sediments ice may be present either in segregated form throughout, or as lenses and wedges. The upper part of the ground in permafrost regions, the active layer, thaws and freezes seasonally. The active layer depth (the maximum thaw depth) depends on air temperature, vegetation cover, soil material, soil moisture (SM) content and the degree of insulation provided by the winter snow cover. In the Arctic, the active layer depth typically varies from less than 50 cm in the coldest regions to 150 cm or more in warmer southern regions. Active layer depths are thicker on south-facing slopes and in coarse-textured soils, and least in peat bogs.

Zones where permafrost may be present account for 24% of the northern hemisphere land surface. However, the actual area underlain by permafrost is smaller if someone takes account for the fractional coverage of different permafrost zones (continuous, 90–100%; discontinuous, 50–90%; sporadic, 10–50%; isolated, less than 10% of the ground underlain by permafrost). On this basis, 12.8–17.8% of the land surface is underlain by permafrost. The area mapped as continuous permafrost represents 47% of the entire permafrost region of the northern hemisphere. In addition, there is frozen ground below much of the Greenland Ice Sheet and the subpolar ice caps.

The distribution of permafrost in the northern hemisphere compiled by the International Permafrost Association (IPA) (Brown et al. 1997), shows that, permafrost is spatially continuous where the mean annual surface air temperature (SAT) is below approximately −7 °C, and is discontinuous for an SAT of −1 °C to −3 °C. The permafrost thins toward the southern borders, near the −1 °C SAT isotherm (Ives et al. 1974). In the region of continuous permafrost, the thickness exceeds 600 m in northeastern Siberia, northern Canada, and Alaska. Extreme depths of 1200–1470 m are found in the Anabar Plateau of north-central Siberia (72–73° N, 113° E) (Tuniel 2002). This is attributed to ice-free conditions through the Quaternary, low geothermal heat fluxes and high bedrock conductivity. Southward from the zone of continuous permafrost in Siberia and Canada, and also in Scandinavia, there is extensive discontinuous permafrost. The typical thickness is in the order of 50 m in the discontinuous zone, but only a few meters where the distribution is sporadic. Exposed treeless ridges and north-facing slopes are underlined by permafrost whereas it may be absent in forest valleys, especially where there are large lakes. In general, the permafrost thins southward and becomes sporadic. Eventually, it is found only in islands with favorable conditions. Some of these permafrost islands may be relics that are unstable in the present climate.

North of the Arctic coastline, there is a considerable zone of submarine permafrost underlying shallow sections of the continental shelf seas (Vigdorchik 1980). This zone is extensive in the Barents, Kara, Laptev, and East Siberian seas, but is narrower in the Chukchi and Beaufort seas. Most submarine permafrost represents relic strata formed during the glacial intervals of the late Pleistocene, when lower sea levels exposed the shelf areas and mean winter temperatures were well below modern values. As sea levels rose during the subsequent Holocene, the upper ice-rich strata were thermally abraded.

The modern occurrence and characteristics of submarine permafrost are determined by a number of factors, including the water mass structure and vertical circulation, the absorption of solar radiation in summer when the coastal seas are ice free, and the corresponding heat loss to the atmosphere in winter through polynyas and leads, geothermal heat flow, the bottom water temperature, salinity and the freezing of bottom sediments by grounded landfast ice and stamukhi (grounded ridges). Seaward of the submarine permafrost, there may also be a narrow zone of seasonally frozen submarine sediments. In the Beaufort Sea, the permafrost upper boundary varies between the seabed and 100–150 m and its base may lie as deep as 600–900 m. In the Laptev Sea region (Hubberten and Romanovskii 2001), permafrost is continuous to the −65 m isobath, and discontinuous to the −100 to −120 m depth, while the maximum submarine permafrost thickness is about 500 m north of Kotelny Island.

The ice content of permafrost is difficult to determine quantitatively, due to its spatial variability and the limited number of boreholes allowing for analysis. Only about 10% of the terrestrial permafrost area is "ice rich." The IPA map shows that ice-rich permafrost, with an ice content greater than 20% by volume in the upper 10–20 m of the ground, is widespread in the coastal Siberian lowlands of Yamal (around the Ob estuary) and east of the Lena Delta in Yakutia.

Surface icings represent a special component of ground ice. They form when surface water from springs and groundwater seepage freezes, due to persistent low temperatures, or when rivers overflow their banks, sometimes as a result of ice jams. They are most extensive in Siberia, but they also occur in Alaska, mainly located to the east of the Colville River on the Arctic coastal plain (Harden et al. 1992). The total area for all of northern Russia is estimated to be 128 000 km^2 (Kotlyakov et al. 1997). The largest single feature in northeastern Siberia (120 km^2) extends for up to 26 km along the valley of the Momy River.

3.1.2 Sea-Ice Cover

The surface of the Arctic Ocean is characterized by its floating cover of sea ice. Sea ice is intimately coupled with the atmospheric energy budget, the atmospheric circulation, the surface energy budget and the hydrologic budget. Accordingly, some basic aspects of sea-ice processes and Arctic oceanography will be introduced here.

On average, the sea-ice cover in the northern hemisphere lying within the Arctic Ocean, ranges in areal extent from about 15 million km^2 in March to about 8 million km^2 in September. There is large annual variability. Mean ice extent for March and September, defined as regions have an ice concentration (i.e. fractional ice cover) of at least 15%. During winter, ice extent in the Arctic Ocean is largely constrained by the surrounding landmasses and in the Greenland, Norwegian, and Barents seas (hereafter termed the Atlantic sector) by relatively warm ocean waters that discourage ice formation. The sea-ice cover is not a featureless slab, but consists of individual ice floes ranging from a few meters to many kilometers across. Floes are separated by roughly linear openings, known as leads, and irregularly shaped openings known as polynyas. Typical winter-ice concentrations over the central Arctic Ocean are 98–99%, compared to 85–95% during late summer. During winter, leads generally quickly refreeze to form areas of new, thin ice. Ice concentrations decrease near the sea-ice open ocean margins. The typical summer sea ice is located in the central Arctic Ocean based on high-resolution visible-band satellite imagery. As is typical in the Arctic, the sea-ice margin ranges from rather sharp to very diffuse.

Apart from areas of landfast ice along coasts (ice locked to the shoreline), the sea-ice cover is in a state of constant motion. This is due to the influence of wind stress on the upper surface of the sea-ice cover, water stress (ocean currents) on the lower surface of the ice, floe-to-floe interactions, and ocean tilt. Ice motion varies strongly from day to day, largely in response to variations in the wind field. Leads and polynyas in the pack ice are largely the result of differential ice motion. Divergent ice motion acts to reduce the ice concentration, while convergent motion increases it. Convergent motion can force one floe to ride over another, a process known as rafting, producing thicker deformed ice characterized by ridges and corresponding underwater keels. Much of the ice cover consists of deformed ice.

Ice growth and melt are determined by deformation and thermodynamic forcing effects together, the latter occurs at both the top and the bottom of the ice cover. During autumn, the marginal seas and openings in the pack ice refreeze, forming new ice that thickens during the winter to become first-year ice. New ice thickens at a rate of between 3 and 10 cm per day for SATs between about −10 °C and −30 °C (Maykut 1986). However, the rate decreases by order of magnitude as the ice thickens from 10 to 60 cm. Nearly half of the ice area consists of deformed ice that persists through the summer, becoming second-year ice and potentially multiyear ice. Ice thickness across the Arctic varies obviously. Multiyear ice is typically 3–5 m thick. Undeformed first-year ice at the end of winter is generally up to 2 m thick. Areas of deformed ice can exceed 10 m in thickness. In general, the thickest ice is found off the Canadian Arctic Archipelago and northern Greenland (mean values of 6–7 m).

Over the period 1975–2000, annual mean Arctic Ocean sea-ice thickness decreased by 1.25 m (i.e. ~31%), with the maximum thickness in 1980 and the minimum in 2000. The sharpest rate of decline occurred in 1990, with a much slower rate by the end of the record. More recently, ice thickness was relatively constant between 2003 and 2007, while undergoing a substantial decrease between 2007 and 2008. The decline in ice freshwater storage is due to a combination of a loss of ice thickness and a loss of ice area. The estimated loss in thickness is in the order of 30% from 1975 to 2000 (Rothrock et al. 2008). Ice area loss is 2% per decade in winter and 9% in summer. Over the period from 1975 to 2000, the total loss in ice freshwater storage would therefore be in the order of 40%. The largest decline over this period is around 25% suggested by model simulation results. The ensemble mean results for different models is nearly half that or a decline of only around 13%. One potential caveat is that the submarine ice thickness data come only from the central basin, whereas the model includes seasonal areas that may have experienced a lesser decline. It is likely that a continuous decline of freshwater storage happened in the ice. The lengthening melt season will result in continued thinning of the ice and a steady decrease in ice extent. Further, the ice is prone to episodic wind events, such as the Arctic oscillation shift around 1990 that flushed old, thick ice out of the Arctic Ocean. The thinning of the ice has led many to refer to the ice pack as "vulnerable" both to steady warming and episodic events.

Most of the Arctic land area is low lying, especially in western and central Siberia, the western part of the Canadian Arctic Archipelago and coastal Alaska. Mountains and high plateaus are prominent over the eastern Canadian Arctic, coastal regions bordering the Greenland Ice Sheet, interior Alaska, northeastern Siberia and Scandinavia. For example, the highest mountain north of the Arctic Circle is Gunnbjorns Fjaeld (86.92° N, 29.87° W), which rises to 3800 m some 80 km inland from the East Greenland coast in Knud Rasmusen Land. In eastern Baffin Island, there are peaks exceeding 2100 m, but in general the dissected plateaus are around 1500–1800 m elevation. On Devon and Ellesmere islands, ice caps rise to 1900–2000 m. In northern Alaska, the more or less zonally oriented Brooks Range rises to 1500–3000 m. The higher peaks in the more massive sub-Arctic Alaska Range typically exceed 3000 m. The Alaska Range contains Denali (Mount McKinley), the highest mountain in North America, with an elevation of 6194 m. There are numerous mountain ranges in northeastern Siberia. East of the Lena River, there are several north–south mountain ranges. From the Arctic Circle at 140–145° E, the Momsky-Chersky ranges, with peaks rising to 2500–3100 m, stretch southeastward. The Okhotsk-Kolyma Mountain ranges of Chukotka rise to 1800–2100 m.

Of the permanent ice masses (glaciers, ice caps, and ice sheets), the Greenland Ice Sheet is the most prominent and is by far the largest ice mass of the Northern rise of about 7.2 m. There is a central dome rising to 3200 m elevation and a subsidiary dome in southern Greenland. About 40% of the ice margin area is subject to annual surface melt on average. Melting does not occur at the highest elevations under present climate conditions. The dimensions of the other Arctic ice caps are insignificant by comparison, although they exert an influence on the local climate conditions.

Contrast to the Antarctic, there are only a few small ice shelves (floating shelves of glacier ice connected to the shore) in the Arctic. The Ward Hunt Shelf extends about 100 km along the north coast of Ellesmere Island (Jeffries 1992). It is up to 20 km wide and occupies three fiords. The shelf is 40–50 m thick and rises 4.5–6.5 m above sea level. Since the early 1900s, its size has been greatly reduced and there have been many earlier events leading to the formation of ice islands that may

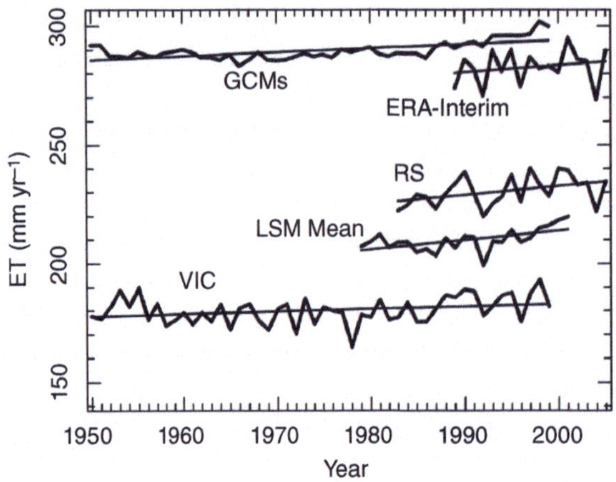

Figure 3.5 Annual evapotranspiration (ET) for the terrestrial region. Time series depicted are from the nine general circulation models (GCMs), the mean among the five land surface models (LSMs), the surface energy balance and remote sensing (RS) method, the variable infiltration capacity (VIC) model, and the ERA-Interim dataset (Slater et al. 2007).

drift around the Arctic Ocean for many years. An example is the ice island which was used as a platform for meteorological surveys.

3.1.3 Land-Surface Evapotranspiration

Surface-based observations of evapotranspiration (ET) across the pan-Arctic are sparse (Figure 3.5). Given the lack of *in situ* observations, the analysis of ET trends involves information from land-surface models and remote sensing data. ET is defined here as the total flux from all sources such as open water evaporation, transpiration from vegetation, and sublimation from snow. Eddy covariance measurements are the primary means of observing turbulent, boundary layer ET fluxes. For regional- and continental-scale studies, models using time-varying climate data (e.g. precipitation and air temperature) are used. Estimates of ET at regional and global scales are also available through satellite remote sensing. These methods are generally based on surface energy balance partitioning among surface-sensible heat, latent heat, and soil heat–heat storage fluxes. Annual ET from variable infiltration capacity (VIC) shows a significant upward trend from 1950 through 1999 of 0.11 mm yr^{-1}. The mean trend (0.40 mm yr^{-1}) among the land surface models (LSMs) also suggests ET intensification (Slater et al. 2007).

3.2 Daily and Seasonal Cycle of Freeze–Thaw in Soil

Soil freeze–thaw is the most prominent phenomenon in frozen ground. Frozen ground occupies more than half of the continents in the northern hemisphere (IPCC, Intergovernmental Panel for Climate Change 2013; Zhang et al. 2003). The soil freeze/thaw phenomenon is the basic characteristic of frozen-ground land processes (Figure 3.6), and it affects the terrestrial hydrological processes and ecological processes, and causes noticeable changes in soil water and heat transport. In the soil freeze–thaw process, soil water movement and heat transport are coupled, which makes the soil water and the heat transport process more complex than that in unfrozen soil. The changes in the soil hydrothermal regime cause the land-surface energy budget and water cycle to vary significantly.

In acting as an impermeable barrier, permafrost plays a major role in the hydrology of the Arctic. Following snowmelt, the ground becomes waterlogged unless there is sufficient topographic relief

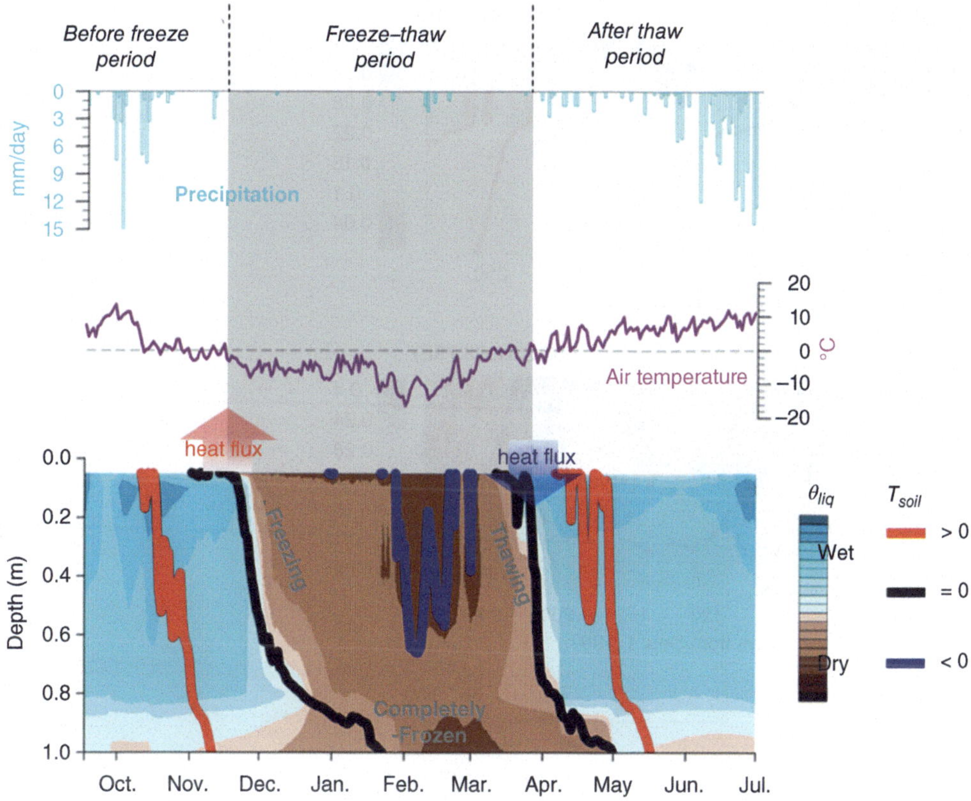

Figure 3.6 Diagram of the soil freeze–thaw process.

for drainage channels to form. Hence, there are numerous shallow lakes. Permafrost is generally absent beneath large deep lakes, especially in the sub-Arctic. In areas with sufficient drainage, the impermeable permafrost barrier fosters rapid channeling of precipitation and meltwater into streams, which impacts strongly on the seasonality of river discharge and ET rates.

The distinct features of the land surface in high-altitude and high-latitude cold regions in spring are snow melting and soil thawing. These processes are the source of increased spring SM content. Studies (Barnett et al. 1989; Qian et al. 2003; Wang et al., 2017) have reported that SM anomalies in spring contain information about snow anomalies from the previous winter/spring. Concurrent precipitation is also a primary factor that influences the variation of SM; the rainy season generally occurs after the increase in SM (Bao et al. 2017; Shang and Wang 2006).

Simulation experiments of the land–atmosphere interaction indicated that, the positive anomaly of precipitation can increase SM and the wet soil can increase the surface ET, which can influence precipitation in turn through the local water cycle or the changes of the large-scale circulation. SM plays an important role in maintaining the subsequent precipitation via the effects of water storage in atmosphere on precipitation cycles. There are responses between soil moisture and climate and widespread negative correlations between SM and temperature. The soil-moisture anomalies have significant influences on the subsequent short-term climate in terms of precipitation and temperature.

The content of SM in freeze–thaw process undergoes "wet-dry-wet" variation (Figure 3.7). The changes of SM are not obvious from December to February, due to the frozen of soil and scarce precipitation in this period. In the early mid-March, as surface temperature rises to higher than 0°C, SM

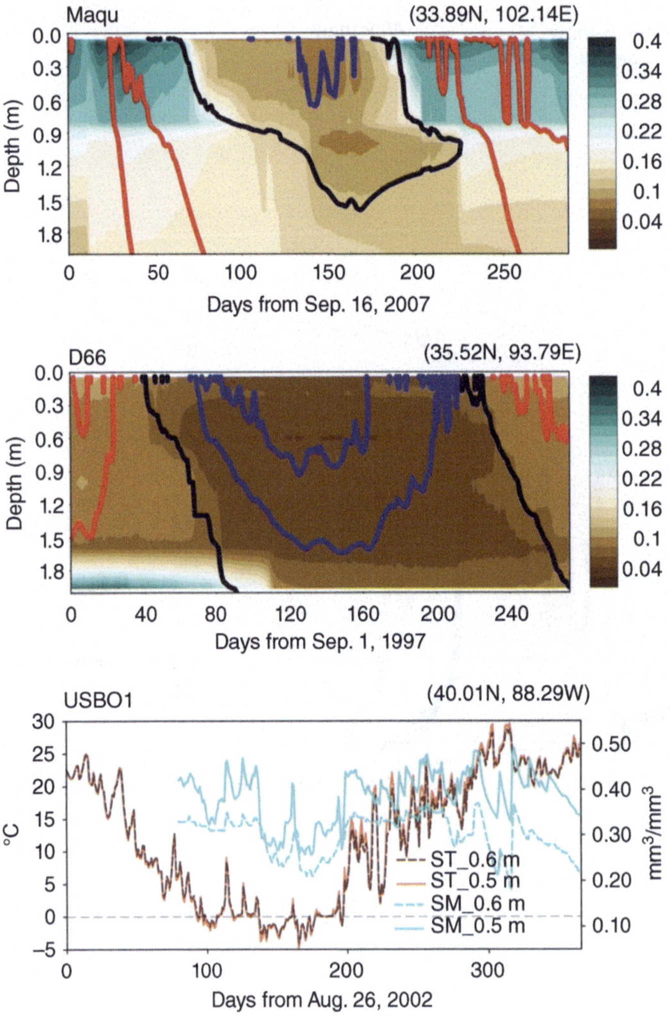

Figure 3.7 Seasonal cycle of soil freeze–thaw process. Section of soil moisture (shading; $mm^3\ mm^{-3}$) and temperature (contour, interval is 5 °C, black line is 0 °C isoline, red and blue lines represent positive and negative values, respectively) in Maqu (top panel) and D66 (middle panel). Series of soil temperature and soil moisture in USBO1 (bottom panel).

rapidly increases, which means that the ground begins to thaw. In April to May after the thawing of soil, with the increase of surface temperature, precipitation appears in succession and then increases, which may reflect the influence of SM on late precipitation. From June to July, with the arrival of the rainy season, the SM increases with the increase of precipitation and decreases with the decrease of precipitation. In early August, the surface temperature rises to the maximum of a year, which may result in the enhancement of the surface evaporation, and precipitation begind to dramatically reduce. This may cause the decrease of SM, and a similar situation appears in mid-September. From mid-August to early November, due to the rapid reduction of precipitation, the effects of precipitation on SM become weak and the effects of surface temperature on SM enhances, as a result, SM increases (decreases) with the decrease (increases) of surface temperature. From mid-to-late November, as the surface temperature drops below 0 °C, the SM significantly reduces, suggesting that the soil begins to freeze.

The performance of reanalysis dataset is the base of land-surface processes studies (e.g. Shang and Wang 2006; Bao et al. 2017) over the Tibetan Plateau (TP). For example, during the frozen period (from January to March and from November to December), SM changes with the fluctuations of surface temperature and is larger than observations, and GLDAS (Global Land Data Assimilation System) data cannot well reflect the thawing process of soil near the surface temperature of 0 °C. The biases of surface temperature are large as well. During non-frozen period (from April to October), the change trend of simulated SM in GLDAS is consistent with observations, but is drier than observation, while the surface temperature is relatively closed to observations. SM mainly changes along with the changes in levels of precipitation. In the middle TP, soil begins to freeze in mid-to-late October, and the SM changes less during the frozen period. In early April, soil begins to thaw. In May, precipitation begins to increase. The changes of SM were mainly affected by precipitation in the wet season (from June to August) and were mainly affected by surface temperature in the dry season (from September to mid-to-late October). During the frozen period (November 2013 to March 2014), the bias of surface temperature in reanalysis datasets such as GLDAS rises to maximum, that is, SM has small fluctuations with the changes of surface temperature and cannot well reflect the thawing process of soil. During the non-frozen period (from July to October and from April to May) the simulated precipitation and SM in GLDAS are obviously lower than observations. During the non-frozen period (from May to October), the change trend of SM in GLDAS is consistent with observations though the simulated SM is larger than observations, and the simulated precipitation is less than observations. During the frozen period (from March to November) in the western TP, the observed SM appears obviously to increase during which the soil begins to thaw and decrease; in other periods, however, changes of SM are marginal; the simulated SM in GLDAS, however, has distinct biases; on one hand, it may be caused by the bias of precipitation forcing, on the other, it may be associated with the biases of surface temperature.

The diurnal freeze–thaw cycles of soil temperature and SM have obvious variations (Figure 3.8). During the soil thawing period, when the soil temperature increases, the SM increases accordingly,

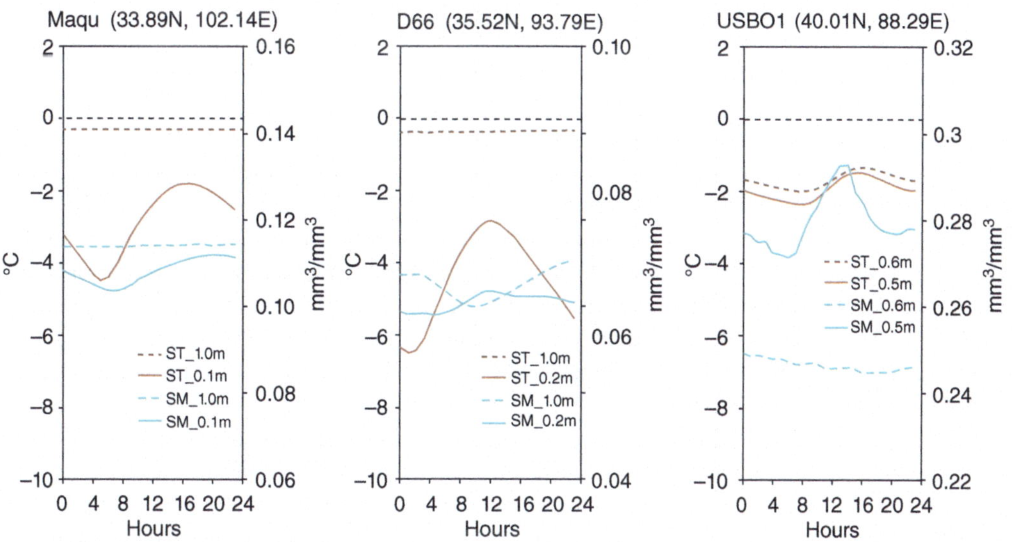

Figure 3.8 Diurnal cycle of the soil freeze–thaw process. Diurnal variations of soil temperature (ST) and soil moisture (SM) from *in situ* observations in Maqu (left panel), D66 (middle panel) and USBO1 (right panel).

and the changes in SM will in turn affect the soil temperature variations, thus affecting the phase change of soil water to soil ice. This process reflects the coupling of soil temperature and moisture in the thawing period.

3.3 Soil Freezing–Thawing and Snow-Melting Processes and their Hydrothermal Effects

In the soil freeze/thaw conversion process, the soil water phase and energy budget change together, which can affect the soil-vegetation-atmosphere system. The freeze–thaw process (FTP) plays a basic role in soil water and heat transport and impacts on the SM and temperature variation. Without soil FTP (i.e. no water phase change occurred), surface soil temperature is about −1.02 °C lower than the freeze–thaw (FT) period, while surface soil temperature is about 0.91 °C higher and surface SM is drier about −0.02 $mm^3\ mm^{-3}$ than after-thaw (AT) period. Soil FTP has a water-storage effect; the storage index (*SI*) can reach 0.95 at surface layer, suggesting that FTP has an effect on restoring SM. Without soil FTP, the surface soil water content at the AT period can be reduced by ~10% (*SI*, which is used to quantify the water-storage effect, is decreased by about −0.05) as a result of enhanced evaporation; the soil-moisture memory was shortened by about −20 days in March. Accordingly, surface-latent heat decreased by −1.07 $W\,m^{-2}$, while the surface-sensible heat increased by 4.72 $W\,m^{-2}$. The effects of FTP on soil water and heat transport at deeper layers are distinguished from that at the surface layer. These imply that the biases of soil temperature and moisture simulation during the FT period could lead to large uncertainties in estimating climate effects of TP. Under climate warming in recent decades, with the delayed beginning of the soil freeze and the advance of ending date of the soil thawing, changes of soil FTP regime would lead to the loss of spring SM.

Accompanied by the phase change of soil water, the soil hydraulic and thermal properties also changed (Figure 3.9), which in turn affects the transport of soil water and heat. Generally, soil ice has higher thermal conductivity and lower heat capacity than soil water. Due to the lack of soil ice and more liquid water during the FT period, the simulated soil thermal conductivity λ was lower and the soil volumetric heat capacity c was higher than that simulated by the control experiment during the FT period. These changes in the soil thermal properties were not beneficial for the simulation of upward heat flux F (the upward direction is defined as positive) and relieved the decreasing of soil temperature during this period. On the contrary, the lack of energy released due to the absence of the soil freezing in the sensitivity experiment (EXP, henceforth) was beneficial for the decreasing of soil temperature during the FT period. During the AT period, because less soil water content was simulated by the sensitivity experiment compared with the control experiment, the soil volumetric heat capacity and thermal conductivity were both lower than those simulated with FTP. Nevertheless, the lack of energy consumed by thawing mainly resulted in the soil being warmer than that simulated by the control experiment during this period. And these changes in the soil temperature profile would further affect the vertical transport of soil heat by conduction.

The changes in the soil liquid water and ice content resulted in the changes in soil hydraulic properties. Soil ice can impede water flow, which is defined as the ice impedance factor (Niu and Yang 2006; Swenson et al. 2012). This factor means that, when ice content θ_{ice} is less, the value of Θ_{ice} is larger, and the impedance effect is lower. With no ice (without FTP), the ice impedance factor Θ_{ice} was larger than that in the simulation with FTP, indicating decreased tortuosity of water flow and increased soil hydraulic conductivity k. These are beneficial for the upward water flux q

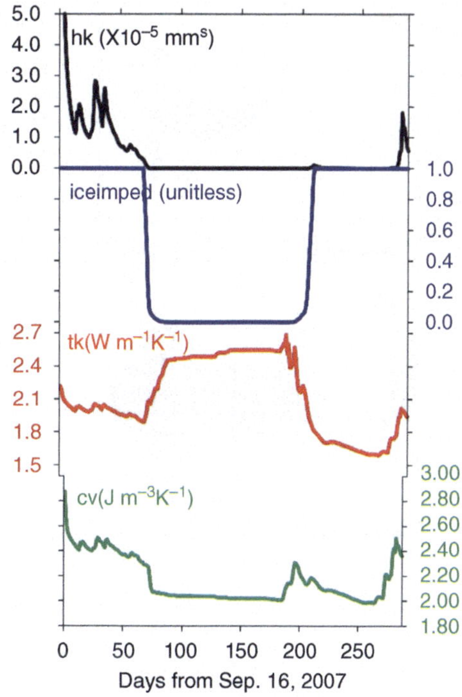

Figure 3.9 Variations of soil hydrothermal properties in the freeze–thaw process. Soil hydraulic conductivity (hk), ice impedance factor (iceimped), soil thermal conductivity (tk) and volumetric heat capacity (cv) simulated by Community Land Model version 4.5 in Maqu (33.89 N, 102.14E).

in the shallow layers (result from several TP sites, for example 0–0.5 m in Maqu and 0–0.1 m in D66, TTH, and ALi), which made these layers drier without FTP compared to that with FTP. In permafrost region, the increase in soil hydraulic conductivity in the deep layers (1.0–2.5 m) led to more upward water flux, which resulted in wetter layers between 0.6 and 1.2 m. As soil is frozen, soil water moves toward the freezing front (Swenson et al. 2012). In seasonally frozen ground, the negative difference in the soil water flux between simulations with FTP and simulations without FTP exhibited no upward water flux toward the freezing front in the absence of FTP, which led to the soil being drier near the freezing front.

The above analysis showed that, in FTP, both the changes in soil thermal properties (thermal conductivity and volumetric heat capacity) and energy in soil water phase influence the variations of soil temperature, but the latter might be dominant, especially during the FT period. The impact of FTP on soil water transport is mainly reflected in the soil hydraulic conductivities changes; experiments without FTP cannot simulate the movement of soil water toward the freezing front.

Decreasing air temperature in winter corresponds to the beginning of soil freezing, and most of the liquid water is frozen into ice, the soil-moisture content (liquid water) decreases and reaches a minimum in winter. As air temperature rises in spring, the ice thaws, which leads to increased SM (liquid water) content. Throughout the FTP, soil water content generally remains unchanged, only its phase is changed, and the soil water is redistributed along with the soil freezing/thawing. Comparing soil-moisture contents during the periods between before freezing (BF) and AT, they are almost unchanged, which implies that soil FTP might act as "water storage."

Soil FTP has a water-storage effect, which is quantified by storage index ($SI = \theta_{BF}/\theta_{AT}$) over TP can reach up to 0.95 (Figure 3.10). Throughout the TP, the *SI* ranges from 0.9 to 1.0 in the 0.2 m layer, regional mean was 0.95, and it is generally larger in the drier soil region than in the wetter soil region, which means that the soil was wetter, and the SI was smaller. The *SI*

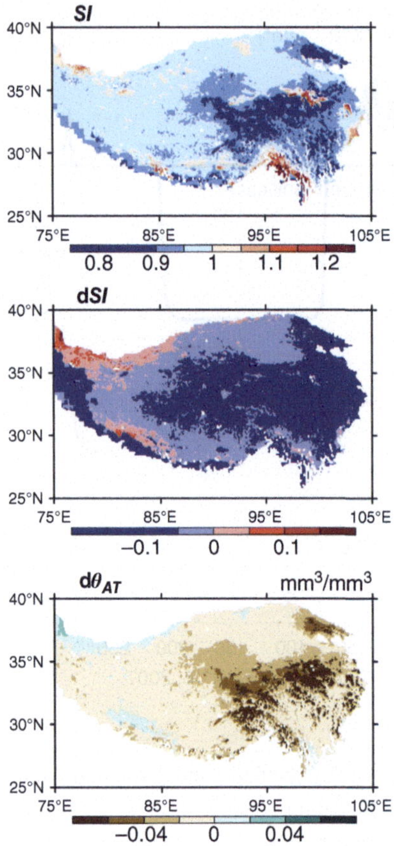

Figure 3.10 Effects of the soil freeze–thaw (FT) process on soil-moisture variations. Distribution of storage index (SI) over the Tibetan Plateau, which represents the water-storage effect of FT process (top panel), changes of SI (d*SI*, middle panel) and changes of soil moisture ($d\theta_{AT}$, bottom panel) after the thawing (AT) period with no FT process.

in the 0.2 m layer was not closely related to the soil temperature. When soil does not undergo FTP, SM in the 0.2 m layer decreases during the AT period (about −0.02 mm^3 mm^{-3} on regional average). Its magnitude can be approximate with the standard deviation of spring SM the *SI* decreased by −0.05 and the decrease of *SI* was greater in the wet region than that in dry region and the warmer soil temperature would enhance the decreasing of *SI*. These further illustrate a water-storage effect of FTP in shallow layers. In the deep (z = 1.0 m) layer, the *SI* was larger than 1.0 (regional mean is 1.02) in most regions over the TP; *SI* was larger with increase of SM and smaller with decrease of soil temperature. These features were distinguished from those at the surface layer (z = 0.2 m), which might be related to the movement of soil water to freezing front. Without soil FTP, the SM during the AT period decreased by about −0.019 mm^3 mm^{-3} and the *SI* also decreased by −0.04, but the reduction of the *SI* in the 1.0 m layer was smaller than that in the 0.2 m layer, and decrease of *SI* was greater when SM was wetter and soil temperature was warmer. The water-storage effect on SM is more distinct in the wetter regions than the drier regions. These can also clearly be seen in the results from the observed sites. The water-storage effect of FTP on SM variation can be understood as soil water variations caused by precipitation during the last summer and autumn preserved in the soil as soil freezing during the winter, and this variation is released along with soil thawing during the spring, which implies that the water-storage effect of FTP might be an important factor influencing the persistence of SM anomalies.

3.4 Impacts of Soil Freeze–Thaw Process and Snow Melting on Surface Diabatic Heating

Land-surface diabatic heating is closely related to the soil hydrothermal conditions. In freeze–thaw process, SM and temperature change rapidly.

In cold region, the land-surface diabatic heating has evident seasonal variations. Duan and Wu (2008) used station observations to analyze the variation of surface diabatic heating on different time scales over the TP. Their results indicated that heating in the central and eastern TP in spring is dominated by surface-sensible heating, while the latent heating of precipitation condensation in summer is larger than surface-sensible heating. Can reanalysis of the data also reflect the seasonal variability of the uninsulated heating of the land surface over the TP? Compare the annual variation of surface-sensible heating and latent heating fluxes in different regions of the TP between observations and reanalysis dataset (i.e. ERA-Interim). The surface-latent heating fluxes in different parts of the TP reached their maximum in different months, with the surface-latent heating reaching its maximum in July over the mid-eastern TP, and in August over the western TP (Figure 3.11). The surface-latent heating flux is related to SM content. Apart from the advent and influence of the rainy season on the TP, the thawing of frozen ground significantly increases the SM, which also has a great influence on the surface-latent heating (Wang, Shi, et al. 2008).

Apart from the significant changes of surface-sensible heating and latent heating in March to May (thawing frozen period), another significant change in the TP is surface snow melting. Surface-sensible heating and latent heating are at their lowest from December to February, which also corresponds to the period when the TP soil is completely frozen, when the exchange of water and energy between the land and atmosphere is very weak. This indicates that the seasonal variation in land-surface diabatic heating of the TP is closely related to the soil FTP.

The increase in SM caused by spring thawing of frozen soil has an important contribution to the changes in surface diabatic heating over the TP (Yang et al. 2002; Wang and Shang 2007). The FTP has a significant effect on soil temperature and moisture. The changes in soil temperature and moisture can cause anomalous changes in surface diabatic heating through the surface energy balance process.

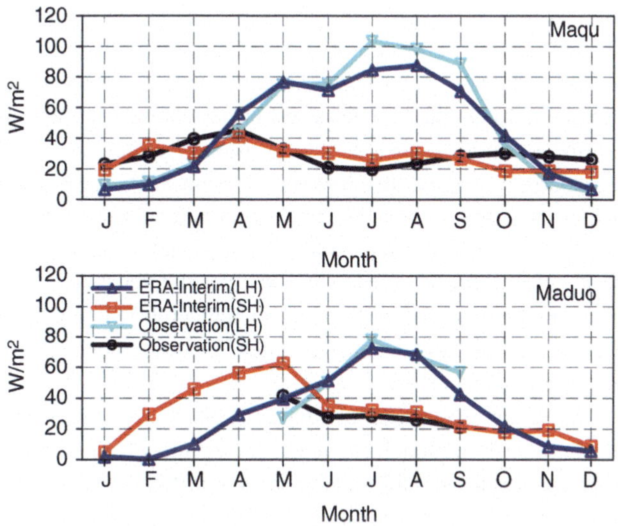

Figure 3.11 Variations of monthly surface-sensible and surface-latent heat fluxes during the freeze–thaw process from observation and ERA-Interim reanalysis dataset in Maqu (seasonally frozen-ground region, top panel) and Maduo (permafrost region, bottom panel) sites in the Tibetan Plateau.

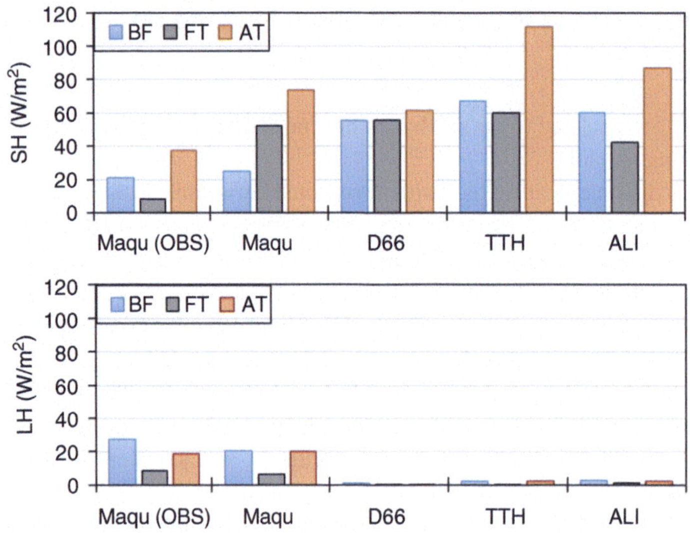

Figure 3.12 Comparison of surface sensible heat (SH) flux (top panel) and surface latent heat (LH) flux (bottom panel) during before freezing (BF) period, freeze-thaw (FT) period and after thawing (AT) period at Maqu, D66, TTH and ALi sites over Tibetan Plateau (TP) from observation (OBS) and ERA-Interim dataset.

The changes of surface-sensible and latent heating fluxes before freezing, during freezing and after thawing periods are different (Figure 3.12). The surface-sensible heating flux from the land-surface model (i.e. Community Land Model version 4.5, CLM4.5) simulations is minimal before freezing and is not consistent with observations. Also, the simulation does not reflect the changing characteristics of surface-sensible heating during freezing and thawing periods, which indicates that the model has large uncertainties in the simulation of surface diabatic heating during the freezing and thawing periods. Results at several sites in the TP (e.g. D66, Tuotuohe, and Ali), the simulated surface-sensible heating fluxes by model are minimum during the freezing period and maximum after the thawing period, and can reproduce the seasonal variation characteristics of surface-sensible heating fluxes in the FTP. The magnitude of surface-latent heating is generally smaller than surface-sensible heating during freezing and thawing periods, indicating that surface diabatic heating effects during freezing and thawing periods are dominated by surface-sensible heating. The observed surface-latent heating flux is the smallest during the freezing period, which is mainly due to the fact that most of the SM during the freezing period freezes into ice, while the unfrozen liquid water is less, and the surface temperature is relatively low. As a result, surface evaporation is weak; the observed maximum surface-latent heating occurs before freezing period, which is different from the change characteristic of surface-sensible heating; the land-surface model can generally reproduce the seasonal variation of the surface-latent heating the during freezing and thawing period, but the simulated values are still different from the observation. The simulated surface-latent heating is smallest during the freeze–thaw periods and larger before the freezing period, and the surface-latent heating after thawing period has similar magnitude to that before freezing period.

The seasonal variation of surface-sensible and latent heating in the FTP can be reproduced in the land-surface model. However, the magnitude of surface-sensible and latent heating simulated by land-surface model still deviates significantly from the observations, especially during the freezing period, and the simulated surface-sensible heating has larger biases than surface-latent heating, which is consistent with the deviation characteristics of the reanalysis dataset. According to the aerodynamic method for calculating surface-sensible heating, the magnitude of surface-sensible heating is related to the thermal resistance coefficient, surface wind speed, and difference between ground temperature and air temperature. The trend of reducing surface-sensible heating on the TP is significantly related to the surface snow cover (Zhu et al. 2018), which affects the surface temperature by changing the surface albedo and decreasing the difference between ground temperature and air temperature, thus affecting the surface-sensible heating. In the freezing and thawing process, the calculated deviations of shallow soil temperature and humidity will cause uncertainties in the surface diabatic heating (Yang et al. 2002; Wang and Shang 2007). In the process of soil freezing and thawing, SM changes significantly, which affects the magnitude of surface albedo, and changes the difference between ground temperature and air temperature through the surface energy balance; the magnitude of the shallow soil temperature also directly affects the magnitude of surface-sensible heating. At present, the simulation of seasonal changes in soil temperature and moisture during freezing and thawing process by the land-surface model is basically consistent with observations, which corresponds to the ability of land-surface model to reproduce the seasonal changes in surface-sensible and latent heating characteristics; however, there are large biases in the simulation of soil temperature and moisture magnitudes by the land-surface model, and the analysis of Yang et al. (2018) indicated that the average biases of land-surface model (e.g. CLM4.5) on the simulation of shallow soil temperature over the TP were 3.0 °C. The average biases of soil-moisture simulation are 0.05 mm^3 $mm-^3$ and these biases can also lead to large biases in the simulation of surface-sensible latent heating magnitude. The above analysis further illustrates that the surface-sensible and latent heating magnitude are significantly affected by the changes of soil temperature and moisture during freezing and thawing periods.

There are significant changes in soil temperature, SM, and surface diabatic heating when the soil freezes completely and begins to thaw to complete thawing. In winter, the soil is completely frozen, SM is low, and the corresponding surface-latent heating is lower than surface-sensible heating. In spring, with the increase of solar short-wave radiation received at the surface, the soil temperature rises, the soil starts to melt, the SM increases rapidly, and the surface-sensible and latent heat increase gradually, but the increase of surface-latent heating is larger than surface-sensible heating, and the magnitude of the surface-latent heating in spring is comparable to surface-sensible heating. Since the beginning of the rainy season on the TP is significantly correlated with the soil thawing and freezing process, and the increase in SM caused by the thawing and freezing process is earlier than the beginning of the rainy season on the TP, therefore, the contribution of SM to precipitation cannot be ignored in the TP. Results at a station on the TP (Wang and Shang 2007) show that, the soil thaw and freeze causes an increase in SM. About 20 days later, precipitation on the TP began, and there was an increase in soil-moisture transport water to the atmosphere through surface ET process. As a result, humidity in atmosphere increased, which provides favorable moisture conditions for trigging precipitation on the TP, and then the arrival of the wet season. Precipitation in turn will increase SM, increasing further the surface-latent heating, reaching the maximum in summer, even though surface-sensible heating is smaller than in spring. Surface diabatic heating in summer in the eastern TP is dominated by surface-latent heating. In autumn and winter before the soil freezing, although SM is relatively abundant, however, with the gradual decrease in soil temperature, surface-latent heating decreases rapidly. The magnitude of surface-sensible heating is comparable, indicating that the changes in surface-latent heating before freezing are closely related to the soil temperature variations. As soil began to freeze, both surface-sensible and latent heating decreased, and reached the lowest values. The above analysis shows that changes in soil temperature and moisture on the TP are associated with the freeze–thaw process, which is an important contribution to the changes in surface-sensible and latent heating on the TP.

The variations of SM and temperature will affect surface diabatic heating (Figure 3.13) and further influence the interaction between land surface and atmosphere. Comparing surface-sensible heating and surface-latent heating during three different periods, both surface-sensible and surface-latent are lowest during the FT period; these are linked to the restrained exchanges of water and heat between land surface and atmosphere. Due to the inexistence of soil water phase change in EXP, more evaporation occurred. The surface-sensible heating decreased and the surface-latent heating increased during the FTP. The changes in surface diabatic heating were generally inversed during the AT period. The surface-sensible heating increased, while the surface-latent heating decreased during this period, reaching 1–10 W m^{-2}, especially at Maqu station, where the soil is wet. The increase of surface-sensible heating and decrease of surface-latent heating accounted for the warmer ground temperature (higher soil temperature in the AT period), the drier surface soil layer (the SM in the 0–0.1 m surface layers decreased), and the reduced surface evaporation.

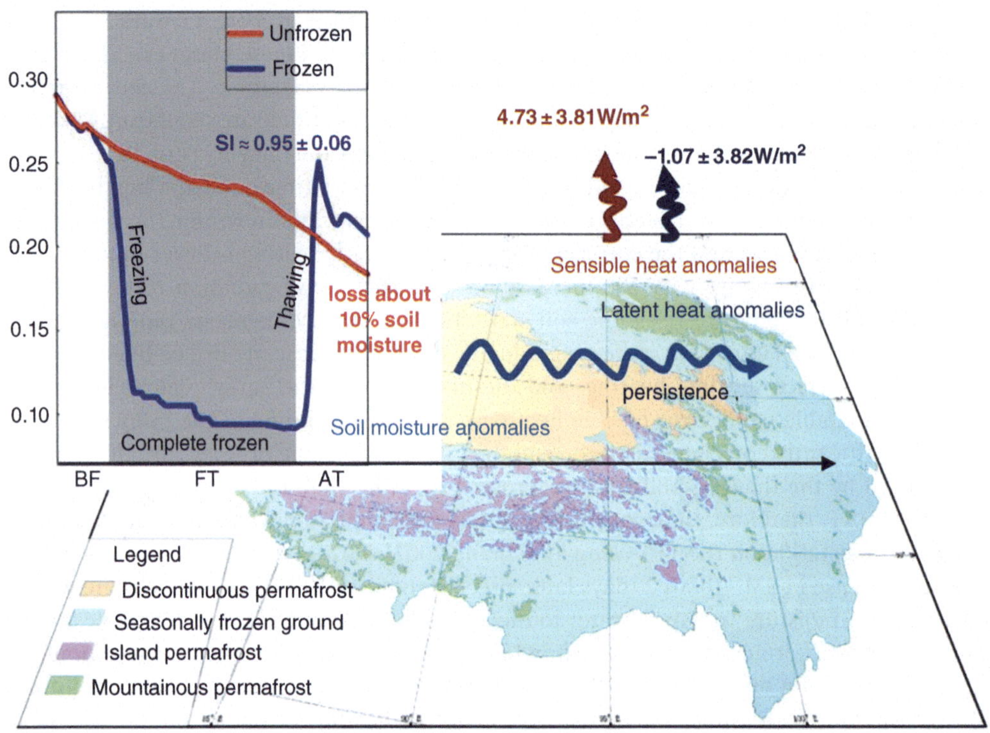

Figure 3.13 A schematic diagram summarizing the impacts of the soil freeze–thaw (FT) process on surface diabatic heating in the Tibetan Plateau (TP). This diagram illustrates that SM varies before freezing (BF), FT, and after-thaw (AT) periods. Soil FT process has a water-storage effect, and the storage index (SI) can reach 0.95 ± 0.06 at surface soil layer. Without soil frozen process, the loss of SM during the AT period would be 10%, which further leads to anomalies of surface-sensible heating by 4.73 ± 3.81 W m^{-2} and anomalies of surface-latent heating by −1.07 ± 3.82 W m^{-2} through persisting soil-moisture anomalies. The map of frozen soil on the TP, showing permafrost and seasonally frozen ground comes from Li and Chen (1996).

On average, the spatial distribution of differences of surface-sensible and surface-latent heating between the simulations with FTP and the simulations without FTP show a decrease of SH about $-2.63\,\mathrm{W\,m^{-2}}$ and an increase of surface-latent heating about $2.92\,\mathrm{W\,m^{-2}}$ in the FT period, and surface-sensible heating increased about $4.72\,\mathrm{W\,m^{-2}}$ and surface-latent heating decreased about $-1.07\,\mathrm{W\,m^{-2}}$ on regional mean at the AT period. These results imply that no or weakened soil FTP would make a positive contribution to surface-sensible heating and a negative contribution to surface-latent heating during AT period in spring, which is different from the FT period. This reflects the soil FTP water-storage effect on SM and buffer effects on soil temperature.

Snow over the TP is another dominant factor that influences the surface diabatic heating through its albedo and hydrological effects (Barnett et al. 1989). The snow cover over the TP has evident spatial and temporal variations (Zhao and Moore 2004; Bo et al. 2014). Studies suggested that variations in the spatial distribution of snow cover over the TP induces the change of energy and water exchanges between land and atmosphere, thereby affecting surface diabatic heating and changing thermal forcing of TP to the atmosphere (Li et al. 2001; Zhang and Tao 2001; Zhu et al. 2007). Due to the albedo cooling effect of snow cover on the TP, the difference between surface and air temperature ($T_s - T_a$) presents a decreasing trend in April. The stronger winter–spring snow cover on the TP would lead to the colder anomalies of surface temperature due to the increase of albedo; both the surface-sensible heating and surface-latent heating decrease before the snow is completely melting. Studies suggested that the hydrological effects are dominant and far more important than the direct thermal or albedo effect (Barnett et al. 2005; Xu and Dirmeyer 2012). The snow melting favors the increase of SM, and results in the increase of surface-latent heating flux, the decrease of surface-sensible heating flux and cooler surface; the surface diabatic heating anomalies persistently affect the atmospheric circulation until the summer (Barnett et al. 1989; Yasunari et al. 1991; Qian et al. 2003; Wang, Yang et al. 2017).

4

Land–Atmospheric Interaction in Cold Region

Interaction between land and atmosphere is an important external forcing for climate apart from the interaction between ocean and atmosphere. However, this interaction has its particularity in cold regions. The interaction between land and atmosphere in cold regions is a slow variation process relative to subtropical and tropical regions. This chapter will introduce the current problem and a new approach for seasonal climate prediction, using signals obtained from the land-surface process in cold regions. This is the key content of this book.

4.1 Spring Predictability Barrier

The prediction of seasonal climates is a challenging issue. Currently, the sea surface temperature (SST) anomalies (e.g. the El Niño/La Niña phenomenon) are regarded as the most significant signal in climate variations, and thus are widely used for climate prediction (e.g. Ropelewski and Halpert 1987; Wang and Eltahir 1999; Wang et al. 2000; Zhang et al. 1999; Dai and Wigley 2000).

Taking the El Niño southern oscillation (ENSO and other ocean signals as essential factors for climate prediction, ENSO's own prediction is firstly involved. The ENSO is one of the most prominent interannual climate phenomena. Advanced and reliable ENSO prediction remains a crucial goal for operational prediction, due to its significant influence on the economy, society, and ecosystem. Despite the development of various dynamical and statistical prediction models in recent decades, the "spring predictability barrier" remains a great challenge for long-lead time (over six months) prediction. Is the "spring barrier" like a brick wall? The answer is clearly no. It is more like a lull or a valley in ENSO forecasting accuracy. In the spring, even making an ENSO forecast for the coming summer is pretty difficult, though model performance is developing continuously (Figure 4.1).

However, hope slowly grows as the "spring barrier" issue is still not deeply understood. In particular, models initiated in May are improving sufficiently to clarify the forthcoming winter variability, which is not too bad. But, still, predictions are far from assured. Using July and August data, about three-quarters of the winter ENSO fluctuations are predicted by the models. Thus, while forecast "surprises" are becoming less frequent, they are still lurking.

Why is the accuracy of the models so bleak after spring? Is there any reason to believe that continuous model development will improve the low predicable skill during the spring? While there are many ideas on why the spring barrier exists, there are no definitive culprits (Webster and Yang 1992; Webster 1995; Torrence and Webster 1998; Duan and Wei 2013).

Climatology in Cold Regions, First Edition. Chenghai Wang.

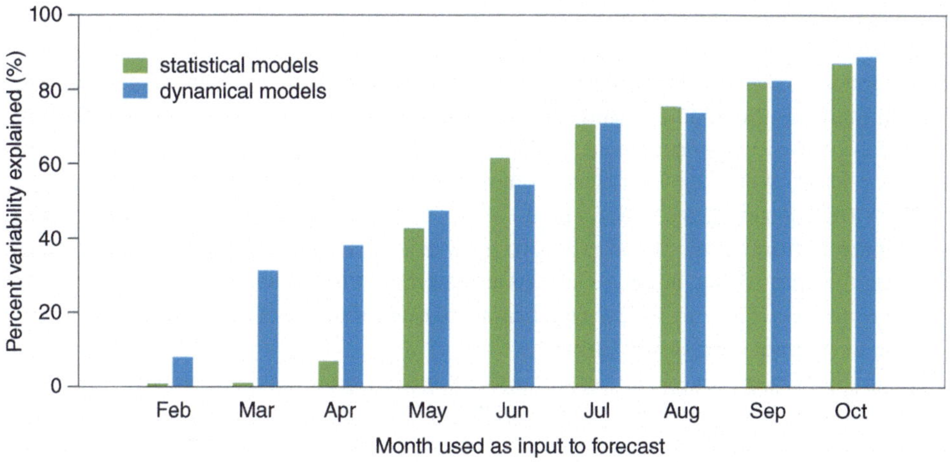

Figure 4.1 The forecast skill (or forecast ability) of model runs based on February–October observations to predict the November–January average value in the Niño-3.4 sea surface temperature region (El Niño southern oscillation). Results shown here are an average correlation coefficient from each of the 20 models between 2002 and 2011. *Source:* Data used from Barnston, Tippet, et al. (2012).

One of the reasons that the "spring barrier" is said to exist is because spring is a transitional period of the year for general circulation, as well as ENSO (signals are low and noise is high). The spring is when ENSO is moving around – often El Niño/La Niña events are decaying after their winter peak, sometimes passing through neutral, before sometimes leading to El Niño/La Niña later in the year. It is harder to predict the beginning or end of an event than to predict an event that is already happening. There is also weaker coupling between the ocean and atmosphere in the spring due to a reduction in the climatological SST gradients in the tropical part of the Pacific Ocean. Evidently, in addition to the locality and teleconnection of ENSO itself, it is not enough to rely only on ocean signals, but must consider land signals. However, for various reasons, these known factors do not fully explain why lower prediction skill exists in spring.

4.2 Signal of Climate Prediction in Land Processes in Cold Regions

In addition to the oceanic "signal," land-surface signals in cold regions (e.g. soil moisture, snow, polar ice) are also a slowly varying signal that can influence the atmosphere for several months (Koster et al. 2004; Wang et al. 2001; Wang et al. 2008; Wang et al. 2003) and can be as other important signals in climate prediction.

Changes in the spatiotemporal distribution of snow cover in Eurasia in response to a changing climate are complex (Brown 2000; Ye and Bao 2001; Liu and Yanai 2002; Brown and Mote 2009; McCabe and Wolock 2010; Brown and Robinson 2011; Brutel-Vuilmet et al. 2013). Snow cover in Eurasia strongly influences the northern hemisphere climate via snow-hydrological and albedo effects on land-surface energy budgets and moisture storage (Barnett et al. 1989; Douville and Royer 1996; Fasullo 2004; Peings and Douville 2010). The different spatiotemporal allocations of Eurasian snow cover have different impacts on the subsequent climate. For example, an analysis of snow-cover data from the National Oceanic and Atmospheric Administration (NOAA) (Bamzai and Shukla 1999) showed that, only winter snow cover in western Eurasia has a significantly negative relationship with summer monsoon precipitation in India, while snow cover in the Tibetan Plateau (TP) has no significant relationship.

The TP is located in the westerly upstream of the East Asian monsoon region. Due to the high topography of the TP in the middle latitude of the northern hemisphere, its dynamic and thermodynamic effects on weather and climate in East Asia have received much attentions (Ye and Gao 1979; Wu et al. 2007, 2009, 2012a,b; Duan and Wu 2008; Liu et al. 2012; Boos and Kuang 2010, 2013; Duan et al. 2013; Wang et al., 2017, 2020; Yang et al., 2017). In addition, there are large amounts of frozen ground and extensive snow cover over the TP, which impacts weather and climate significantly. Wang, Shi et al. (2003a) suggested that hydrothermal changes caused by the freeze–thaw process (FTP) are an external heat force that influences East Asian climate and is closely related to summer rain belts in China.

The effect of snow cover over the TP (as part of the Eurasian snow cover) on the the East Asian monsoon (Zwiers 1993; Wu et al. 2012a,b), the variability of rain belts (Wang et al. 2000, 2008; Wu and Qian 2003), and the associated anomaly of large-scale circulation (Wang et al. 2003) are distinguished from those in other regions of Eurasia (Yasunari et al. 1991; Zhang, Li, and Wang 2004). Furthermore, snow cover on the TP has undergone significant spatiotemporal changes in past decades (Qin et al. 2006; Wang et al. 2009a; Bo et al. 2014).

Impact of snow anomalies on climate has been always a hot issue. Earlier, the range and thickness of snow in the Himalayas was revealed to be negatively correlated with summer precipitation in northwest India (Blanford 1884; Walker 1910). The establishment of the Asian summer monsoon (ASM) depends on the influence of the TP through its dynamic and thermodynamic effects on atmospheric circulation (Hahn and Manabe 1975; Ose 1996; Zhang et al. 2004; Liang et al. 2005). A significantly stable relationship between TP snow cover and summer precipitation over the Yangtze River Basin and south China has been clearly established through diagnosis and numerical analysis (Chen 1997, 1998, 2001; Chen and Wu 2000; Zhang et al. 2000; Qin et al. 2006; Ding et al. 2010). Snowpack pollution would result in the increase of TP diabatic heating and enhancement of the East Asian monsoon, which further causes the anomalously wet, dry, and slightly wet patterns over southern China, the Yangtze River basin, and northern China, respectively. The winter–spring snow cover in the TP influences the spring precipitation and the pre-rainy season over northwestern and eastern China (Xu and Li 1994; Wang et al. 2000). In addition, changes in snow cover in the TP have a close relationship with the extent of the Meiyu front (Wu and Qian 2003). Research into the relationship between winter–spring snow in the TP and spring–summer precipitation in China suggested that a positive correlation exists between winter snow in the southern TP (STP) and spring precipitation in southern China, and a negative correlation in the Huaihe Basin. There is positive also correlation between winter snow over the northern TP (NTP) and summer precipitation in the Yangtze-Huaihe River Basin. There are different impacts of winter–spring snow in different regions of the TP on spring and summer precipitation in eastern China (Wang, Yang et al. 2017). The possible physical mechanism of the model result was explained by the impact of TP snow anomalies on the establishment of ASM. The impact of anomalies of ground temperature in spring over different regions of the TP showed that ground temperature anomalies in May over northeastern and northwestern TP have different effects on general circulation in summer in East Asia. The clouds, moisture recycling, and precipitation over the TP have distinct spatial differences (Wang and Guo 2012; Guo and Wang 2014; Wang and Wang 2015; Wang et al. 2015). Observations showed the prominent spatiotemporal distribution characteristics of TP snow cover (Bo et al. 2014) and diabatic heating (Zhao and Qian 2007) on the TP. Thus, the impacts of snow cover on climate in eastern China vary in different parts of the TP.

Based on dataset derived from SMMR, SSM/I, and AMSR-E passive microwave remote sensing data (Che, Xin et al. 2008), analysis of the distribution of snow depth in the cold season over the STP and NTP indicated that the snow over the TP is distributed mainly over the southeastern TP, Himalaya, Kunlun, and the Qilian Mountains. The standard deviations (interannual variations) of snow depth over the whole TP are considerable. The seasonal variations of winter–spring snow cover over the TP contain melting snow to some extent. Soil-moisture anomalies may contain the information about snow anomalies in the preceding winter and spring through the snow melting process (Barnett et al. 1989; Yasunari et al. 1991; Qian et al. 2003). Due to the complex underlying surface (e.g. topography, vegetation type), there should be a great difference in spatial variations of snow cover. The interannual and interdecadal variations of snow depth over STP and NTP have distinct differences (Figure 4.2). The difference in snow depth between NTP and STP has significant correlation with the East Asian Summer Monsoon Index (EASMI). This indicates that the impacts of the spatiotemporal anomalies of snow over the TP on the Asian Summer Monsoon (ASM) system are significant.

Snow cover has been used as an important indicator of summer precipitation in eastern China in operational forecasting (Chen 1997, 1998, 2001; Chen and Wu 2000; Wang et al. 2017, 2020, 2021). Although the majority of meteorological observational stations have been established on the eastern and southeastern TP, snow cover/depth data, however, are incomplete and rare; in addition, due to the uneven distribution of *in situ* observational stations and the various ways in which parameters such as snow cover and snow depth are measured, the snow cover/depth data from different source is inconsistent. Snow cover in winter over the TP has a stronger relationship with summer precipitation than spring precipitation in western China (Wang et al. 2000), and has dominant influence on precipitation over south China and the middle and lower reaches of the Yangtze River. Using NOAA/National Environmental Satellite, Data, and Information Service satellite snow data, Chen et al. (2000) suggested that TP winter snow is positively related to summer precipitation over the middle and lower reaches of the Yangtze River, especially the northern Jiangnan region. The relationship between the spatiotemporal distributions of TP winter–spring snow cover and summer precipitation in eastern China is still an open issue. Therefore, the uncertainties in the relationship between snow cover over the TP and summer precipitation in eastern China, as well as the underlying mechanisms are worth further exploring.

Variations in the spatial distribution of snow cover over the TP will induces the variation of energy and water exchanges between land and atmosphere, thereby affecting surface diabatic heating, and further changing the TP thermal forcing on the atmosphere (Li et al. 2001; Zhang and Tao 2001; Zhu et al. 2007; Wang et al. 2017). The anomalies of TP thermal forcing could greatly influence subtropical westerly and easterly winds on the NTP and STP, respectively. Numerical simulation results demonstrated that TP thermal forcing accelerates the higher latitude westerly on its west and lower latitude and the easterly on its east in summer, forming anticyclonic circulation on both sides (Liu et al. 2007; Wu et al. 2014; Wang et al. 2017). The different heating pattern induces changes in the meridional temperature and pressure gradients across two sides of the TP, thus changing the westerly and easterly stream over the region (Xu et al. 2002; Wu et al. 2014). Furthermore, the thermal forcing on the northern and southern branches of the westerlies caused by snow-cover anomalies in the northern and southern region of the TP differ from each other. Observations showed that summer precipitation in eastern China is connected with the activity of the 500 hPa trough in the Bay of Bengal, which extends from the southern branch of the westerlies caused by TP blocking (Ye and Gao 1979;

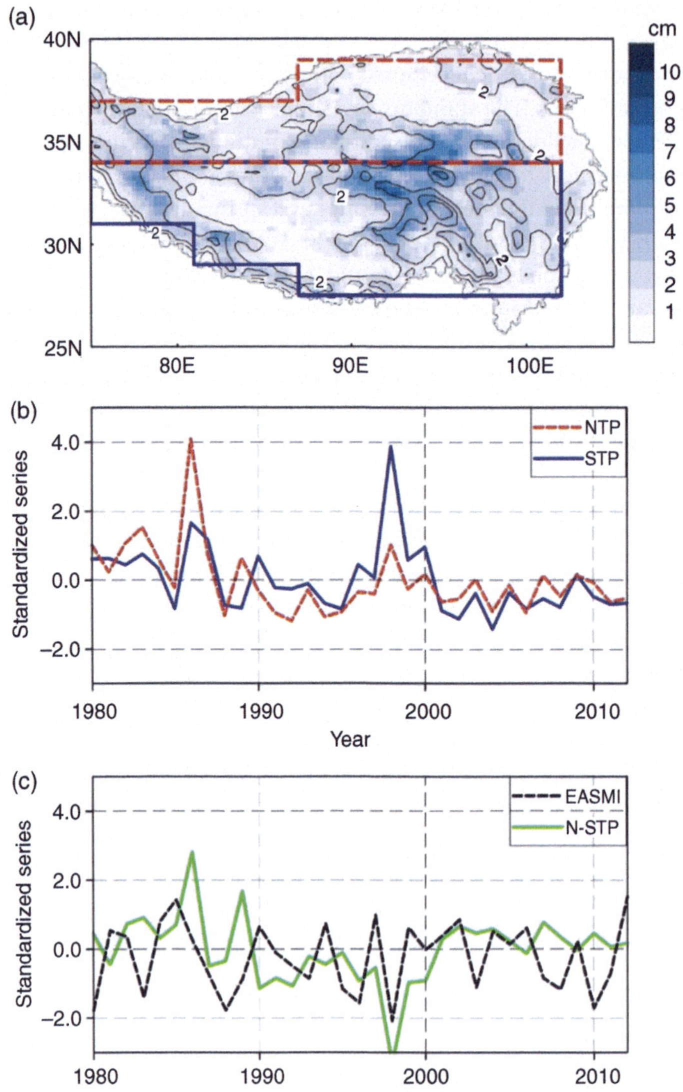

Figure 4.2 (a) Distribution of stand deviation (color) and climatological mean (1981–2010; contour, interval is 4 cm) of snow depth in the Tibetan Plateau (TP) in the cold seasons (defined as the duration from November to April the following year) over the TP; regions surrounded by the blue broken lines and red solid lines indicate the southern TP (STP) and the northern TP (NTP), respectively. (b) Evaluations of standardized series of snow depth averaged over STP and NTP. (c) Evaluations of standardized series of snow depth differences between STP and NTP (N-STP) and East Asian Summer Monsoon Index (EASMI); defined by Li and Zeng 2001). The snow depth dataset is derived from SMMR, SSM/I, and AMSR-E passive microwave remote sensing data. *Source:* Data from Che et al. (2005).

Benn 1998; Liu and Yin 2001; Wu and Qian 2003; Wang et al. 2017), and is also connected with the activity of the subtropical jet.

The observed relationship between regional snow cover in the TP and summer precipitation in eastern China shows three main spatial collocations. With heavier snow cover on the STP, the subsequent summer precipitation increases in the upper reaches of the Yangtze River and decreases

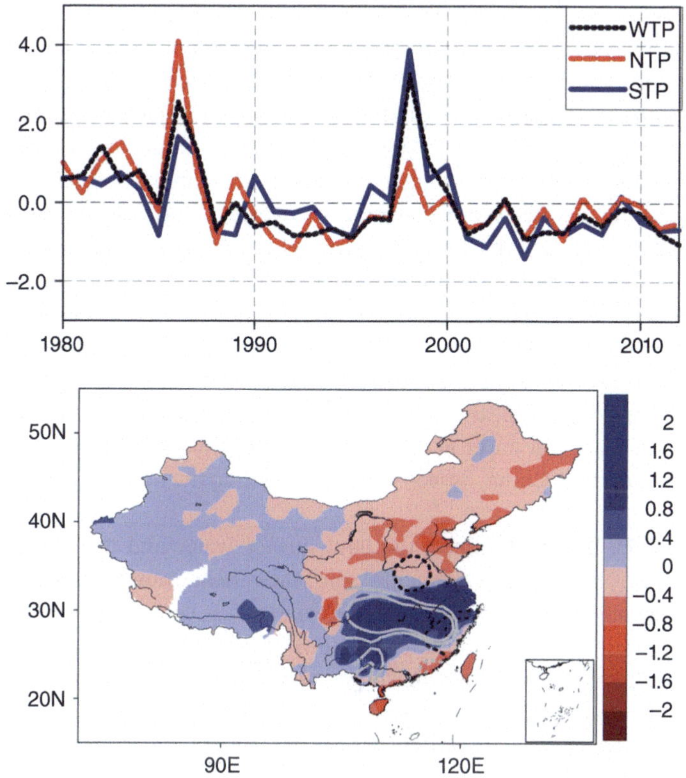

Figure 4.3 Evaluations of standardized series of snow depth average over the southern, western, and northern regions of the Tibetan Plateau (TP) (hereinafter STP, WTP, and NTP, respectively; top panel). Spatial pattern of the leading mode of summer (JJA) precipitation in China, gray and black contours represent precipitation anomalies (bottom panel; compared to 1981–2010 climatology) in 1998 (heavier snow in the southern TP) and 1986 (heavier snow in the northern TP), respectively.

in south China. Heavier snow cover on the western TP leads to excessive summer precipitation in northeastern and southern China. Increased snow cover on the NTP is usually accompanied by the enhanced summer precipitation over the Yangtze-Huaihe River valley and the weakened summer precipitation over northeastern China (Figure 4.3).

The possible mechanism by which increased snow cover in different regions of the TP affects summer precipitation in eastern China has been explored (e.g. Chen 1997, 1998, 2001; Chen and Wu 2000; Qian et al. 2003; Wu and Qian 2003; Zhang et al. 2003; Wang and Yang 2017). With heavier snow cover on the STP, the cooling effects result in a decrease of air temperature gradient and negative anomalies of the westerly jet on the north side of the TP at 200 hPa, which causes a southward shift of the westerly jet exit region and an anomalous cyclone. With heavier snow cover in the western and NTP, the westerly jet moves northward at the exit region, accompanied by an anomalous anticyclone. Diabatic heating anomalies caused by snow cover anomalies in different regions of the TP also lead to anomalous westerlies at 200 hPa along the whole TP, which influence the convergence zone where the southern branch and the northern branch of the westerly flow converge in eastern China, as well as the westerly flow on the south side of the TP at 500 hPa. Along with anomalies of the westerly jet, the negative anomalies of geopotential height in eastern China influence the western Pacific subtropical high activity at 500 hPa, compared with studies that emphasized the role of an anomalous anticyclone over the Philippines (Wang et al. 2008; Wu et al. 2005). Corresponding to the heavier snow cover in the STP, the anomalous anticyclone at 850 hPa over the eastern China Sea will transport more water vapor to the Yangtze River Valley and less to northern China, which results in excessive precipitation in the upper Yangtze River and deficient precipitation toward the south. The difference between snow cover anomalies over the STP and over the NTP are that the latter results in a stronger anomalous anticyclone, located farther north (over Japan), leading to more water vapor convergence over northern China and a northward shift of summer precipitation in eastern China. The atmospheric circulation anomalies in eastern China result from the anomalous cold in the north (caused by heavier snow cover on the STP), which gives rise to the enhanced subtropical westerly jet in the exit region, while heavier snow cover in the NTP weakens the subtropical westerly jet. The relationship between snow-cover anomalies in different regions of the TP and large-scale summer precipitation in eastern China were investigated. However, the underlying physical mechanism needs further investigation. The present results provide new evidence that snow cover in different regions of the TP can be used as a predictor signal for precipitation in eastern China, as well as in investigating the development of the summer monsoon in East Asia.

Soil freeze–thaw (FT) across the TP distinctly affects the land-surface water and the energy budget by changing the soil water phase and influencing soil water–heat transport (Wang et al. 2003a,b; Guo et al. 2011; Wang and Yang 2018). Soil-moisture anomalies caused by the FTP and the snowmelt strongly affect the summer precipitation in eastern China (Wang et al. 2003a,b, 2017; Shang and Wang 2006). Accompanied by soil thawing and snow melting during spring, there are significant temporal and spatial variations of soil moisture across the TP. FTP over the TP has significant impacts on the surrounding general circulations, and is closely related to summer precipitation in eastern China. The variation of the South China Sea summer monsoon is negatively correlated with the anomalies of soil frozen periods over the TP (Figure 4.4). Soil FT played a "water-storage" role that lengthened the soil-moisture memory (SMM) and affected the soil-moisture content during spring.

The relationship between early period soil-moisture anomalies on the TP and summer precipitation in eastern China shows that, when spring (March, April, May soil-moisture anomalies in the eastern TP were negative, summer precipitation in south China and the Yellow River Basin was deficient, whereas it was excessive in the Yangtze River Basin and northeast China (Figure 4.5). Soil-moisture anomalies in the eastern TP during the preceding December, January, February and the preceding September, October, November also showed similar effects on summer precipitation in eastern China. Comparing correlation patterns of summer precipitation with soil moisture during preceding seasons, an interesting feature is identified: PDJF and PSON soil-moisture anomalies

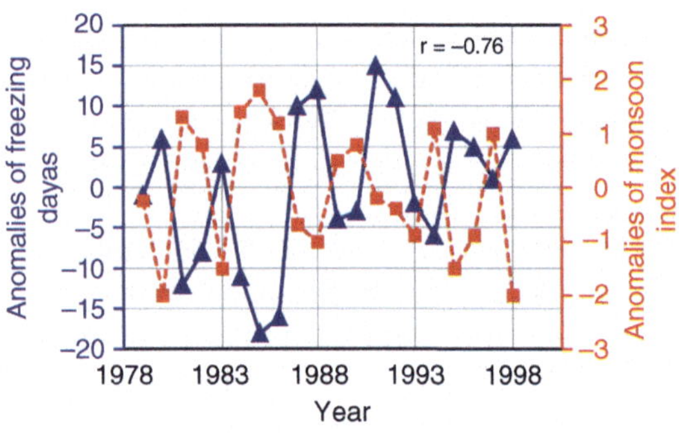

Figure 4.4 Relation between soil freeze–thaw process and the Asian summer monsoon. Changes in the anomalies of soil freezing days are indicated by the blue solid line and South China Sea summer monsoon index by the broken red line.

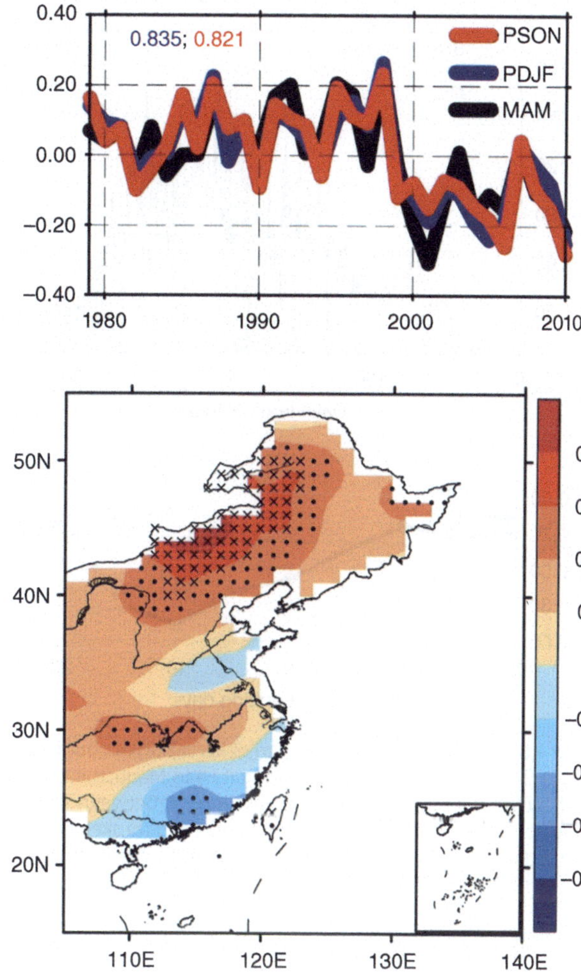

Figure 4.5 Series of soil moisture (principal component of the empirical orthogonal function) in the preceding autumn (PSON), preceding winter (PDJF), and spring (MAM); values colored in blue and red are the correlation coefficients between PDJF or PSON and MAM, respectively (upper). Distribution of correlation between series of soil moisture in PSON and summer precipitation in EC, black dots and crosses represent grid values significant at 80% and 90% confidence levels under the student's t test, respectively (bottom).

showed significant relationships with summer precipitation in eastern China, similar to the effects of spring soil moisture. Additionally, reanalysis data from the Climate Prediction Centre and MERRA-2 also showed similar correlation patterns. It appears that soil-moisture anomalies in the preceding autumn can be maintained in the soil for two seasons, with their effects being shown in the spring. These results imply that the impacts of spring soil moisture on substantial climate (precipitation and air temperature) might include the effects of soil moisture during the preceding winter or autumn.

Along with the soil FTP, the soil moisture, soil temperature and surface diabatic heating vary dramatically. The soil FTP has significant effects on local climate (i.e. precipitation). With the thawing of frozen ground, soil-moisture content and surface-latent heating (LH) increase (Figure 4.6a). In particular, the increase of LH has considerable importance with surface-sensible heating (SH) in spring; about a month later, the LH continues to increase and becomes dominant, while the SH decreases, which is related to the coming of the rainy season in the eastern TP. The increase of soil moisture caused by the thawing of frozen ground after about 20 days, results in the onset of the subsequent rainy season in the TP due to the supplement of abundant water vapor from surface evaporation (Figure 4.6b). This implies that the thawing of frozen ground has an important role in the transition from dry season to wet season, as well as the seasonal variations of

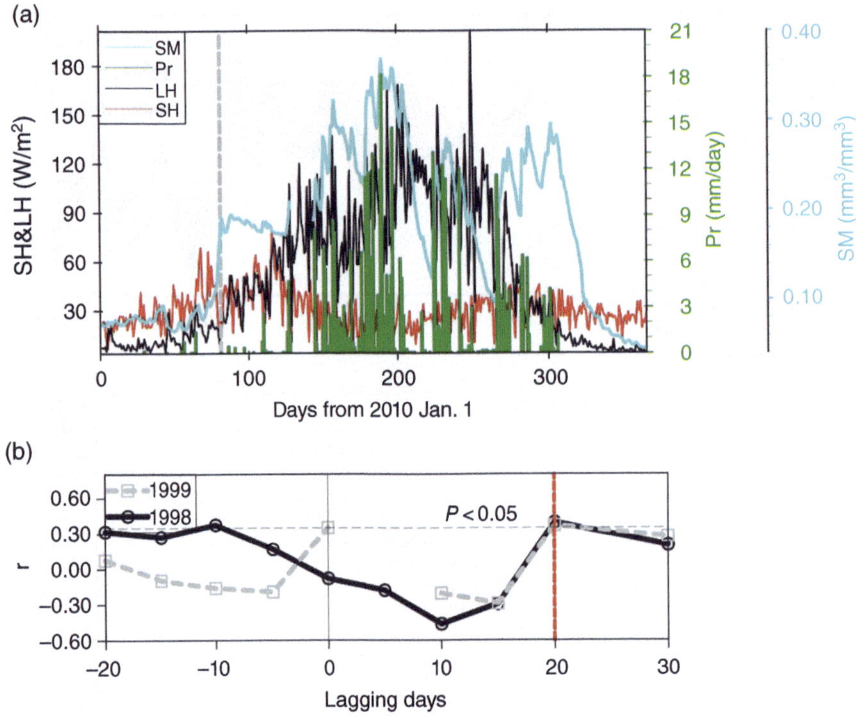

Figure 4.6 (a) Evolution of the *in situ* observation of soil moisture (SM), precipitation (Pr), surface-sensible heating (SH), and surface-latent heating (LH) in Maqu, the broken gray line indicates the date of complete soil thaw (days = 82). (b) Lagging correlations (*r*) between soil moisture and precipitation in TTH (tuotuohe) during spring in 1998 (black solid line) and 1999 (broken gray line); the broken red line represents the lagging days (=20) by which r in both 1998 and 1999 pass the student-*t* significant test ($p<0.05$). *Source:* This figure is based on the data in Wang and Shang (2007).

surface diabatic heating over the TP. The impacts of soil moisture and freeze–thaw on the coming of the wet season in the TP can also be seen in the features of clouds and convection during pre-onset periods of ASM. The frequency of deep convective cloud increases rapidly after the onset of the South Asian summer monsoon, which is attributed to the soil moisture–FTP and the increase of surface diabatic heating (Wang et al. 2009).

4.3 Improvement of Climate Prediction by Assimilating Land-Surface Signals in Cold Regions

It is well known that land-surface processes have substantial impacts on short-term climate variability (Charney 1975; Yeh et al. 1984; Manabe and Wetherald 1987; Tzeng and Lee 2001; Lorenz et al. 2012). Land-surface processes provide the low-level boundary conditions for atmospheric movement, which interact with the atmosphere through momentum, heat and moisture exchange at the same time, leading to a great influence on climate.

Soil moisture has dramatic changes with season. Soil moisture plays a crucial role in storing precipitation and has the ability to "remember" wet or dry conditions. These storage anomalies are slowly dissipated and last for several weeks or even months (Koster and Suarez 2001; Dong et al. 2006; Seneviratne et al. 2010; Bao et al. 2017). Climate prediction via a climate model has received widespread attention over the past decades (Small et al. 1999; Nobre et al. 2001; Leung et al. 2003; Liang et al. 2007; Wang et al. 2011; Yuan and Liang 2011). Accurate atmospheric forcing and soil-moisture initialization are vital for predicting seasonal climates. However, current land-surface models (LSMs) coupled with the climate model have uncertainty in reproducing the observed monthly variations of soil moisture (Ni-Meister et al. 2006 Xia et al. 2015), and studies have proposed a straightforward approach such as the second Global Soil Wetness Project to improve the performance of the climate model (Nijssen et al. 2001; Dirmeyer et al. 2006; Srivastava et al. 2015). The basic idea is that the land-surface state is typically obtained from an uncoupled LSM that is directly driven by model-based atmospheric forcing. This method strongly depends on the quality of atmospheric forcing data. However, the great uncertainties from atmospheric modeling seem to be inevitable. This implies that soil-moisture data from the model are usually not reliable as they involve poor model initialization techniques and forcing errors. A feasible approach to constrain initial soil moisture of the model is data assimilation, which has been regarded as an effective technique that can be used to curb unrealistic states that are related to the initial land-surface conditions.

Land data assimilation have mainly used on offline LSM (uncoupled with an atmospheric model), and observations have mainly originated from remotely sensed soil moisture (Lin et al. 2013). For example, assimilating scanning multifrequency microwave radiometer soil-moisture data into a LSM suggested that, improvement of the model depends on the quality of the remote sensing data. Incorporating AMSR-E soil-moisture retrievals into offline LSM via an EnKF approach suggested that, the statistical improvement in soil moisture and surface fluxes are potential limitations with sometimes neutral or degrading effects.

Soil moisture largely controls the exchange of water and energy between land surface and atmosphere though evaporation and plant transpiration (Seneviratne et al. 2010). However, because of the errors of the retrieval algorithm and bias correction, current global remotely sensed soil-moisture data have large biases compared to *in situ* soil-moisture measurements (Reichle et al. 2004, 2007; Pu and Tao 2004; Liu et al. 2005; Ni-Meister et al. 2005; Albergel et al. 2012; Zhang et al. 2015; Colliander et al. 2017). Unlike conventional meteorological observations, *in situ* soil moisture is difficult to obtain for data assimilation because of the scarcity and nonuniformity in its spatial distribution (Pleim and Xiu 2003). An indirect soil nudging (ISN) approach, which adjusts soil moisture and temperature according to the *in situ* meteorological observations, has been adopted recently to investigate the effect of indirect soil-moisture assimilation on climate prediction (Wang and Cui 2011; Shen et al. 2022). This method consists of two parts: one uses the model bias of 2-m air temperature and 2-m relative humidity (RH) data to dynamically adjust the soil moisture (Pleim and Xiu 2003), and the other adjusts deep soil temperature with the deviation of 2-m air temperature only at night because the surface forcing is weak at that time. Deep soil temperature nudging of ISN could effectively improve the reliability of deep soil moisture. Therefore, deep soil temperature nudging combined with soil-moisture nudging can be applied to track realistic changes in soil moisture. Several comparatively strong coupling sites exist in South America, Central Asia, and China. This means that land-surface processes in these regions in atmospheric models are a necessary external forcing while implementing climate prediction. The ISN scheme (Pleim et al. 2009) includes soil-moisture nudging term in the prognostic equation, the basic expression of the relaxation term is:

$$\frac{\partial w_g}{\partial t} = \alpha_1\left(T^a - T^f\right) + \alpha_2\left(RH^a - RH^f\right) \tag{4.1}$$

$$\frac{\partial w_2}{\partial t} = \beta_1\left(T^a - T^f\right) + \beta_2\left(RH^a - RH^f\right) \tag{4.2}$$

where w_g represents soil moisture in the first soil layer. w_2 represents soil moisture in the deep soil layer. T and RH represent temperature and RH at 2 m above ground level. Superscript a and f represent the observed and the simulated variable, respectively. α_1, α_2, β_1, β_2 represent the Newton relaxation coefficients, which are determined by model parameters such as vegetation cover, soil texture, and so on.

Soil FT process and snow in cold regions can be the effective signal for climate prediction. Could assimilating land-surface information (i.e. soil moisture) improve the simulation of subsequent precipitation? To answer this question, several experiments were designed by Wang and Cui (2018), that are, (i) grid nudging (GN): assimilating atmospheric variables above the Planetary Boundary Layer (PBL) with the GN method in all domains and the assimilation data coming from Final Analysis (FNL); (ii) grid nudging plus indirect soil nudging (GN_SOIL): the same as GN, but assimilating soil moisture and soil temperature with the ISN scheme. Their results indicated that the root mean square error (*RMSE*) of precipitation increases shown in the annual mean precipitation. Two particular areas with the maximum precipitation *RMSE* are the Yangtze River and the Pearl River Basins. The precipitation *RMSE* in other places is basically below 25 mm day^{-1}; compared with the GN experiment, GN_SOIL mainly reduces the precipitation *RMSE* in northern China (Figure 4.7).

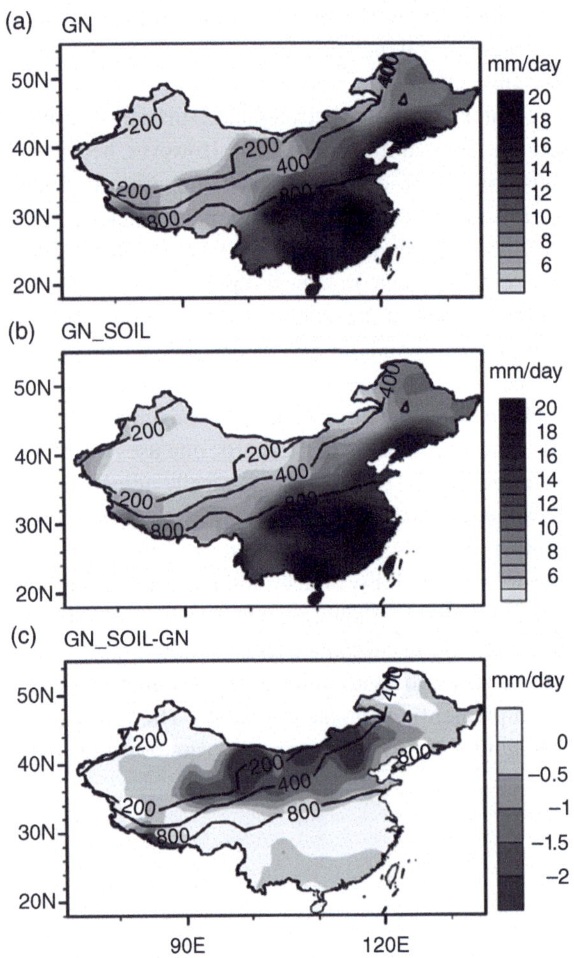

Figure 4.7 Distribution of summer (June, July, August) precipitation root mean square error (RMSE) (mm day^{-1}) in the (a) Grid nudging (GN) and (b) GN_SOIL experiments. (c) The RMSE deviation between GN_SOIL and GN (GN_SOIL minus GN). Contour lines are annual average precipitation (mm yr21) over 30 years (1981–2010) of precipitation data from the Global Precipitation Climatology Project.

Wang and Cui (2018) also investigated the benefit of assimilating spring soil moisture in the TP for the subsequent summer precipitation simulation in eastern China. Their results showed that the impacts of assimilating spring soil moisture in the TP with ISN scheme on the simulation of summer precipitation in China demonstrates that using the ISN method can reduce the spring soil-moisture bias over the TP, which further decreases the *RMSE* of summer precipitation simulation, especially in south China, north China, northeast China, and the middle and lower Yangtze River. The assimilation of spring soil moisture with the ISN method evidently modulates the distribution and intensity of surface diabatic heating over the TP, which leads to changes in the general circulation anomaly over East Asia. Specifically, the upward movement over the TP gets stronger and leads to the position of 200 hPa westerly jet shifts to the west and is closer to ERA5 reanalysis. Due to the influence of the upper layer, the location of the 500 hPa western subtropical high is also corrected, which results in a change in the distribution of water vapor in eastern China. As a result, the precipitation simulation biases are alleviated. It suggests that improving soil-moisture accuracy caused by the snow melting and frozen soil thawing, can be an effective approach to improving the seasonal prediction skills of precipitation in China. Therefore, assimilating land-surface information in cold region can effectively improve the simulation of the subsequent summer precipitation, which has great potential for applications in climate prediction.

4.4 Soil-Moisture Memory in Cold Regions

Persistence of soil-moisture anomalies has received much attention in recent years, with most studies focusing on soil-moisture anomalies at seasonal-scale or sub-seasonal scale (e.g. Koster and Suarez 2001; Seneviratne, Corti, et al. 2010). Considering the water storage effect of soil FT process (Yang and Wang 2019), frozen ground in cold regions can produce a soil-moisture anomaly that persists longer. Soil-moisture anomaly persistence can be quantified by soil moisture meter (SMM), which is defined as the average time required for lagged autocorrelation of soil moisture dropping below the 99% confidence level (Dirmeyer et al. 2008). This lag is backward in time and the memory implies how long ago the anomaly arose. For example, a 120-day memory for March means that soil anomalies in November can persist to March.

Investigating the persistence of soil-moisture anomalies showed that the memory of soil moisture across most regions of the TP in March is generally longer than 120 days, with memory reaching 180 days in the middle and northern regions of the TP, where low soil-moisture content, low precipitation, and low permafrost are distributed. SMM was short on the TP's southern edges (especially in the Himalayas) as these regions had abundant precipitation. The spatial distribution characteristics of SMM in April and May were similar to those in March, whereas the western region and southern edges of the TP had soil-moisture memories that became shorter, which could be associated with the beginning of the rainy season and snowmelt (Wei et al. 2003). Backward lagged days reached approximately 150, 180, and 210 days for soil moisture in March, April, and May, respectively, after which soil moisture autocorrelation drops below the 99% confidence level. This implies that spring soil-moisture anomalies are connected with previous states and that soil moisture anomalies in the previous September, October, November and December, January, February can persist into March, April, May.

The water-storage effect of the FT process (discussed in Chapter 2) on soil-moisture variation can be understood as soil water variations caused by precipitation during the last summer and autumn preserved in the soil while frozen during the winter, and this variation is released along with the soil thawing during the spring, which implies that the water-storage effect of FT process might be an important factor influencing the persistence of soil-moisture anomalies. The persistence of soil-moisture anomalies is quantified as the SMM (Koster and Suarez 2001). From the spatial distribution of SMM in spring (Figure 4.8), spatial variations of SMM for each month are characterized by long memory length in the northwest regions of the TP, where soil moisture content and amount of precipitation are also low (Maussion et al. 2013), and short memory length exists in the southeast regions because of the abundant soil moisture and precipitation. During spring, the SMM in the 0.2 m layer in March was longest (>120 days in most regions of the TP), about one season (an average of 136 days), while memory was shortest in May (<60 days) in most regions of the southeastern TP, within a season (an average of 78 days), which might be due to the influence of rainy season in the TP from May to September (Wei, Huang, and Dong 2003). The SMM in a 1.0 m layer was steady from March to May, which is generally beyond one season (greater than 150 days). This illustrates that soil moisture anomalies in winter or autumn of the previous year can persist to spring.

The FTP has significant impacts on SMM. If the FTP did not occur, the SMM in the 0.2 layer would decrease in March, which was consistent with the decrease in the storage index. The reduction of SSM was about 10–30 days (Figure 4.9), which implies that without the soil FTP, soil moisture anomalies from the previous autumn cannot persist to spring. The decrease in SMM still exists in April, which means that the impact of the FTP on soil moisture during the after thawing (AT) period can persist for two months. But memory in May increased without the soil FTP. The impact of the soil FTP on the SMM in the 1.0 m layer was generally contrary to those in the 0.2 m

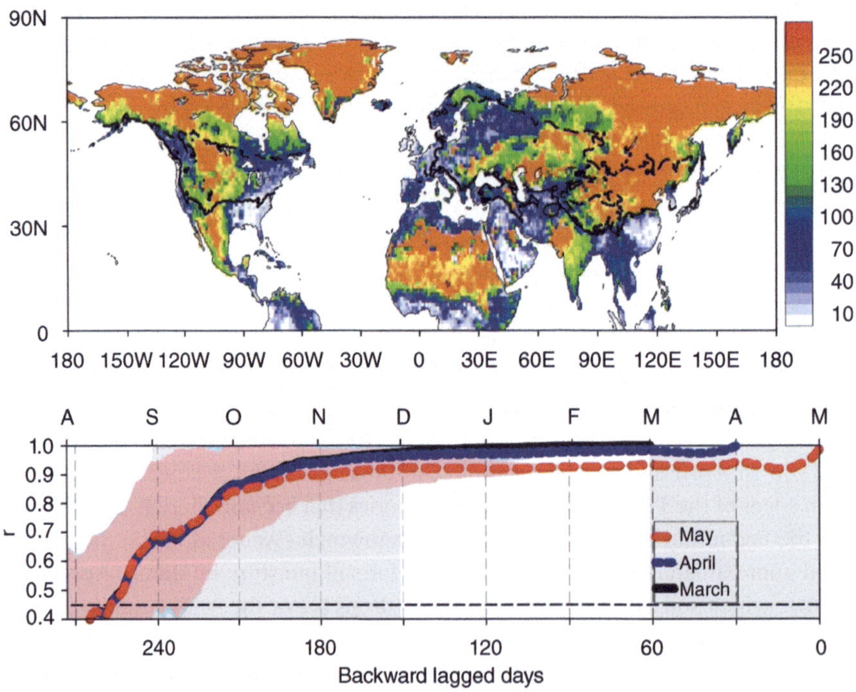

Figure 4.8 Soil moisture memory (unit: days) in spring (March–April–May average) in the northern hemisphere, which is calculated from the Global Land Data Assimilation System daily dataset (top panel). Evolution of backward lagged autocorrelation for soul moisture in March, April, and May (bottom panel; the shading for mean ± 1 standard deviation), the broken line represents the 99% significance level.

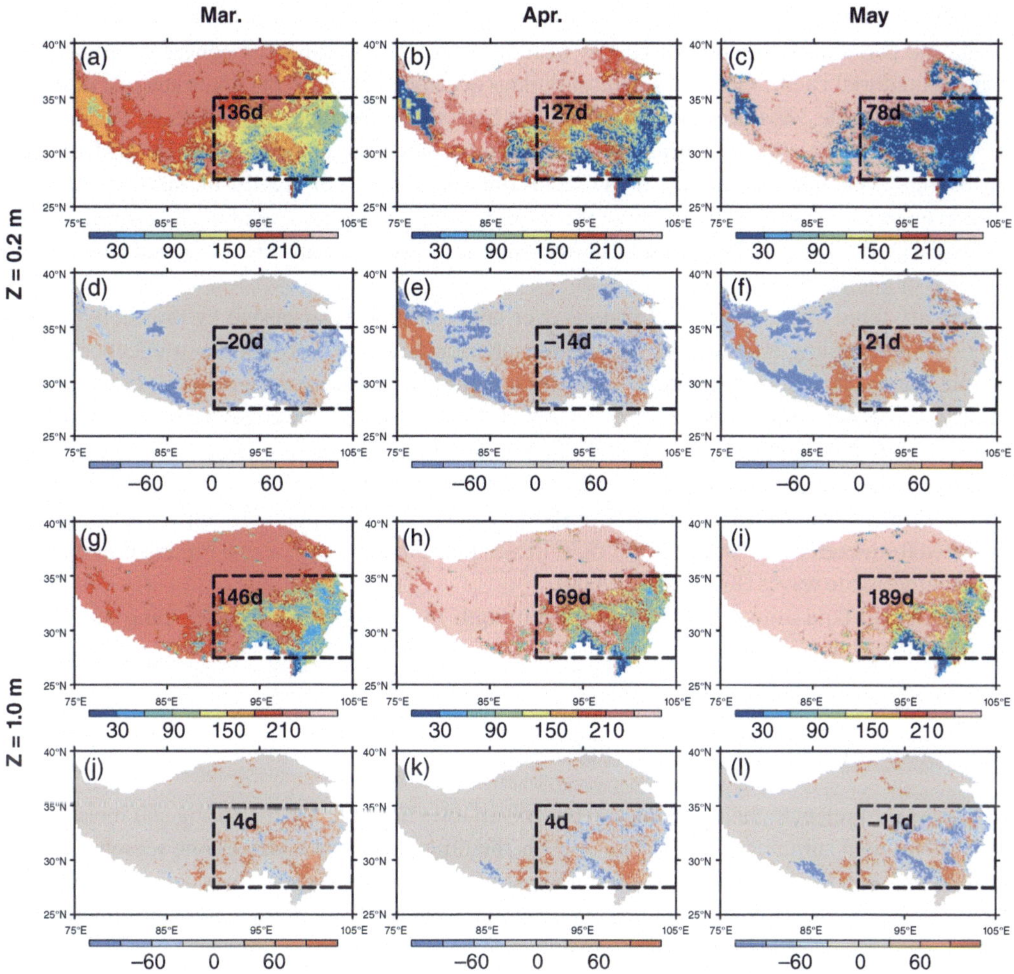

Figure 4.9 Spatial distribution of soil moisture memory (days) during March, April, and May from (a–c) the simulation with the freeze–thaw (FT) process and (d–f) the difference between simulation without FT process and simulation with FT process for the 0.20 m layer. (g–i) and (j–l) are the same as (a–c) and (d–f) except for the 1.0 m layer. The values in each panel are the average of the SSM in the rectangle region.

layer. Without the soil FTP, there would be an increase of SMM March, and this memory increase could also last to April, which was mainly attributed to the removing effects of the FTP on soil water transport in deep layers. However, SMM in May decreased because of a positive contribution of precipitation to the increase of soil moisture, and the effect of precipitation is not beneficial to the persistence of the soil moisture anomaly.

4.5 Soil-Moisture Climate Coupling in Cold Regions

SST anomalies can explain approximately 40% of the variance in summer precipitation anomalies globally, especially in tropical oceanic regions in the northern hemisphere; SST anomalies are usually used for climate predictions for a year or more in advance and have been employed in climate prediction operations by many organizations around the world. The atmosphere responds to SST anomalies, particularly in certain regions with "hot spots" of ocean–atmosphere coupling. The eastern equatorial Pacific Ocean is the most famous oceanic "hot spot," playing a key role in the El Niño-La Niña cycle. Nevertheless, several studies have suggested that the impacts of SST anomalies (e.g. El Niño/La Niña) are limited in mid-high latitude inland regions that are far from the ocean. Some modeling studies suggested that in continental mid-latitudes during summer, oceanic impacts on precipitation are weaker than the impacts of soil moisture. Therefore, determining how to obtain a prediction signal from land-atmosphere interactions remains uncertain and has become the focus of attention.

There are three main "hot spots" regions over land globally, as suggested by the Global Land-Atmosphere Coupling Experiment. Land–atmosphere interaction has significant effects on both the interannual variability in summer precipitation and 2-m temperature as a result of increasing greenhouse gas concentrations. In general, state-of-the-art atmospheric general circulation models do not present any significant soil moisture-precipitation or soil moisture-temperature interaction in Europe, which might partly cause precipitation biases in atmospheric general circulation simulations. However, the distributions of "hot spots" described by the previous studies have become separated from each other. On the scale view, these "hot spots" act as external forcing that impacts only weather-scale waves (Koster et al. 2004).

The seasonal land-atmosphere interaction strength index on seasonal scale (Yang and Wang 2022) is defined as:

$$k = \frac{cov\left(S\left(t-\tau\right),P\left(t\right)\right)}{cov\left(S\left(t-\tau\right),S\left(t\right)\right)} \tag{4.3}$$

where $cov(S, P)$ refers to the covariance between variables S and P, S is the spring soil moisture, P is the subsequent summer precipitation, and τ is the time interval, which was one season.

The zones with the strongest seasonal soil moisture-precipitation interactions coincide well with the position of the seasonal freeze–thaw region (SFTR) (Figure 4.10). In particular, the effect of soil moisture on precipitation is manifested not only in local areas but also in remote areas. Considering that the seasonal FTP is an important mechanism for transmitting soil water in different phases and reassigning soil water to different soil layers, land-surface diabatic heating, including the sensible heat flux and LH flux, is closely related to soil moisture (expressed as soil liquid water), and changes in the zone of land–atmosphere interaction (ZLAI) would lead to soil moisture and land-surface diabatic heating anomalies in these regions. As soil moisture is contained in the soil

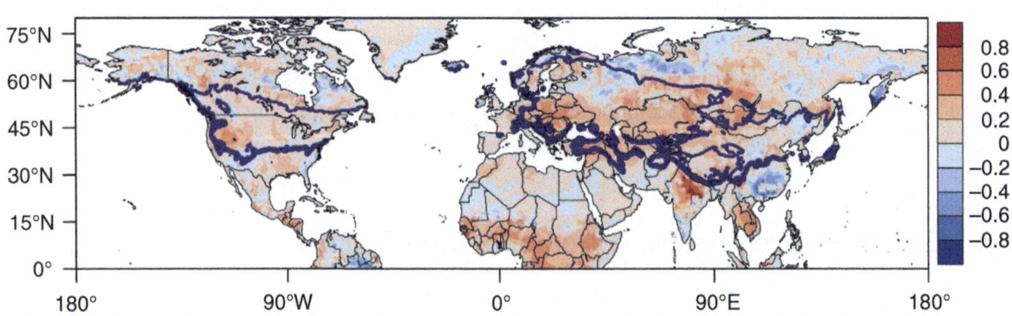

Figure 4.10 Soil moisture–precipitation coupling in the northern hemisphere. Distribution of coupling index between spring soil moisture and summer precipitation. The blue soil line denotes the range of seasonally frozen ground.

column, evaporation from the surface requires a relatively long time, and the land-atmosphere exchange also persists for a long time and can extend to the summer. Evaporation rather than precipitation varies with soil moisture, which persists to the summer, and can affect the subsequent precipitation. Evaporation is greatly influenced by soil moisture, which is caused by the thawing of frozen soil. Meanwhile, large interannual variability in soil moisture also occurs in the ZLAI. The effects of the ZLAI can be illustrated as follows: as an air mass passes over mid-latitudes across the ocean (Pacific and Atlantic), the interaction between the ocean and atmosphere would change the thermodynamic regime of the air mass; as an air mass crosses over land, the persistent interaction between the land and atmosphere will change the thermodynamic regime of the air mass again; the impacts that originate from both land and ocean would form quasi-stationary external forcing over the mid-latitude atmosphere. In theory, impacts from the ocean mainly focus on changes in the atmospheric thermodynamic regime from autumn to winter and persist to influence summer precipitation, while impacts from the ZLAI influence the general circulation in the summer through land-atmosphere interactions that occur in spring.

Land heating in the non-monsoon region of Eurasia is a dominant factor that influences the summer precipitation anomalies in Asia and North America is more significant than oceanic forcing. The variability in the ZLAI (or SFTR) can better illustrate the impacts of changes in land heating on summer precipitation in the northern hemisphere. The soil moisture and ocean information (i.e. ENSO) have different effects on boreal summer precipitation. If we combine the effects of soil moisture and the ocean, the pattern of the combined partial correlations is very similar to the EOF (Empirical Orthogonal Function) pattern of summer precipitation in the northern hemisphere (NH). This result means that the combined effects of soil moisture and the ocean can explain most of the variability in summer precipitation.

On a seasonal scale, the performance of summer precipitation simulations in models might partly depend on their capability to reproduce the seasonal land-atmosphere interaction strength in the ZLAI. Compared with the "El Niño/La Niña" phenomenon in the eastern tropical Pacific Ocean, the ZLAI has considerable effects on boreal summer precipitation, especially in the mid-latitude inland regions, and seasonal climate predictions can be improved by focusing on this region.

In the spring, which is the transition period from dry to wet season over the TP, snowmelt and the thawing of frozen soil on the land surface lead to the rapid changes in soil water and heat transport, and enhance the energy and water exchange between the land surface and atmosphere (Wang et al. 2008). Analysis from *in situ* observations suggested the increase in soil moisture caused by the spring soil thawing process plays an important role in the occurrence of subsequent precipitation in the early summer in the TP (Wang and Shang 2007). However, due to the larger spatial heterogeneity, physical processes linked to the effects of soil moisture on subsequent precipitation in whole TP are still unclear.

In most regions of the TP, the coupling relationship between spring soil moisture and summer precipitation is positive, especially in the southeastern and northern regions of the TP, with coupling coefficients of 0.2–0.4, indicating that when the spring soil moisture increases, it leads to an increase in late-summer precipitation. In the western part of the TP (75–85°E), the coupling relationship between spring soil moisture and summer precipitation is basically negative, with coupling coefficients about −0.2, indicating that an increase in spring soil moisture in this region will instead lead to a decrease in summer precipitation over the TP (Figure 4.11). The zonal variations of coupling between spring soil moisture and summer precipitation show features of increasing and then decreasing from east to west, with the coupling coefficients reaching its largest value near 92°E. The meridional variations of coupling between spring soil moisture and summer

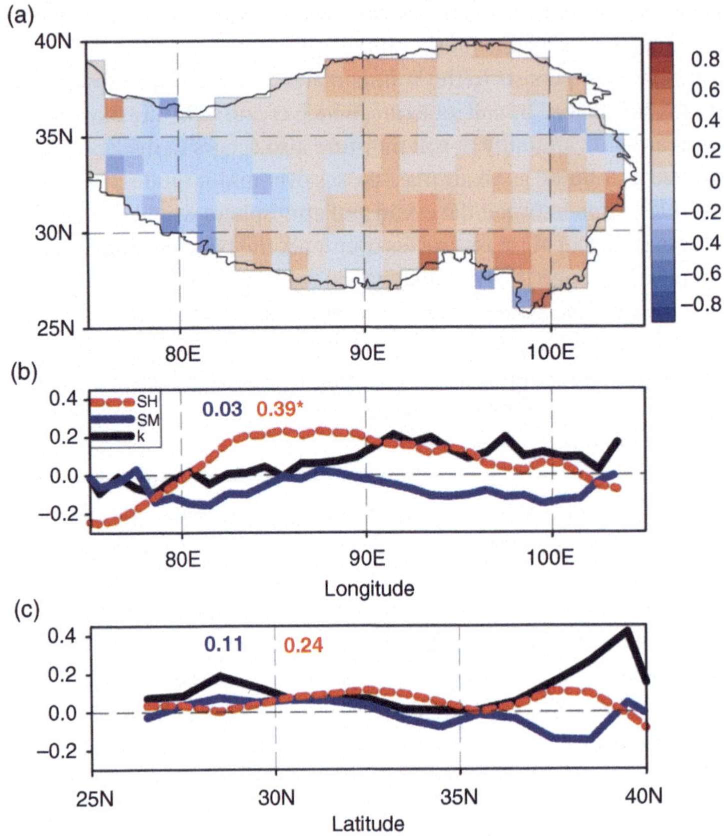

Figure 4.11 Distribution of coupling index *k* between spring (MAM) soil moisture and summer (JJA) precipitation over the Tibetan Plateau (a) Spatial distribution, (b) meridional mean, (c) zonal mean. The blue and red values in the figure represent the correlation coefficients of soil moisture (SM; average of 0–1.0 m soil layer), surface-SH and the coupling coefficient *k*. Asterisks (*) represents the correlation coefficient which passed the 90% significance student-*t* test. The coupling coefficient is the average of multiple sets of data (Global Precipitation Climatology Centre) precipitation and Global Land Data Assimilation System (GLDAS) soil moisture, ERA-interim precipitation and soil moisture, and NCEP-II precipitation and soil moisture), surface-sensible heat for ERA-interim data, soil moisture for GLDAS data.

precipitation show decreasing first and then increasing from south to north, with large values in the southern and northern edges of the TP, while the coupling strength is small in the middle region (about 33°N). The coupling relationship between spring soil moisture and summer precipitation in the TP has obvious differences between zonal and meridional variations, which may be related to soil hydrothermal conditions, atmospheric boundary layer processes, and large-scale circulation (Cook, Bonan, et al. 2006; Seneviratne et al. 2010; Xu and Dirmeyer 2011).

The coupling between soil moisture and precipitation is related to soil moisture and surface energy budget in local feedback (Figure 4.12). It has been shown that the soil moisture and precipitation coupling hotspot areas are mainly concentrated in the transition zones between wet and dry climates (Koster et al. 2004). This is mainly due to the fact that, in wet climates, although soil moisture is abundant, the surface evaporation is mainly influenced by the surface net radiative energy and is not sensitive to soil moisture variations, so the effect of soil moisture variations on precipitation is relatively small; in dry climates, although surface evaporation is sensitive to soil moisture variations, surface evaporation is small due to the dry soil and cannot have a significant effect on precipitation. In the transition zone between dry and wet climate, where surface evaporation is suitably high and still sensitive to soil moisture variations, while surface net short-wave radiation can also satisfy the atmospheric motion conditions for precipitation formation, such as the boundary layer humidity and convective trigger convection (Sud, Chao et al. 1993).

The relationships between meridional and zonal averages of soil moisture, surface-sensible heat, and spring soil moisture-summer precipitation coupling coefficients were unclear. The coupling coefficient was not significantly correlated with soil moisture in zonal variation, but was significantly positively correlated with surface sensible heat ($r = 0.39$; $p < 0.1$), indicating that the zonal variation of spring soil moisture-summer precipitation coupling on the TP from east to west was mainly influenced by surface-sensible heat, especially in the east-central region (90–105°E) of the TP. The connections of coupling coefficient with the soil moisture and surface-sensible heat have obvious meridional variations. In the southern region (27–35°N) of the TP, the variation of coupling coefficient is consistent with the variation of soil moisture. This might be due to the larger surface net short-wave radiation over the STP during spring, satisfying the conditions of abundant surface evaporation and convection required for precipitation occurrence. In the northern region (35–40°N) of the TP, the variation of coupling coefficient is consistent with the surface-sensible heat, which is mainly due to the lower surface temperature in the NTP where the altitude is higher

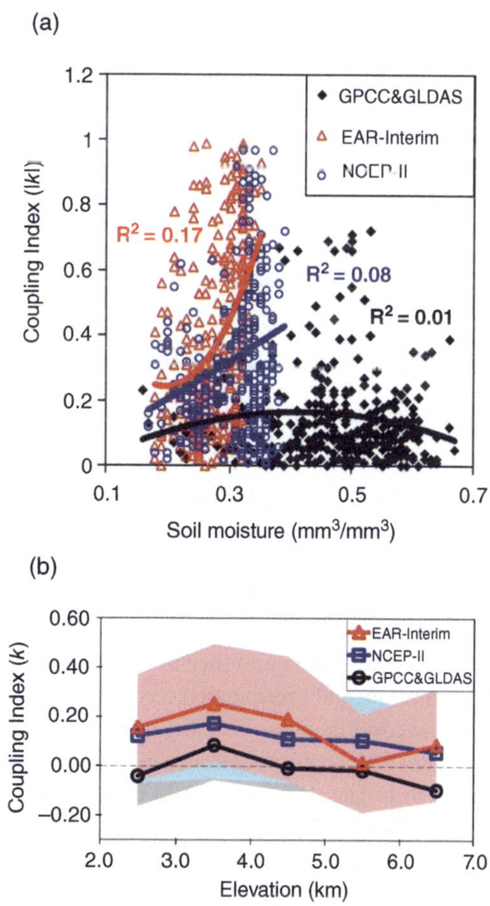

Figure 4.12 Evaluation characteristics of the coupling relation between soil moisture and precipitation. (a) Relation between soil moisture (mm^3/mm^3) and coupling index (|k|) for three reanalysis datasets (Global Precipitation Climatology Centre precipitation and Global Land Data Assimilation System soil moisture, ERA-interim precipitation and soil moisture, and NCEP-II precipitation and soil moisture). The absolute value of the coupling explains the strength of coupling varying with soil moisture. (b) Coupling Index (k) vs. elevation bands with 1 km width, the shading indicated with ± means 0.5 standard deviation.

(>4000 m) than the STP. The increase of SH will benefit the surface evaporation and convection and favor the precipitation formation.

The coupling coefficients basically increase with the increase of soil moisture content. When the soil moisture is greater than 0.4 mm^3/mm^3, the coupling coefficient decreases as the soil moisture increases. Generally, the soil moisture in most regions of the TP is 0.3–0.4 mm^3/mm^3, which belongs to the dry-wet transition zone type. There should be a strong coupling relationship between soil moisture and precipitation.

It has been shown that the TP warming and variations of SH have elevation-dependent characteristics (Gao, Chen, et al. 2018; Zhu, Huang, et al. 2018). The coupling relationship between spring soil moisture and summer precipitation on the TP calculated using different datasets is relatively consistent. The coupling coefficient reaches the maximum value at 3–4 km elevation, and the average value is about 0.2. In regions with elevations of less than 4 km, the coupling coefficient gradually increases with the altitude; in regions with elevations greater than 4 km, the coupling coefficient gradually decreases with the altitude; in regions with altitude greater than 6 km, the coupling coefficient reaches the minimum, and this means the effects of spring soil moisture variations on subsequent precipitation is smaller in the higher elevation regions. The strength of coupling between spring soil moisture and summer precipitation is generally elevation-dependent, which is supposed to be related to the frozen soil thawing in spring. In the lower altitude regions (eastern TP), soil thawing in spring makes soil moisture increase. With the increase of solar radiation received at the surface, increase in surface evaporation is conducive to precipitation occurrence. The eastern region of TP is also the area with large interannual variability of spring soil moisture (Yang and Wang 2019b). In regions of the TP with altitude greater than 4 km, there are large amounts of permafrost. The soil thaws relatively late and most of the soil is still frozen in spring (Yang and Wang 2019a). Soil moisture (liquid water content) and surface evaporation are relatively small, and the effect of soil moisture variation on precipitation is also small. In spring, surface-sensible heat in the western TP is greater than that in the eastern TP, while LH is inverse (Duan and Wu 2008). This spatial difference in partition of surface energy also reflects the effects of soil moisture on surface evaporation as well as precipitation.

4.6 Snow–Climate Coupling in Cold Regions

Snow–atmosphere coupling strength has attracted much attention. The spatial and temporal distribution of coupling strength during the snow depletion phase (March through June) over the northern hemisphere (Xu and Dirmeyer 2012) indicated that snow coupling strength evolves in three distinct stages: before snowmelt, during snowmelt, and after snowmelt. The coupling strength is strongest during snowmelt. The coupling strength after snowmelt, mainly due to hydrological effect (soil moisture impact), is stronger than coupling strength due to the albedo effect (radiative impact) before snowmelt.

Snow albedo effect and the snow-hydrological effect that operates through soil moisture, evapotranspiration, and precipitation feedback impacts on the atmosphere. The albedo effect is governed by snow-cover fraction, while the hydrological effect is controlled by anomalies in snow water equivalent. Realistic snow cover from satellite estimates is prescribed and compared with model-generated values to isolate the snow albedo effect. The snow albedo effect is found to be active before, and especially during the snowmelt period, and regions of strong albedo-driven coupling move northward during spring, with the retreating edge of the snowpack in the northern hemisphere. The snow-hydrological effect appears first during the snowmelt period and can persist for months afterward.

5

Models and Modeling in Cold Regions

The maximum uncertainty in climate modeling comes from the description and parameterization in land-surface process in cold regions. The Climate and Cryosphere Project was launched to develop the land-surface parametrization and improve model's performance. This chapter will discuss the cold regional model and modeling in land-surface process from three aspects, including performance of land-surface model simulation, improvement of snow and freeze–thaw parameterizations, and how to assimilate soil moisture during the transition season to improve the seasonal climate prediction.

5.1 Snow and Freeze–Thaw Parameterizations

The Community Land Model version 4.5 (CLM4.5), which was released by the National Center for Atmospheric Research, is the land component of the Community Earth System Model. Compared with previous versions of the CLM (e.g. CLM3.5, CLM4.0), improvements in CLM4.5 include, for instance, the new hydraulic properties of frozen soil (Swenson et al. 2012) and fractional snow-covered area parameterizations (Swenson and Lawrence 2012), improved simulation of the terrestrial hydrological cycle in cold regions, and so on. Apart from the introduction of CLM4.5 component, a set of offline simulations with CLM4.5 will also be introduced in this chapter.

In CLM4.5, the vertical transport of soil moisture is described using a modified Richards equation (Zeng and Decker 2009). The heat transport is calculated using a one-dimensional heat conduction equation. The temperature profile is calculated first without phase change and then readjusted to the phase change.

The soil temperatures are evaluated to determine whether a phase change will take place as follows:

$$T > T_f \text{ and } w_{ice} > 0 \text{ thawing} \tag{5.1}$$

$$T \left\langle T_f \text{ and } W_{liq} \right\rangle W_{liq,\max} \text{ freezing} \tag{5.2}$$

where T is the soil temperature, w_{ice} and w_{liq} are the mass of ice and liquid water (kg m^{-2}) in each soil layer, respectively, T_f is the freezing temperature (K), and $w_{liq,\max}$ is the maximum liquid water when the soil temperature is below the freezing temperature.

Climatology in Cold Regions, First Edition. Chenghai Wang.

The phase change is determined by soil temperature. The maximum liquid water (Niu and Yang 2006) is estimated as follows:

$$w_{liq,\max,i} = \Delta z_i \theta_{sat,i} \left[\frac{10^3 L_f \left(T_f - T_i\right)}{g T_i \psi_{sat,i}} \right]^{-1/B_i} \tag{5.3}$$

The above equation is the function of soil temperature T_i (unit is K), where Δz_i (mm) is the thickness of the soil layer i, $\theta_{sat,\,i}$ ($mm^3\ mm^{-3}$) and $\psi_{sat,\,i}$ (mm) are the saturation water content and matric potential, respectively, L_f (unit is $J\,kg^{-1}$) is the latent heat of fusion and, g and B_i are the gravitational acceleration ($m\,s^{-2}$) and Clapp and Hornberger exponent, respectively. Combined with the freezing criteria, the soil liquid water content during the completely frozen stage can also be determined by using the soil temperature. Thus, variations in soil moisture are controlled by the soil temperature.

In CLM4.5, when the freezing criterion is met, the soil ice is calculated as follows:

$$w_{ice,i}^{n+1} = \begin{cases} \min\left(w_{liq,i}^n + w_{ice,i}^n - w_{liq,\max,i}^n, w_{ice,i}^n - H_m\right) & w_{liq,i}^n + w_{ice,i}^n \geq w_{liq,\max,i}^n \\ 0 & w_{liq,i}^n + w_{ice,i}^n < w_{liq,\max,i}^n \end{cases} \tag{5.4}$$

where $w_{liq,i}^n$ and $w_{ice,i}^n$ are the soil liquid water and ice at current step *n*, respectively, $w_{liq,\max,i}^n$ is the maximum liquid water, $w_{ice,i}^{n+1}$ is the soil ice at next step $n+1$, and H_m is the hypothetical ice mass when the soil temperature decreases to the freezing point. Details are shown in the technical description of CLM 4.5 in Chapter 6 (Oleson et al. 2013). The soil liquid water $w_{liq,i}^{n+1}$ at the next step n + 1 is readjusted as follows:

$$w_{liq,i}^{n+1} = w_{liq,i}^n + w_{ice,i}^n - w_{ice,i}^{n+1} \geq 0 \tag{5.5}$$

In cold regions, the soil temperature during the frozen season allows freezing of soil water, whereas soil water content is limited. Therefore, the term $w_{ice,i}^n - H_m$ is always larger than the term $w_{liq,i}^n + w_{ice,i}^n - w_{liq,\max,i}^n$ in Eq. (5.4); the variations in soil ice and liquid water are determined by the combined effects of the total water content (sum of soil liquid water and ice) and soil temperature. Thus, the calculation of soil liquid water can be rewritten as follows:

$$w_{liq,i}^{n+1} = w_{liq,\max,i}^n \tag{5.6}$$

In CLM4.5, when the thawing criterion is met, the soil ice is calculated as follows:

$$w_{ice,i}^{n+1} = w_{ice,i}^n - H_m \geq 0 \tag{5.7}$$

Soil liquid content is calculated using Eq. (5.5). It can be seen that soil temperature plays a pre-judgment and dominant role in the freezing–thawing process (FTP).

The soil temperature is adjusted in phase change as:

$$T_i^{n+1} = T_i^n - \frac{L_f}{c_i \Delta z_i}\left(w_{ice,i}^{n+1} - w_{ice,i}^n\right) \tag{5.8}$$

where T_i^n is the soil temperature calculated by solving the heat conduction equation, c_i is the volumetric heat capacity ($J\,m^{-3}\,K^{-1}$).

Table 5.1 The root mean square errors of soil temperature (ST) and soil moisture (SM) between regional simulation and observation in the freeze–thaw process.

	ST (°C)		SM ($mm^3\ mm^{-3}$)	
	Shallow layer	Deep layer	Shallow layer	Deep layer
Maqu	2.36	1.21	0.047	0.022
D66	2.21	2.12	0.038	0.050
TTH	2.74	3.12	0.079	0.107
D110	3.42	2.64	0.084	0.062
Noda	1.92	1.03	0.054	0.072
Amdo	3.36	1.72	0.063	0.021
MS3608	2.43	1.43	0.070	0.033
MS3637	2.15	1.38	0.041	0.035
Naqu-BJ	5.73	4.79	0.018	0.055
ALi	3.37	1.88	0.030	0.023
Mean	2.97	2.13	0.052	0.048

The performance of CLM4.5 in simulating soil moisture and temperature during the FT process has been evaluated in many studies. Yang et al. (2008) showed that the error of daily soil temperature is within 3°C and error of daily soil moisture is within 0.10 $mm^3\ mm^{-3}$ in FTP (Table 5.1), averaged at 10 *in situ* sites in the whole Tibetan Plateau. Single-point simulations showed that model biases partly attribute to uncertainties of initial condition and soil category data. Studies also indicated that biases of simulated soil moisture and temperature by CLM4.5 come from the biases of atmospheric forcing dataset, especially during the unfrozen period (Bao et al. 2017; Li et al. 2018). Excluding impacts of model setting, large biases of soil moisture still exist during soil thawing (Figure 5.1), and CLM4.5 fails to simulate the diurnal cycle of soil moisture in this period.

In the FTP, the coupling between soil water and heat transport is crucial for the variations of soil temperature and soil moisture (Swenson et al. 2012; Zhang, Sun et al. 2007). The variations of soil moisture in the freezing process simulated by CLM4.5 were consistent with the observation; in the thawing process, even soil temperature was below 0°C. With the increase of soil temperature, the observed soil moisture in shallow layers ($z = 0.1$ m or 0.2 m) increases slowly at first, and when soil temperature nears 0C, the soil thaws rapidly, while the simulated soil moisture from current freeze–thaw (FT) parameterization cannot capture this feature.

The relationship between soil temperature and soil moisture is very important in the FTP (Xu et al. 2001; Yu and Wang 2012) and is utilized in the parameterization scheme of land surface models. This relationship in soil freezing and thawing process was compared between simulation and observation. According to the observed soil temperature and moisture, as the soil temperature decreases below the freezing point, a rapid reduction in soil moisture also occurs when soil temperature is still near 0°C; as the soil temperature continues to decrease, the soil moisture slowly decreases, which is like a power function. Simulations from current FT parameterizations can reproduce these variations in soil moisture along with decreasing soil temperature during the freezing period. It implied that the parameterization scheme (Eq. 5.3) in CLM4.5 can generally describe the relationship between soil temperature and moisture during freezing.

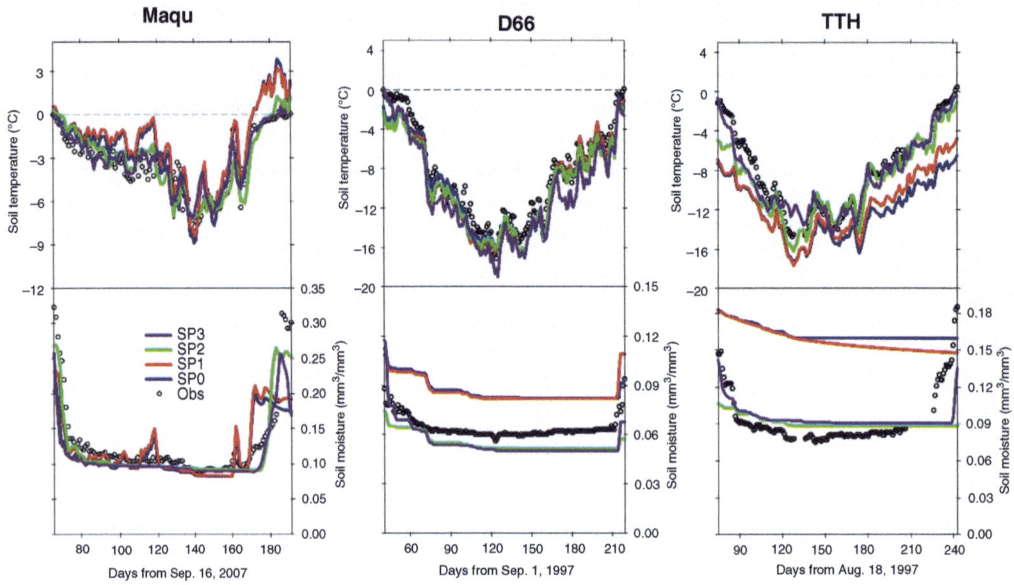

Figure 5.1 Comparison of daily soil temperature (upper row; °C) and soil moisture (lower row; $mm^3 \cdot mm^{-3}$) between observation and simulation by CLM4.5 in the FT process.

Compared to the freezing process, the thawing process is relatively rapid. Soil moisture varies with soil temperature in the thawing process. With rising temperature, observed soil moisture increases slowly at first, then with the soil temperature increasing to nearly 0°C, the soil thaws rapidly and soil moisture increases; however, the simulated soil moisture from current FT parameterizations shows a sudden increase when soil temperature reaches 0°C, and its variations were steeper than the observation, the simulated relationship between soil temperature and moisture did not coincide with the observations in the thawing process. This implies that the current frozen-thawing parameterization scheme in CLM4.5 may be problematic in describing the soil thawing process.

The soil ice prominently affects the soil water and heat transport. Many studies (e.g. Decker and Zeng 2006; Koren et al. 1999; Niu and Yang 2006; Zheng et al. 2017) have discussed the effects of soil ice on relationship between soil liquid water and soil temperature (Eq. 5.3). The relationship of soil hydrological and thermal states proposed by Koren et al. (1999) and Decker and Zeng (2006) can be expressed as:

$$\frac{g\psi_{sat}}{L_f}\left(1+c_k\theta_{ice}\right)^2\left(\frac{\theta_{liq}}{\theta_{sat}}\right)^{-b} - \frac{T-T_f}{T}=0 \qquad (5.9)$$

where $c_k = 8$. This significantly improved soil moisture simulation.

The term $(1+c_k\theta_{ice})^2$ in Eq. (5.9) might produce more liquid water and less ice faction, when set $c_k = 0$ (i.e. does not consider the ice effects in Eq. 5.3), then it is convenient for equation solution without iteration. The experiments with $c_k = 0$ and $c_k = 8$ were performed for testing the relationship from Koren et al. (1999) and Decker and Zeng (2006) against Eq. (5.3), the results were similar. However, Eq. (5.9) still failed to reproduce the daily variations of soil moisture during soil thawing process. These mean that how to represent the relationship between soil moisture and soil temperature still needs to be explored for improving simulation of the soil thawing process. Considering that some basic parameterizations in CLM4.5 have been validated and improved from observations, such as hydraulic conductivity formulation (Swenson et al. 2012), this chapter will focus on introducing how to determine the phase change occurring, and how to optimize simulation of soil moisture when the temperature is around 0 °C.

The freezing temperature used to determine the occurrence of phase change should not be a constant (e.g. Zhang, Sun, et al. 2007; Zheng, van der Velde, et al. 2017), which can be calculated as follows:

$$T_v = \frac{10^3 L_f T_f}{10^3 L_f - g \cdot \psi_{sat} \left(\frac{\theta_{liq}}{\theta_{sat}} \right)^{-B}} \tag{5.10}$$

the values in this equation were previously defined in Eq. (5.3).

Furthermore, Eq. (5.10) is definite as virtual soil temperature is used to establish a new the criterion of phase change:

$$T > T_v \text{ and } w_{ice} > 0 \text{ thawing} \tag{5.11}$$

$$T < T_v \text{ and } w_{liq} > w_{liq,\max} \text{ freezing} \tag{5.12}$$

The phase-change efficiency (ε) is also introduced (Masson et al. 2013):

$$\varepsilon = \begin{cases} \theta_{liq} / \theta_{sat} & (Freezing \quad process) \\ \theta_{ice} / \theta_{sat} & (Thawing \quad process) \end{cases} \tag{5.13}$$

This principle assumes that when liquid soil moisture is relatively large, more energy is used for freezing, compared to a more dry soil with the same available energy for freezing. The same basic idea holds for soil ice melting.

In the phase-change process, the hypothetical ice mass (H_m) when the energy is used for the soil water freezing or thawing in the process that soil temperature decreases to the freezing point is calculated as (Masson et al. 2013):

$$H_m = -\frac{c_i \Delta z_i}{L_f} \left(T_{v,i} - T_i \right) \cdot \varepsilon_i \tag{5.14}$$

where the c_i (J m^{-3} K^{-1}) is the volumetric heat capacity of soil for layer i. The detailed calculations of soil liquid water and ice and adjustment of soil temperature are described below.

Comparing simulation of daily soil temperature and moisture among simulations from current FT parameterization and modified FT parameterization, in shallow soil layers, experimenting with the modified phase-change parameterization improved the simulation of soil moisture in the FTP, especially in the thawing process, which was more consistent with the observed soil moisture variation feature (Figure 5.2). The relationship between soil moisture and soil temperature variation simulated from modified FT parameterization is confirmed by the observation (Figure 5.3). Also, the simulaitons from modified FT parameterization reproduced the diurnal variations of soil moisture, which cannot be simulated from current FT parameterization, and content of soil moisture is closer to the observation (Figure 5.4). These improvements are partly attributed to the introduction of virtual soil temperature (T_v) and the modified criterion of the

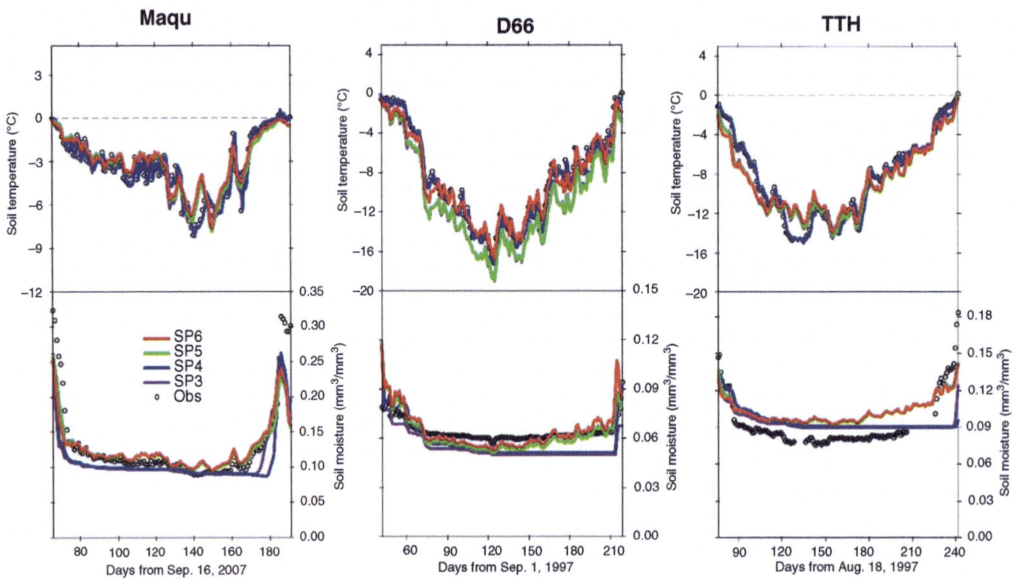

Figure 5.2 Comparison of daily soil temperature (upper row; °C) and soil moisture (lower row; $mm^3 \cdot mm^{-3}$) in Maqu (left panel), D66 (middle panel), and TTH (right panel) between observation (Obs) and simulation by original scheme and modified schemes in the FT process.

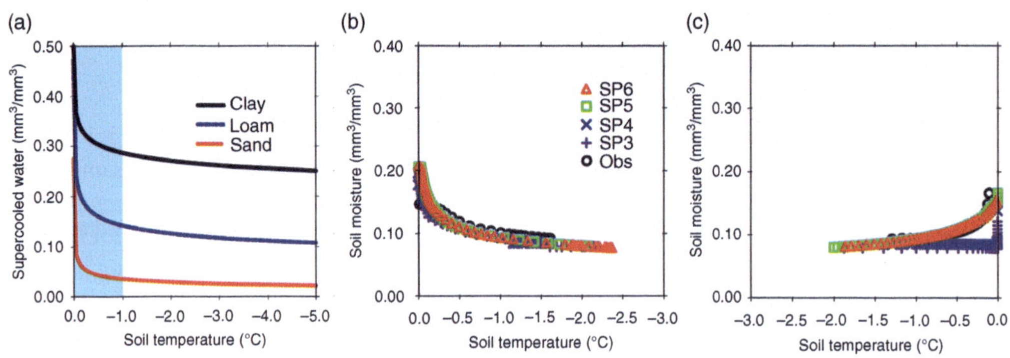

Figure 5.3 Coupling relations between soil temperature and soil moisture during soil freezing and soil thawing: (a) the variations in supercooled water ($\theta_{liq,max}$) with soil temperature for clay, loam, and sand when soil temperature is below 0 °C; (b) variations in soil moisture with soil temperature from simulations and observations (Obs) in freezing processes; (c) variations in soil moisture with soil temperature from simulations and observations (Obs) in the thawing processes.

phase change (Eqs. 5.11–5.12). These change the simple criterion that soil is frozen when the temperature is below constant freezing point (0°C) and soil thawing with the temperature above 0°C. The introduction of phase-change efficiency (ε) makes the freezing and thawing process more fit for observation during soil freezing and thawing. This means the phase-change efficiency or rate varies with changes of soil ice and liquid water content during different periods in soil FT. However, the degenerations of soil temperature simulations existed during the soil thawing period from modified FT parameterization, which was partly due to the energy consumed in soil thawing and changes of volumetric heat capacity and thermal conductivity along with transformation from ice to liquid water.

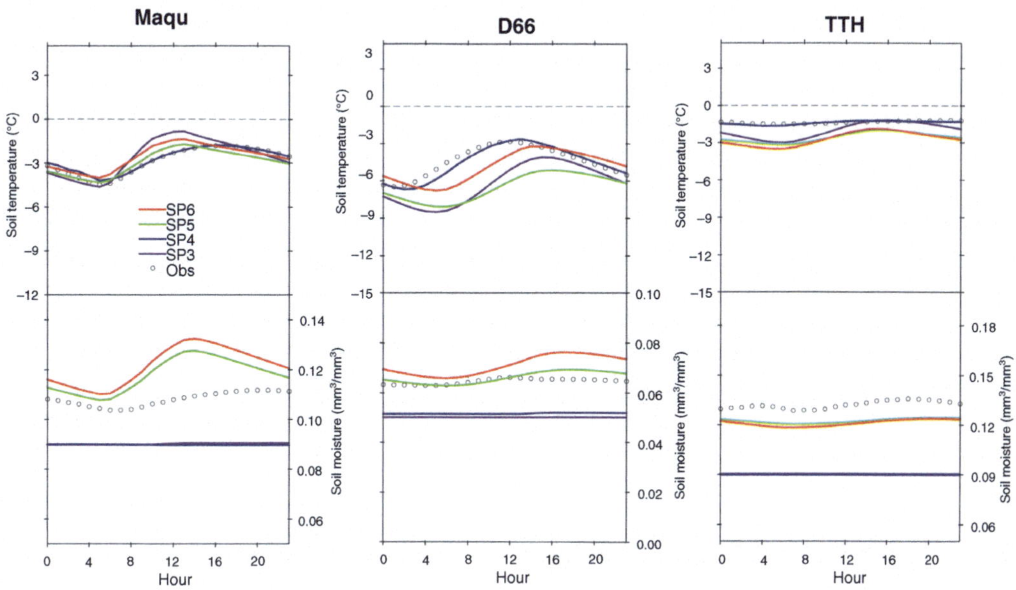

Figure 5.4 Diurnal cycles of soil temperature (upper row; °C) and soil moisture (lower row; $mm^3 \cdot mm^{-3}$) at Maqu (left panel), D66 (middle panel), and TTH (right panel) in the soil thawing period, which is averaged for 10 days before the date of complete soil thawing.

In CLM4.5, the calculation of soil-heat conduction process is separated from the phase-change process. The soil temperature is first calculated by neglecting the phase change, and then the temperature is adjusted from the energy consumed in soil thawing or released in freezing. This procedure may lead to inaccuracies in the simulation of soil temperature (Nicolsky et al. 2007). A possible impact of the phase-change process on soil-heat conduction might not be considered, which may be the cause of the cold bias of soil temperature during the soil thawing period. Thus, it assumes that the extra soil heat is transported by conduction due to soil freezing or thawing, which would further readjust the soil temperature:

$$T_i^{n+1'} = T_i^{n+1} + \frac{\Delta t}{c_i \Delta z_i}\left[\alpha\left(F_i^n - F_{i-1}^n\right) + (1-\alpha)\left(F_i^{n+1} - F_{i-1}^{n+1}\right)\right] \cdot \eta \qquad i = 2, \ldots, N_{levsoi} \tag{5.15}$$

where T_i^{n+1} is the temperature which has been adjusted in the phase-change process (Eq. 5.8), Δt is the time integration step, $N_{levsoi} = 10$, $\alpha = 0.5$, F_i ($W\,m^{-2}$) is the heat flux from layer i to layer $i+1$:

$$F_i = -\lambda\left[z_{h,i}\right]\left(\frac{T_i - T_{i+1}}{z_{i+1} - z_i}\right) \tag{5.16}$$

where $\lambda[z_{h,i}]$ is the thermal conductivity at the interface, z_i and z_{i+1} are the soil node depth of layer i and $i+1$. The detail can be seen in the technical description of CLM 4.5 (Oleson et al. 2013).

The η is the gain factor defined as:

$$\eta = \begin{cases} \left(\theta_{liq} / \theta_{sat}\right) \cdot \psi_1 & (Freezing \quad process) \\ \left(\theta_{ice} / \theta_{sat}\right) \cdot \psi_2 & (Thawing \quad process) \end{cases} \tag{5.17}$$

where the ψ_1, ψ_2 are the empirical coefficients. Simulations in Maqu: $\psi_1 = 1$ and $\psi_2 = 1$; simulations in D66: $\psi_1 = 1$ and $\psi_2 = 4$; simulations in TTH (Tuotuohe observational site): $\psi_1 = 1.5$ and $\psi_2 = 4$.

By considering the inferred impacts of phase change on soil heat conduction, compared to the simulations including T_v and ε, the cold biases of simulated soil temperature during the thawing period were reduced in simulations from modified FT parameterization (Eq. 5.15–5.17). Simulations of soil temperature and moisture in the FTP had some improvement; while some incidental degeneration in simulations of soil temperature and moisture also exist, which can be explained as: (i) the variation of soil temperature interacts with the soil moisture, and improvement in the simulation of soil temperature (or moisture) may have positive or negative impacts on the other one; (ii) since vertical transports of soil heat in different soil layers and different periods (freezing or thawing) are complex, modified parameterizations in describing the inferred impacts of phase change on soil-heat conduction may partly include the process of soil-heat transport. Nevertheless, the performances of CLM4.5 with these modifications are generally better than its original parameterization.

5.2 Hydrothermal Coupling Scheme in the Land-Surface Model

In CLM4.5, the terrestrial hydrological processes include interception of precipitation by the vegetation canopy, throughfall, surface and subsurface runoff, infiltration, evaporation, effects of snow, and soil water content evolution. The soil energy budget is governed by heat flux between soil layers and phase-change processes.

Vertical soil moisture transport is described within CLM4.5 by the modified Richards equation (Zeng and Decker 2009):

$$\frac{\partial \theta}{\partial t} = \frac{\partial}{\partial z}\left[k\left(\frac{\partial(\psi - \psi_E)}{\partial z}\right)\right] - Q \tag{5.18}$$

where θ is the volumetric soil water content (mm^3 mm^{-3}), k is the hydraulic conductivity (mm s^{-1}), Ψ is the soil matric potential (mm), Ψ_E is the equilibrium soil matric potential (mm), and Q is a soil moisture sink term (s^{-1}). The first term on the right-hand side of Eq. (5.17) represents the water flux gradient in the vertical direction, with the result that this model of water flow only considers the effects of liquid-phase movement due to the vertical hydraulic potential gradient.

The heat conduction equation, in one-dimensional form, is

$$c\frac{\partial T}{\partial t} = \frac{\partial}{\partial z}\left[\lambda \frac{\partial T}{\partial z}\right] \tag{5.19}$$

where c is the volumetric heat capacity (J m^{-3} K^{-1}) of the soil or snow, T is the soil temperature (K), and λ is the thermal conductivity (W m^{-1} K^{-1}). CLM4.5 solves this equation numerically to calculate the soil, snow, and surface water temperatures for a 15-layer soil column with up to five overlying layers of snow and a single surface water layer. Boundary conditions include a heat flux in the top soil, snow, and surface water layers from the overlying atmosphere and an assumption of zero heat flux at the bottom of the soil column. The calculation of phase changes within the soil column is separated from the thermal diffusion process; the temperature profile is calculated first, without accounting for phase changes, and then, these temperatures are used to model phase-change processes, which alter the final temperatures (Niu and Yang 2006).

Apparently, Eqs. (5.17) and (5.19) illustrate that the water and heat transport scheme used in CLM4.5 is isothermal (Figure 5.5): they assume that the flow of liquid water is not directly influenced by soil temperature, and the calculation of soil temperature only considers heat flux by conduction. In addition, water vapor flux is neglected in the current scheme. Incorporating a fully coupled water and heat transport scheme into CLM4.5 can examine and understand the impacts of these physical mechanisms, and make CLM4.5 more complete theoretically.

To do this, a fully coupled heat transport and variably saturated water flow parameterization scheme (Fully Coupled Scheme (FCS), Wang and Yang 2018), which was equipped to simulate conditions both above and below 0°C, was adapted from the scheme proposed by Hansson et al. (2004). The variably saturated water flow for temperatures above and below 0 °C is described using the modified Richards equation:

$$\frac{\partial \theta_L}{\partial t} + \frac{\rho_{ice}}{\rho_w}\frac{\partial \theta_{ice}}{\partial t} = \frac{\partial}{\partial z}\left[K_{L\psi}\frac{\partial(\psi - \psi_E)}{\partial z} + K_{LT}\frac{\partial T}{\partial z} + K_{v\psi}\frac{\partial \psi}{\partial z} + K_{vT}\frac{\partial T}{\partial z}\right] - S \tag{5.20}$$

where θ_L is the volumetric liquid water content (mm^3 mm^{-3}); θ_{ice} is the volumetric water ice content (mm^3 mm^{-3}); Ψ is the soil matric potential (mm); Ψ_E is the equilibrium soil matric potential (mm); T is the soil temperature (K); ρ_w is the density of liquid water (kg m^{-3}); ρ_{ice} is the density of water ice (kg m^{-3}); t is the time (s); z is the soil depth (mm); $K_L\Psi$ (mm s^{-1}) and K_{LT} (mm^2 K^{-1} s^{-1}) are the isothermal and thermal hydraulic conductivities for liquid-phase fluxes driven by gradients in Ψ and T, respectively; $K_v\Psi$ (mm s^{-1}) and K_{vT} (mm^2 K^{-1} s^{-1}) are isothermal and thermal water vapor hydraulic conductivities, respectively; and S is a sink term (s^{-1}) which usually accounts for root-water uptake. The methods for calculating all four hydraulic conductivities in Eq. (5.20) are presented in Hansson et al. (2004). To solve the modified Richards equation conveniently, the terms related to ice and phase changes are calculated only after the changes of total water content and preliminary soil temperatures have been solved. At this point, the changes of soil

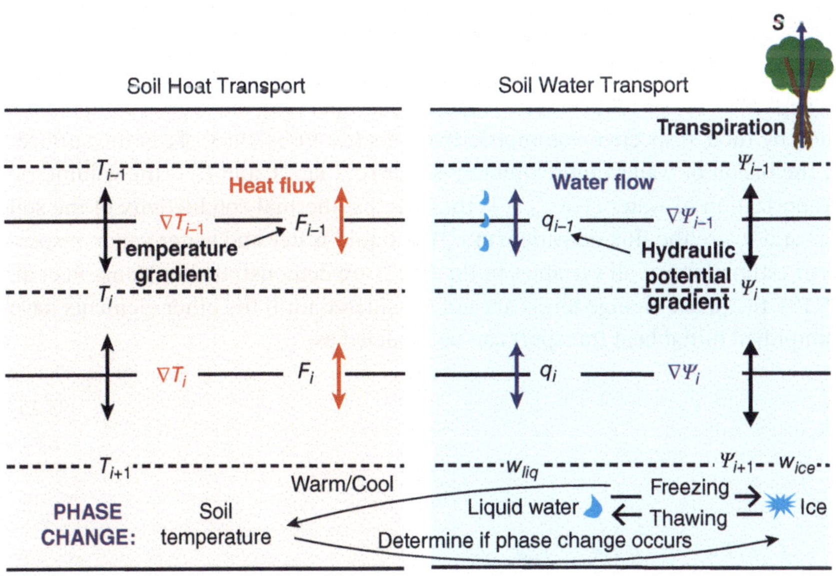

Figure 5.5 Schematic diagram of water–heat transport processes during soil freeze–thaw (FT) in the land-surface model.

liquid water and ice are calculated, and the soil temperature is readjusted based on phase-change thermodynamics.

The modified Richards equation can be written as

$$\frac{\partial \theta}{\partial t} = -\frac{\partial q}{\partial z} - S \tag{5.21a}$$

$$\text{where } q = -\left[\underset{\substack{\uparrow \\ W\_1}}{K_{L\psi}\frac{\partial(\psi - \psi_E)}{\partial z}} + \underset{\substack{\uparrow \\ W\_2}}{K_{LT}\frac{\partial T}{\partial z}} + \underset{\substack{\uparrow \\ W\_3}}{K_{v\psi}\frac{\partial \psi}{\partial z}} + \underset{\substack{\uparrow \\ W\_4}}{K_{vT}\frac{\partial T}{\partial z}}\right] \tag{5.21b}$$

and where θ is the total (liquid and solid) volumetric soil water content (mm^3 mm^{-3}) and q is the water flux. The first, second, third, and fourth terms of Eq. (5.21b) are represented by *W_1*, *W_2*, *W_3*, and *W_4*, respectively.

The governing equation for heat transport in soil is

$$C_p\frac{\partial T}{\partial t} - L_f\rho_{ice}\frac{\partial \theta_{ice}}{\partial t} = \frac{\partial}{\partial z}\left[\lambda\frac{\partial T}{\partial z}\right] - C_w\frac{\partial q_L T}{\partial z} - C_v\frac{\partial q_v T}{\partial z} - L_0\frac{\partial q_v}{\partial z} \tag{5.22}$$

where

$$q_L = -\left[K_{L\psi}\frac{\partial(\psi - \psi_E)}{\partial z} + K_{LT}\frac{\partial T}{\partial z}\right] \tag{5.23}$$

and

$$q_v = -\left[K_{v\psi}\frac{\partial \psi}{\partial z} + K_{vT}\frac{\partial T}{\partial z}\right] \tag{5.24}$$

and where C_p is the volumetric heat capacity of the soil (J m^{-3} K^{-1}), which is defined as the sum of the volumetric heat capacities of the solid, organic, liquid (C_w), vapor (C_v), and frozen components of the soil multiplied by their respective volumetric fractions (de Vries 1963). L_f is the surface-latent heat (LH) of the fusion of water (approximately 3.34×10^5 J kg^{-1}), and L_0 is the volumetric latent heat of the vaporization of water (J m^{-3}). λ is the effective thermal conductivity of the soil (W m^{-1} K^{-1}), and q_L and q_v are the flux densities (m s^{-1}) of liquid water and water vapor, respectively. Calculations or estimations of all variables in Eq. (5.22) are demonstrated in Hansson et al. (2004). As in Eq. (5.19), the phase change terms are not considered until the other elements have been solved. This simplified initial heat transport can be modeled as

$$C_p\frac{\partial T}{\partial t} = -\frac{\partial F}{\partial z} \tag{5.25}$$

where

$$F = \underset{\substack{\uparrow \\ H\_1}}{-\lambda\frac{\partial T}{\partial z}} + \underset{\substack{\uparrow \\ H\_2}}{C_w q_L T} + \underset{\substack{\uparrow \\ H\_3}}{C_v q_v T} + \underset{\substack{\uparrow \\ H\_4}}{L_0 q_v} \tag{5.26}$$

The fully coupled water and heat transport formulation defined in Eqs. (5.19) and (5.22) is more complete than CLM4.5 (Eqs. 5.17 and 5.18) in both form and theory. The water flux in FCS is driven not only by the hydraulic potential gradient but also by the temperature distribution within the soil. Most notably, the water vapor flux is considered as a part of vertical water transport. Soil hydraulic and thermal properties in above equations can be found in Appendix A.

The soil water flux in FCS includes four different processes (Figure 5.6): liquid fluxes due to the hydraulic potential gradient (*W_1*) and the temperature gradient (*W_2*), and water vapor fluxes due to the hydraulic potential (*W_3*) and temperature gradient (*W_4*). FCS describes soil-heat flux by accounting for conduction (*H_1*), convection of sensible heat (SH) with flowing water (*H_2*), transfer of SH by diffusion of water vapor (*H_3*), and transfer of LH by diffusion of water vapor (*H_4*). The energy loss associated with root-water uptake is ignored. Positive heat and water fluxes are defined as moving upward.

Numerical solutions to the current scheme are achieved by the same approaches used in the current formulation in CLM 4.5. Eq. (5.17) is solved by implicit expression, whereas Eq. (5.18) is solved using the Crank–Nicolson method. The soil layer structure is identical to the structure used in CLM 4.5 (Oleson et al. 2013): the soil column is divided into 15 layers, with soil temperature calculated in each layer, and soil moisture is only calculated in the upper 10 layers. The details of numerical discretization of the governing equations and the associated solving approaches can been found in Appendix B.

The upper boundary condition for the modified Richards equation is the infiltration rate, and the bottom boundary condition depends on the depth of the water table. Heat transport is

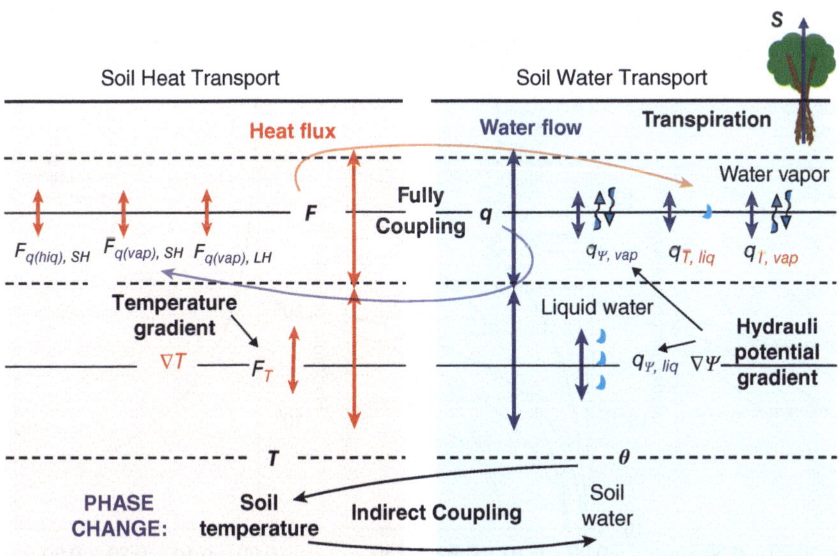

Figure 5.6 Schematic illustration of soil-heat transport (left) and soil water transport (right). F_T represents soil heat flux by conduction, $F_{q(liq),\,SH}$ represents convection of SH with flowing liquid water, $F_{q(vap),SH}$ represents transfer of sensible heat by diffusion of water vapor, $F_{q(vap),LH}$ represents transfer of LH by diffusion of water vapor, F is the total heat flux ($F = F_T + F_{q(liq),SH} + F_{q(vap),SH} + F_{q(vap),LH}$). $q\mathbf{\Psi}_{,liq}$ and $q\mathbf{\Psi}_{,vap}$ represent liquid water and vapor flux due to hydraulic potential ($\mathbf{\Psi}$) gradient, respectively; $q_{T,liq}$ and $q_{T,vap}$ represent liquid water and vapor flux due to soil temperature (T) gradient, respectively; q is the total water flux ($q = q\mathbf{\Psi}_{,liq} + q\mathbf{\Psi}_{,vap} + q_{T,liq} + q_{T,vap}$). S is the sink term including transpiration. Red subscripts mean the impact of heat transport on water flux, blue subscripts indicate the impact of water transport on heat flux.

simulated using a variable heat flux upper boundary condition, which models the heat flux from the overlying atmosphere. The heat flux at the bottom of the soil column is assumed to be negligible and is therefore set to zero. Additional details can be found in Sections 6 and 7 of the technical description of CLM4.5 (Oleson et al. 2013).

The soil moisture content and soil texture effects on hydraulic conductivity in different temperature conditions, Figure 5.7 shows the comparison of the variations of hydraulic conductivity in different soil temperature conditions. Results show that, the effects of water vapor on soil water–heat transport might be larger in higher temperature conditions than in lower temperature conditions. This can well explain the more evident effects of water vapor in arid regions than incold regions. The total effects of water vapor on soil water and heat transport (upward or downward transport) are determined by the dominance between soil hydraulic potential gradient and temperature gradient. In arid regions, due to the large vertical variations of soil water in surface layers, the hydraulic potential gradient is very large and dominant in diffusion of soil water vapor. The water fluxes by diffusion of soil water vapor are generally upward, while the opposite occurs in cold regions. This can explain the different effects of water vapor on soil water–heat transport in arid and cold regions (Figure 5.7) and the soil hydraulic potential increases with the decreasing of soil water content. It should be noted that, although the magnitude of vapor thermal hydraulic conductivity (K_{vT}) is larger than the vapor isothermal hydraulic conductivity ($K_v\Psi$), the soil hydraulic potential gradient is much larger than the soil temperature gradient. The effects of water vapor diffusion caused by hydraulic potential and temperature gradient are considerable and are both important. Along with the water vapor diffusion, the LH fluxes have influence on soil-heat transport.

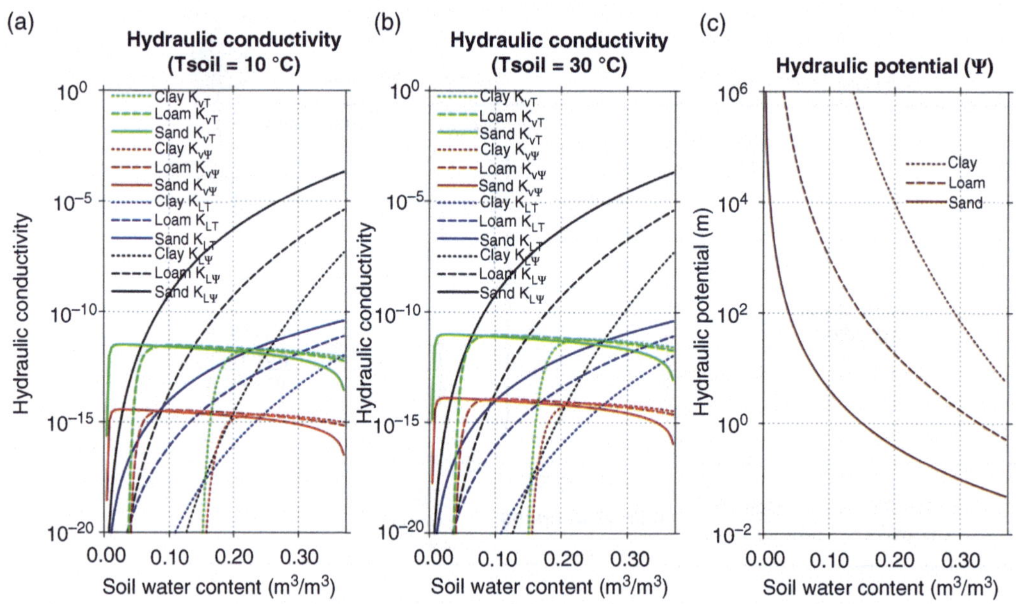

Figure 5.7 Impacts of soil water content and soil textures on hydraulic conductivities in the FCS scheme. (a) Variations of liquid water isothermal hydraulic conductivity ($K_L\mathbf{\Psi}$; m s^{-1}; black line) and thermal hydraulic conductivity (K_{LT}; m^2 K^{-1} s^{-1}; blue line), water vapor isothermal hydraulic conductivity ($K_v\mathbf{\Psi}$; m s^{-1}; red line), and thermal hydraulic conductivity (K_{vT}; m^2 K^{-1} s^{-1}; green line) for three soil textures (sand, loam, and clay) as a function of soil water content (m^3 m^{-3}) on condition that soil temperature is 10 °C. (b) It is the same as (a), except for the condition that soil temperature is 30 °C. (c) Variations of soil hydraulic potential ($\mathbf{\Psi}$; m) for three textures as a function of soil water content (m^3 m^{-3}).

5.3 Simulation of Land-Surface Process in Cold Regions

The soil temperature and volumetric water content simulated by FCS for different soil layers were compared to CLM simulation and corresponding observation data in five sites. The fully coupled water–heat transport scheme (FCS) can generally improve the simulation of soil temperature. Compared to the observations at D66 and the TTH site (Figure 5.8), both CLM and FCS reproduced well the variations of soil temperature in surface layer ($z = 0.04$ m). However, the soil temperatures simulated by CLM were lower than the observed values during the freezing period in deep layers of soil ($z = 1.0$ m). The FCS-simulated soil temperatures were better than the CLM and matched the observation well. The soil temperature simulated by FCS were also closer to the observed values and better than the CLM in surface layers ($z = 0.05$ m) at Maqu. The simulated soil temperature in deeper layers at Maqu ($z = 1.0$ m) and USBO1 ($z = 0.6$ m), in regions where is the humid climate, seemed to have slight improvements or some degradation.

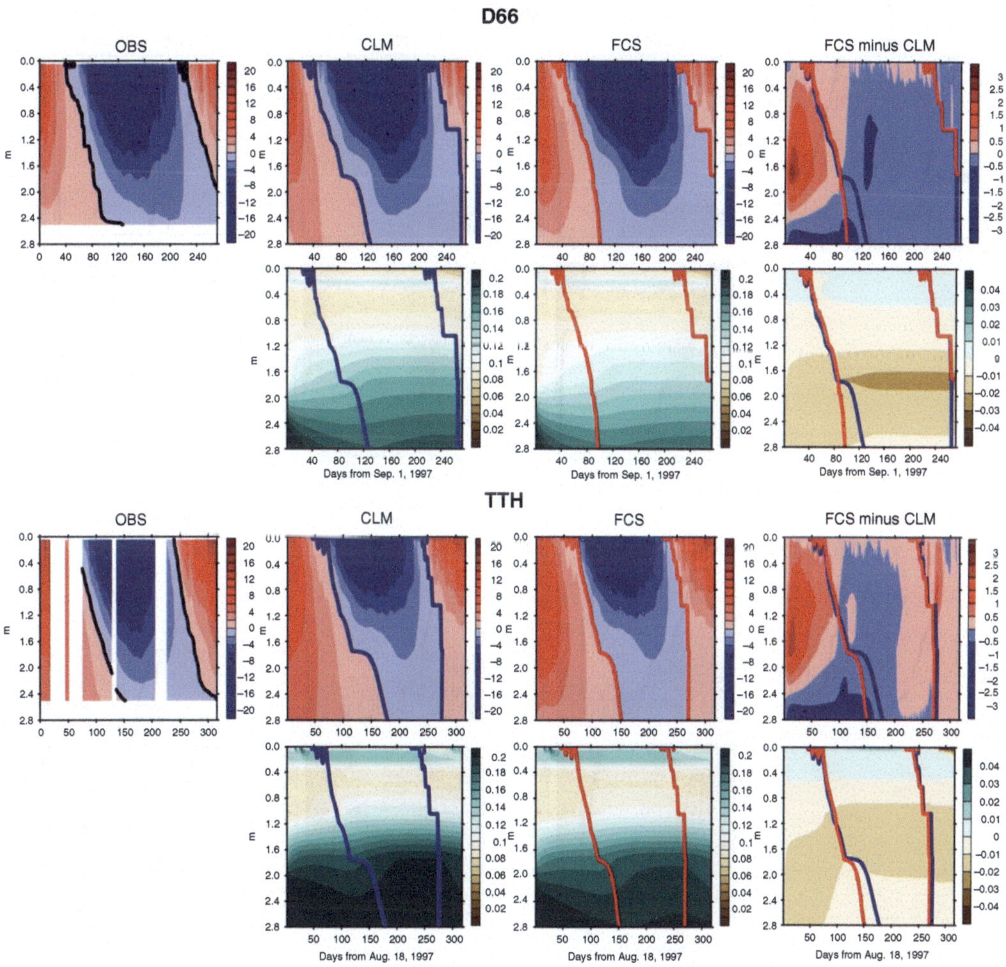

Figure 5.8 Vertical sections of daily soil temperature (°C) (first and third rows), and soil volumetric water (SVW; mm^3 mm^{-3}) (second and fourth rows) from observation (OBS), simulation of CLM and FCS, and the difference (FCS minus CLM) between CLM and FCS. The solid line represents zero degree isotherm, which is displayed by black, blue, and red solid lines for OBS, CLM, and FCS, respectively.

Due to lack of observed soil total volumetric water content, the simulated total volumetric water content is compared between FSC (fully coupled water-heat transport scheme) and CLM4.5, which shows that there are prominent regional differences between FCS and CLM4.5. At sites in cold regions (D66, TTH, and Maqu), the total volumetric water content of surface layers during the unfrozen period simulated by FCS were generally wetter than the CLM4.5. This phenomenon might be related to the vapor contribution in land surface.

Comparisons were also made of soil moisture and ice content between simulations and observations. FCS showed some considerable improvements in simulating soil moisture (Figure 5.9), especially in TTH and D66 sites. Compared to the observed soil moisture, at the TTH site, the FCS biases of soil moisture were less than the CLM during while the soil was unfrozen. The soil moisture at deeper layers (z = 1.0 m) in FCS at Maqu was also better than CLM. From the root mean square error (RMSE) for soil temperature and soil moisture FCS and CLM, generally, FCS improved the simulation of soil temperature and soil moisture at most of the sites.

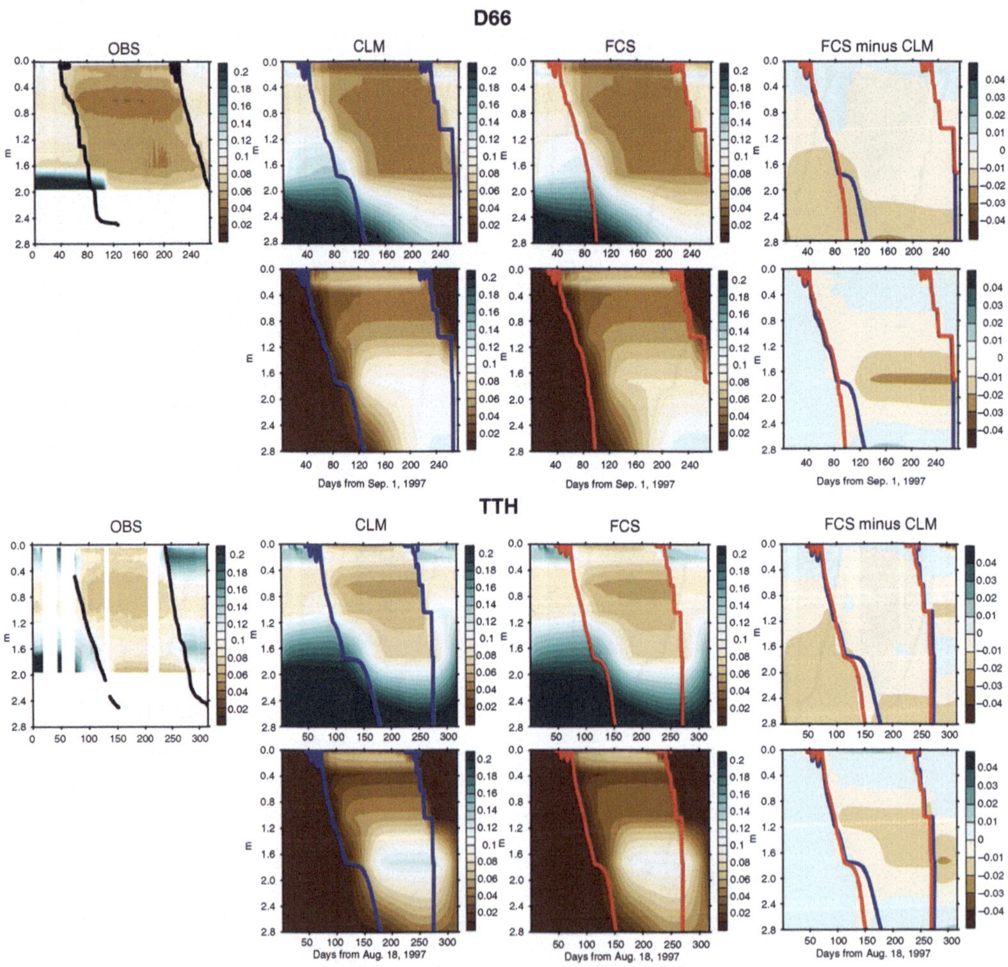

Figure 5.9 Vertical sections of daily soil moisture (expressed as liquid water; mm^3 mm^{-3}) (first and third rows), and soil ice (mm^3 mm^{-3}) (second and fourth rows) from observation (OBS), CLM and FCS simulation, and the difference (FCS minus CLM) between CLM and FCS. The solid line represents zero degree isotherm, which is displayed by black, blue, and red solid lines for OBS, CLM, and FCS, respectively.

The performances of FCS in the vertical sections of soil temperature and total volumetric soil water content simulated by CLM and FCS and observations, and illustrates the differences between them are shown in Figure 5.8. Compared to the observed soil temperature profile, in surface layers (0.0–0.1 m), the soil temperature simulated by FCS was more close to observation, and which was slightly warmer than CLM at D66 and TTH sites. The soil volumetric water (SVW) (no observation) in surface layers simulated by FCS was wetter than the CLM. At D66 and TTH, soil temperatures simulated by FCS were warmer than CLM in subsurface layers (0.1–0.5 m) during the freezing period and colder during the completely frozen and thawing period; similar soil temperature differences can also be seen in deeper soil layers (1.0–2.0 m). In general, observed soil temperature decreased with deeper soil depth, and total vertical distribution characteristics of soil temperature reproduced by FCS were closer to the observation and better than CLM. The total volumetric soil water content simulated by FCS was wetter than that simulated by CLM in subsurface layers, and was drier than the CLM at deeper soil layers (1.0–2.0 m). At the arid Naiman site, compared to CLM, FCS generally simulated the drier and colder surface layers; the soil temperature simulated by FCS in subsurface layers was closer to observation than the CLM, especially during warm season. Besides, the soil water of subsurface layers in FCS was drier than that in CLM.

The changes in soil temperature would affect the simulation of soil liquid and ice (Swenson et al. 2012). Compared to the observed soil moisture (expressed as liquid water), soil moisture simulated by FCS was closer to the observation. It was wetter than that simulated by CLM for layers between 0 and 0.2 m at D66. FCS reduced the drier biases during the completely frozen period. At the TTH site, in the observed soil moisture profile, the soil moisture during the unfrozen period showed "dry-wet-dry-wet" vertical distribution from surface to bottom. The profile of soil moisture difference between FCS and CLM (fourth column in Figure 5.9) indicated that the pattern of soil moisture vertical profile reproduced by FCS was more consistent with the observation. At the Naiman site, the CLM overestimated the soil moisture for layers between 0 and 0.5 m, and the FCS reduced this wetter bias, especially in surface layers (0–0.1 m), which should be the result that FCS considers vapor diffusion at surface.

To examine the performances and universality of new scheme (FCS), comparison between the Global Land Data Assimilation System (GLDAS) and FCS, CLM simulations showed that simulations from both FCS and CLM were close to the GLDAS. However, soil temperature simulated by FCS was a little warmer and SVW was wetter than the CLM in cold regions, while FCS simulations showed colder and drier soil than the CLM in arid regions. The differences of soil temperature and SVW were prominent in the regions where soil is dry (Figure 5.6). These are consistent with the results from single-point simulations in observed sites.

Furthermore, the comparisons of the simulated soil temperature and SVW from FCS and CLM against the GLDAS dataset over four different climate zones of cold, arid, midlatitude, and tropical regions (Yang, Z.L. et al. 2011) are shown. In cold regions, the differences between FCS and CLM were little. The variations of SVW simulated by FCS were closer to the GLDAS than the CLM ($Corr_{CLM} = 0.517$, $Corr_{FCS} = 0.521$, $P < 0.05$), while the RMSE of SVW in FCS was almost same as the CLM ($RMSE_{CLM} = 0.149\,mm^3\,mm^{-3}$, $RMSE_{FCS} = 0.150\,mm^3\,mm^{-3}$); improvement of FCS in the wetter bias of CLM in cold regions was not prominent. In arid regions, FCS reduced the soil temperature warm biases ($RMSE_{CLM}$ = 1.249 °C, $RMSE_{FCS}$ = 1.120 °C) and SVW wet biases ($RMSE_{CLM} = 0.014\,mm^3\,mm^{-3}$, $RMSE_{FCS} = 0.012\,mm^3\,mm^{-3}$) from CLM and improved the simulations, especially during the warm seasone. This would be due to FCS considering the soil water vapor diffusion process, which is important for near-surface soil water and energy budget (Bittelli et al. 2008; Novak 2010). In midlatitude, FCS-simulated soil temperature was generally better than the CLM ($RMSE_{CLM}$ = 1.525 °C, $RMSE_{FCS}$ = 1.484 °C), and FCS better described the SVW variations ($Corr_{CLM} = 0.733$, $Corr_{FCS} = 0.738$, $P<0.05$), while improvements of SVW magnitude was not evident in FCS ($RMSE_{CLM} = 0.030\,mm^3\,mm^{-3}$, $RMSE_{FCS} = 0.032\,mm^3\,mm^{-3}$). In tropical regions, both CLM and FCS-simulated soil temperature well and SVW.

The performances of FCS at both single-point and global scales showed that, the simulations of soil moisture and temperature by FCS had some improvements, and the differences between FCS and CLM in simulating soil water seemed more prominent in arid and midlatitude regions than the other regions. These illustrate the theory optimized in the land-surface model can improve the detailed simulations for soil water–heat transport.

A prominent function in FCS is the mutual influences between soil moisture and heat transport, which not only produces mutual impacts between hydrological movement and thermal regime, but is also beneficial to understanding well the detailed physics process in soil moisture and heat transportation.

The vertical sections of the heat fluxes (Eq. 5.25) and water fluxes (Eq. 5.21b) are compared. The first heat flux term (*H_1*) is a basic term in the heat conduction equation, which dominates heat flux forced by conduction with temperature gradient that includes in FCS and CLM, controls soil-heat transport, with an upward heat flux during the freezing period and a downward heat flux after the spring thaw. The second term (*H_2*) represents the convection of SH by flowing water, and it had considerable magnitude when water flow occurred. The third term (*H_3*), which indicates the transfer of SH by diffusion of water vapor, made little contribution to the soil-heat flux. The fourth term (*H_4*) represents the transfer of LH by diffusion of water vapor, which also had some contributions to soil-heat flux, especially surface layers (Table 5.2).

The influences between soil moisture and heat transport at surface layers is mutual, the evolutions of heat fluxes (*H_1*, *H_2*, *H_3* and *H_4*) in Eq. (5.25) at four surface layers ($z = 0.007, 0.028, 0.062, 0.119\,m$) are compared. As for LH transfer by diffusion of water vapor (*H_4*), except for the first layer (layer1), it resulted in downward heat fluxes at the soil surface layers in D66 and TTH. In cold region where soil is dry (e.g. Naiman), soil LH transport by diffusion of water vapor was generally upward. These illustrate the influences of soil water flow on soil-heat transport, and make considerable contributions to heat fluxes compared to the dominant term (*H_1*). The results from the model simulations were consistent with the magnitude analysis for each heat flux term.

Table 5.2 The magnitude of variables in the heat conduction equation (Eq. 5.25).

	Magnitude	Unit
λ	10^0	$W{\cdot}m^{-1}K^{-1}$
$\partial T/\partial z$	$10^1/10^1–10^1/10^{-2}$	$K{\cdot}m^{-1}$
T	10^2	K
q_l	10^{-8}	$m{\cdot}s^{-1}$
q_v	$10^{-12}–10^{-9}$	$m{\cdot}s^{-1}$
L_0	10^9	$J{\cdot}s^{-1}$
H_1	$10^0–10^3$	$W{\cdot}m^{-2}$
H_2	$10^{-3}–10^0$	$W{\cdot}m^{-2}$
H_3	$10^{-7}–10^{-4}$	$W{\cdot}m^{-2}$
H_4	$10^{-3}–10^0$	$W{\cdot}m^{-2}$

Improvements of FCS in simulating soil temperature were mainly attributes to surface LH and surface SH transport caused by diffusion of water vapor and water flow in the calculation of soil-heat fluxes. During freezing period, there is the downward heat flux at the surface layer. The soil temperatures at the surface layers (0.0–0.1 m) simulated by FCS in cold regions (D66 and TTH) were warmer than those simulated by CLM and closer to the observation. Similarly, the upward heat flux caused by soil water vapor diffusion at surface layers (0.0–0.1 m) reduced the soil temperature warmer biases in Naiman. Prominent FCS improved the soil temperatures in subsurface layers (0.1–0.5 m) and reduced the cold bias during the freezing period in CLM, which was mainly attributed to the additional soil water presented during this period. Meanwhile, FCS improved the soil temperatures in deep layers (1.5–2.8 m). The profile of soil temperature simulated by FCS was generally closer to the observed vertical distribution, that was due to the large upward SH flux (*H_2*) that accompanied the upward migration of water from the saturated zone to deep unsaturated layers.

Water flux was predominately governed by the first term of the water flux formula (*W_1*), which is the liquid water flux induced by the hydraulic potential gradient. During the non-frozen period, the water fluxes in the surface layers (0.0–0.1 m) were upward, induced by surface evaporation; due to the variations of soil hydraulic potential gradient, the soil water in subsurface layers was transported downward, and the soil water fluxes in deep layers (1.0–2.8 m) were transported upward. This is the basic characteristic and physical process of soil water transport and has been well reproduced by most of current land-surface models. FCS even provides some detailed processes that influence water transport. The second term (*W_2*) in the new scheme (FCS), which represents the liquid water flux induced by the temperature gradient, made almost no contribution to the water flux, even in surface layers. The third term (*W_3*) and fourth term (*W_4*) are the water vapor flux caused by the hydraulic potential gradient and the temperature gradient, contributed the movement of water at the surface (Table 5.3).

The temporal variations of water fluxes (*W_1*, *W_2*, *W_3* and *W_4*) in Eq. (5.21b) at four surface layers are analyzed. Water vapor fluxes caused by the hydraulic potential gradient (*W_3*) were transported upward, while water vapor fluxes caused by the temperature gradient (*W_4*) were transported downward. Considering the synergistic effects of water vapor transport by hydraulic potential gradient and temperature gradient, the total water vapor flux (*W_3*+ *W_4*) at the surface layers (layer 2–4;

Table 5.3 The magnitude of variables in the modified Richards equations (Eq. 5.21b).

	Magnitude	Unit
$\partial\Psi/\partial z$	$10^{-1}/10^{1}–10^{3}/10^{-2}$	$m{\cdot}m^{-1}$
$\partial T/\partial z$	$10^{1}/10^{1}–10^{1}/10^{-2}$	$K{\cdot}m^{-1}$
$K_L\Psi$	$10^{-13}–10^{-6}$	$m{\cdot}s^{-1}$
K_{LT}	$10^{-18}–10^{-11}$	$m^2{\cdot}K^{-1}{\cdot}s^{-1}$
$K_v\Psi$	10^{-14}	$m{\cdot}s^{-1}$
K_{vT}	10^{-12}	$m^2{\cdot}K^{-1}{\cdot}s^{-1}$
W_1	10^{-8}	$m{\cdot}s^{-1}$
W_2	$10^{-15}–10^{-11}$	$m{\cdot}s^{-1}$
W_3	$10^{-16}–10^{-9}$	$m{\cdot}s^{-1}$
W_3	$10^{-12}–10^{-9}$	$m{\cdot}s^{-1}$

z = 0.028–0.119 m) were transported downward; thus, the first layer (layer 1) was drier and lower three layers (layers 2–4) were wetter than the CLM in D66 and TTH; while the total water vapor fluxes at layers 1–4 were transported upward in Naiman, which led to the entire drier surface layers compared to the CLM. The water flux from vapor diffusion in arid regions (Naiman; about 10–20 $\times 10^{-7}$ mm s^{-1}) were larger than those in the cold region where soil is relatively wet (D66 and TTH; about 5–10 $\times 10^{-7}$ mm s^{-1}). These model results are consistent with the magnitude analysis for each term of water flux. These soil water fluxes resulted in drier surface layers in D66 and TTH, which was accompanied by the wetter subsurface layers, and drier deep layers. The differences in the simulated soil temperatures and soil water contents between FCS and CLM4.5 in the surface layers confirmed the importance of soil water vapor diffusion in simulating surface soil water and heat transport.

Land characteristics are mainly constituted by the variation of soil texture and soil water content (wetness or dryness). The mutual influence between soil water and heat transport have distinct regional characteristics. In dry-soil conditions ($\theta < 0.2\,m^3\,m^{-3}$), the vapor isothermal and thermal hydraulic conductivity ($K_v\Psi$ and K_{vT}) have the considerable magnitude with the liquid water isothermal hydraulic conductivity ($K_L\Psi$), and in the extremely dry condition ($\theta < 0.1\,m^3\,m^{-3}$), the effects of water vapor on soil water transport are larger than the effects of liquid water. Generally, layers near the surface remain dry (excluding rainfall condition); thus, the effects of water vapor on soil water and heat transport are more prominent in near surface layers than the deeper layers.

Altering the soil temperature gradient and hydraulic potential gradient, FCS further influences the simulation of soil temperature and moisture in turn. The heat flux by conduction (*H_1*), simulated by FCS, was generally weaker than the conduction simulated by CLM4.5 at surface layers in cold regions D66 and TTH. This was the result of the decreases in the temperature gradients as subsurface layers were warmer during freezing periods and colder during thawing periods in D66 and TTH. This negative feedback partly retarded the cooling effects of LH flux by diffusion of water vapor. In Naiman, the cooling in surface layers would enhance the temperature difference between soil and air, and the downward heat fluxes transported into surface layers would consequently increase. The upward water fluxes induced by the hydraulic potential (*W_1*) near the surface simulated by FCS, are weak compared to CLM4.5. Drier surface layers and upward water flux anomalies in the subsurface layers contributed to this phenomenon. In deep layers (1.0–2.5 m), the upward water fluxes simulated by FCS were weaker than those from CLM. FCS-simulated drier middle layers, caused by a decreased hydraulic potential gradient between drier deep layers (about 1.0 m) and middle layers (about 0.5 m). FCS simulated more warming prior to soil freezing, compared to the simulations from CLM. Soil water contents simulated by FCS were lower than the values from CLM4.5. Owing to the decrease of total volumetric water content, soil liquid water simulated by FCS also decreased, and soil ice was less abundant due to warmer soil temperatures during freezing conditions. Although the magnitudes of the added terms in FCS (*W_2*, *W_3*, and *W_4* or *H_2*, *H_3*, and *H_4*) are small relative to the first terms (*W_1* or *H_1*), the addition of these physical processes had considerable direct impact and feedback on the simulations of soil water and heat transport.

The changes in simulating soil moisture and temperature would affect the surface adiabatic heating. As the differences in the simulated surface diabatic heating between FCS and CLM show, the simulated surface SH fluxes by FCS decreased, while the simulated surface LH fluxes increased. In cold regions where soil is wet (D66 and TTH), the increase of LH was linked to the enhanced ground evaporation from FCS. In cold regions where soil is dry (Naiman), the decrease of SH was related to the lower-ground temperature simulated by FCS. The increase of LH resulted from more upward water flux by soil water vapor diffusion in surface layers. The changes of SH and LH can reach about 10^{-1}–10^{0} W m^{-2}. These changes would dramatically impact surface air temperature (e.g. Ban-Weiss et al. 2011).

6

Hydrological Processes in Cold Regions

Cold regions are the sources of rivers and freshwater. Glacier and snow melt are the predominant water resource in cold regions. Solid precipitation formed in the cold seasons is the water resource in warm seasons for cold regions and the adjacent areas. The hydrological processes of cold regions are usually closely related to their underlying surface and environment, which also determine the partitioning of runoff and infiltration. According to water form, the hydrological processes can be generally divided into glacier melting water, snow melting water, frozen soil thawing, river and lake ice, and sea ice hydrological process. The main characteristic of cold region hydrology is that water is stored in solid form and released in liquid. The ice-water inter-conversion process and its impacts on resources and the environment are the focus of people's attention.

Most of this this chapter will take China as an example to present the hydrological phenomena of glaciers, frozen soil, snow, sea ice, river ice and lake ice, as well as the hydrological processes of the water bodies in cold regions, such as mountain rivers, alpine lakes, and wetlands. In addition, the hydrological mechanism and change process of the elements in the cryosphere, as well as the effects of cryospheric changes in the cold regions and beyond, will also be discussed.

6.1 Glacier Hydrology

As an essential part of terrestrial freshwater resources, glaciers play an important role in the development and utilization of water resources in inland region, especially in the arid areas. For instance, in the 1980 and 1990s, glacier water resources assessment in China through the comprehensive glacier meltwater runoff modulus method, flow rate and temperature relations, comparative observation experiments, etc., the representative areas expanded to mountains and countries, the estimated Chinese glacier total annual runoff is $563.3 \times 10^8\,m^3$ (Yang 1991) and $604.7 \times 10^8\,m^3$ (Kang 2000), respectively. According to the simulation results of the response of the glacier system to climate change, the annual total runoff of glaciers in China is $615.8 \times 10^8\,m^3$ (Xie et al. 2006). According to an improved monthly degree-day factor model, the annual mean runoff of glaciers in China is $629.56 \times 10^8\,m^3$ (Zhang et al. 2011), and is about 2.2% of the national river runoff (Hydrology Bureau of the Ministry of Water Resources and Power 1987), which is more than the annual mean runoff of the Yellow River into the sea, and equivalent to the 10.5% of river runoff of Gansu, Qinghai, Xinjiang, and Xizang provinces ($5760 \times 10^8\,m^3$) in western China.

For most glaciers, runoff occurs during the so-called warm season, when the daily mean temperature is above 0 °C (Figure 6.1). In other words, the melting process presents apparent seasonal

Climatology in Cold Regions, First Edition. Chenghai Wang.

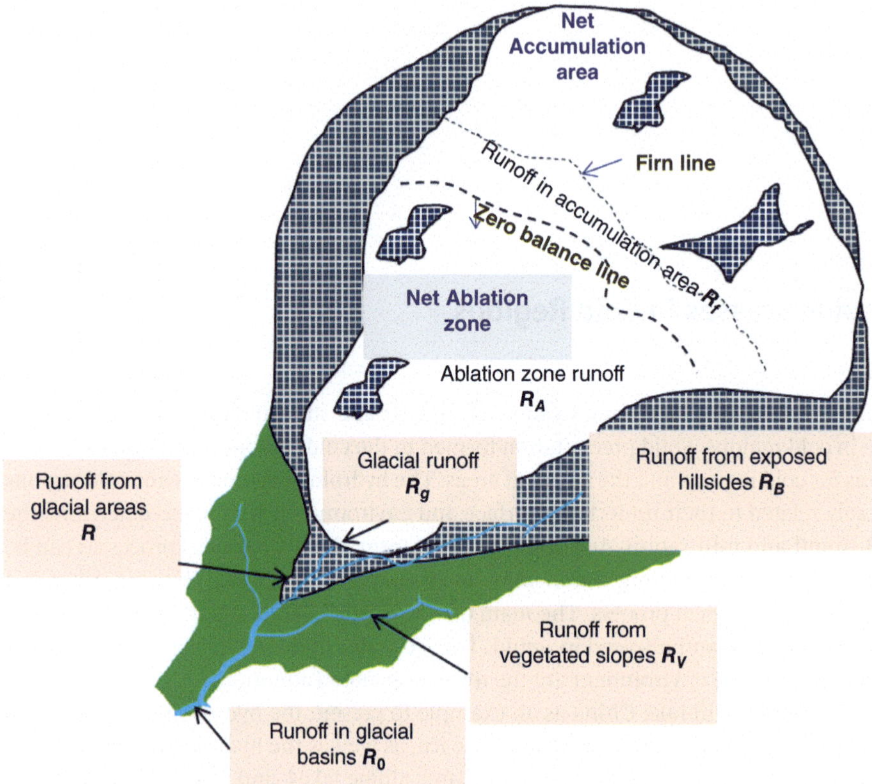

Figure 6.1 Runoff components of continental glacial watershed (Ding 2017).

variation. Glaciers in the northern hemisphere melt from June to August, while melting from December to February of the following year in the southern hemisphere. The characteristics of meltwater runoff are also different for different types of glaciers. The variation of runoff from continental glaciers is most significant, but the distribution is not uniform. The melting period is short (May–September), and the meltwater is highly concentrated in June–August. The annual variation of marine glacier runoff is slight, and the distribution is relatively uniform. The ablation period is relatively long (April–October), and the ablation is distinct from May to September. Glacial meltwater has undergone a complex water migration process from generation to outflow (Figure 6.2). Meltwater can reach the end of a glacier by three routes: surface, in-ice, and sub-ice, which in turn, communicate with each other to form a complex glacier drainage system. The drainage system and confluence process of different types and sizes of glaciers also differ significantly.

6.2 Snowmelt Runoff

Snow cover is a layer of snow that exists on the earth's surface for less than a year. Snow cover is a critical and unique hydrological element on the land surface. In winter in the northern hemisphere, 40% of the land surface is covered by snow. As an unique underlying surface, it has the characteristics of high albedo and low thermal conductivity, which has an important impact on the weather and climate system, thus is a factor that cannot be ignored. Meanwhile, snow is a vital water resource in middle and high latitudes. Snow hydrological processes include two aspects:

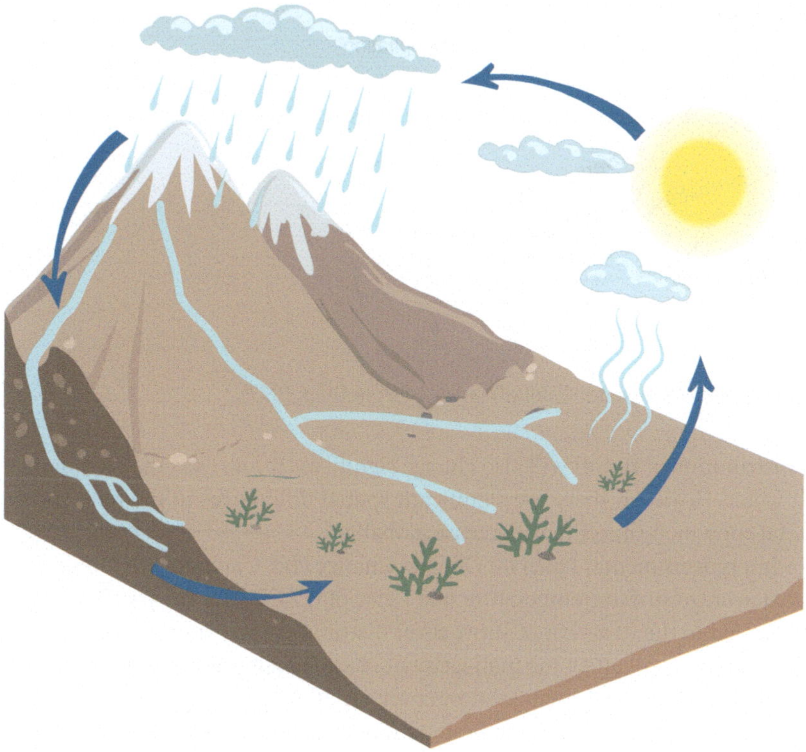

Figure 6.2 Water cycle in glacial areas.

infiltration and runoff. In spring, when the surface temperature rises (T > 0°C), snow melting infiltration enters the soil and increases soil moisture and matric potential. Snowmelt runoff accounts for 50–70% of annual surface runoff. Even in alpine areas at lower latitudes, snowmelt water accounts for a significant proportion of channel runoff. Therefore, snow cover and its melting process impact the global energy and water cycle.

The size and duration of snowmelt runoff are mainly related to accumulation, the strength of ablation, characteristics of the underlying surface and surface evaporation. In the Tianshan and Altai Mountains of China and other regions, the snowmelt runoff is strong in spring due to a large accumulation in winter, while in the Qilian Mountains, the snowmelt runoff process is weak. The snow accumulation in high-altitude areas is more distinct than that in low-altitude areas; accordingly, the snowmelt runoff is more significant in high-altitude areas than in low-altitude areas. In permafrost regions, because of the poor infiltration capacity of meltwater, snowmelt is more likely to produce runoff, while in soils with well-developed vegetation and humus layers, snowmelt runoff is mostly collected in the form of soil flow. Before the end of snow melting, the upper soil is frozen, and the infiltration capacity of snow meltwater in the soil is limited. Once the meltwater reaches the surface, it will rapidly converge at the bottom of the snow layer and form surface runoff. After the snowmelt water reaches the surface, it can enter the river channel through the following four confluence modes: (i) direct runoff; (ii) the slope surface flow; (iii) the runoff in soil; (iv) the underground runoff (Figure 6.3) (Ding, 2017).

With global warming, it has become an indisputable fact that the snowmelt period is significantly earlier. Due to the early melting of a large amount of snow, the snowmelt runoff increased significantly in the early period and decreased correspondingly in the late period; thus, the snowmelt runoff also decreased. This will change the annual distribution of snowmelt runoff in the

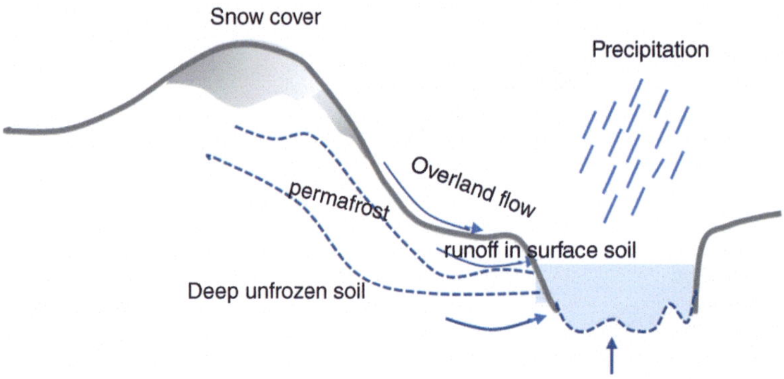

Figure 6.3 The confluence of snow meltwater (rainfall) (adapted from Ding, 2017).

basin, and the annual distribution of runoff will also change significantly for rivers that are mainly supplied by snowmelt water. However, there are significant spatial differences in the total snowmelt runoff, and the total snowmelt runoff may decrease globally.

It was highlighted in Intergovernmental Panel on Climate Change (IPCC) reports (Solomon and Qin 2007) that the global average surface temperature had a rise of 0.74°C over the past 100 years. Global climate change has led to the variation of snow cover distribution and snowmelt runoff on different scales, and this variation will affect the management of snowmelt water resources. Rising temperature has caused redistribution and trended snow cover variations at watershed scales in the northern hemisphere, High Asia, and mid-latitude mountains (McCabe and Wolock 2009). Water resources supply according to climate change is the most critical issue in recent studies, which indicated that the discharge regime in snowmelt-dominated river basins is most sensitive to temperature increases in the snow season (Rauscher et al. 2008; Adam et al. 2009; Day 2009; Stewart et al. 2004). Spring snowmelt runoff was earlier observed in many watersheds worldwide (Bates et al. 2008; Lemke et al. 2007). With more than one-sixth of the Earth's population relying on glaciers and seasonal snow packs for their water supply, it is still necessary to diagnose and predict the consequences of these hydrological changes for future water availability (Barnett et al. 2005).

According to the research of Wang et al. (2010), take the Heihe Basin (Hexi corridor in Gansu province of China), second inland river in northwest China, as an example; the impact of climate change on snowmelt runoff in mountainous regions is distinct. From observation of 46 *in situ* stations, the annual mean air temperature has increased about 7–14°C during 1956–1995; the maximum rises in January and February, and air temperature rising in winter is most prominent; meanwhile, snowfall also increased, and snowfall increasing due to air temperature rising can reach by 7–10%.

Based on the Snowmelt Runoff Model (SRM), Wang et al. (2010) analyzed the responses of snowmelt runoff to different climate change scenarios: increases in air temperature at +2°C, +4°C, +6°C thresholds and heavier precipitation multiplied by 1.5 and 2 times. They found the relation between snowmelt and temperature in the period from March 5 to June 19, 2004, namely, with increases of air temperature (+2°C, +4°C, +6°C), the time of peak snowmelt runoff moved ahead and the discharge increased as well (Figure 6.4a). The first peak discharge happened on 30 April 2004 when there was no increase in air temperature. With an increase of +2°C, the peak value increased from 102.8 $m^3 s^{-1}$ to 197.5 $m^3 s^{-1}$ and the start time of snowmelt runoff shifted unobtrusively; with an increase of +4°C, the peak discharge increased to 167.1 $m^3 s^{-1}$ and for the first time it moved ahead to 24 April; with an increase of +2°C, the peak value of snowmelt runoff increased to 240.0 $m^3 s^{-1}$ and for the first time of it moved ahead to 21 April.

Wang et al. (2010) reported that earlier snowmelt runoff and larger discharge are typical features of responses of snowmelt runoff to air temperature increase. The lag effect of advanced snowmelt is that the discharge of the later snow season becomes less and then discharge decreases correspondingly. With more temperature increases, the discharge in the later snow season decreased more. The other after effect of it is that the rainy season comes early because precipitation happens at the same time as rain in higher temperature conditions, so the snowmelt confluence processes were also affected. In the rainy season, the influence of temperature increase on runoff is small (Figure 6.4a). Moreover, the mean daily runoff of the whole snow season is 47.8, 39.4, 35.0, 36.1 $m^3 s^{-1}$, corresponding to four scenarios simulated by the SRM model, respectively: present, +2°C, +4°C, +6°C. As a result, the influence of temperature rising on the average discharge of the snow season is not apparent.

Assuming three scenarios of precipitation, increases will be 1.5 times and 2 times greater than original precipitation, and precipitation will be multiplied by twice and with a +4°C increase of air temperature (Figure 6.4b). The discharge process showed a marked increase with the increase of precipitation. The discharge was enlarged, but there is almost no influence on the occurrence of snowmelt runoff, and the difference between precipitation fluctuation trends of different increased precipitation scenarios is not obvious. With a comparison among three different scenarios, there is no noticeable discharge change before the snow melting occurred in winter, while the discharge has been enlarged when the snowmelt season comes, and this amplification effect is more evident in the mixed rain and snow season and the rainy season. The increased precipitation affected mainly the

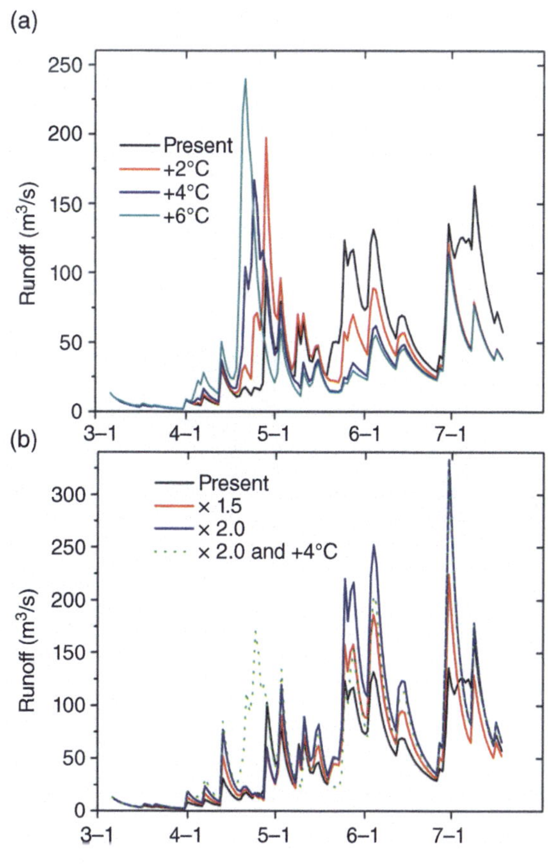

Figure 6.4 Responses of snowmelt runoff to air temperature increases, +2°C, +4°C, +6°C (a). Responses to three different scenarios: precipitation multiplied by 1.5 and 2 times, and precipitation multiplied by two times and with a +4°C increase of air temperature (b) (Wang et al. 2010).

expansion of the discharge. The average discharge increased from 47.8 to 52.6 $m^3 s^{-1}$ and 67.9 $m^3 s^{-1}$, with amplification of 10% and 42%, respectively, when precipitation increases 1.5—2 times.

Another climate change scenario simulated by the SRM is precipitation multiplied twice and with a +4 °C increase in air temperature (Figure 6.4b). This case could be considered as the combination of air temperature and precipitation increases. The snowmelt runoff has a substantial advance and the peak discharge has dramatically increased, but the discharge in the later snow season is less than in the scenario of precipitation multiplied only by two. The average discharge is 67.0 $m^3 s^{-1}$, close to 67.9 $m^3 s^{-1}$ that the discharge of the scenario with single precipitation increases. Further, the total runoff is mainly affected by precipitation other than the air temperature.

The relation between snowmelt and temperature in the Heihe corridor (Basin) indicates that air temperature has increased gradually in the past 30 years (1956–1986) in the upper Heihe River Basin, and the increasing trend was slow from 1956 to 1980 but accelerated after 1980. The air temperature increase is about 2.1, 2.6, and 2.9°C at different weather stations in the basin. Annual precipitation has also increased; the most substantial rising precipitation was from 1977 to 1986. In recent years, the increasing snowfall trend has been less pronounced but with a significant fluctuation. The comparison of monthly discharges showed that snowmelt runoff had increased obviously from 1970 to the present while the autumn discharge hydrograph shifted ahead and its volume decreased. Snowmelt runoff occurred earlier and discharges become more significant as the air temperature increased, and the influence of temperature rising to the average discharge of snow season is not apparent. With increased precipitation, the discharge hydrograph showed noticeable increases in the volumes of peak discharges, but there is insignificant influence on the occurring date of peak discharge. As a result, the total discharge is mainly affected by increased precipitation other than air temperature.

Another example is the Spiti River, a high-altitude Himalayan river located in the western Himalayan region, where snow and glacier melt runoff also have significant contributions to its total streamflow (Figure 6.5). Although snow water equivalent reduces with an increase in air temperature, no significant change is found in the snow water equivalent of some regions in the Spiti Basin by the projected increase in air temperature (1–3°C). An increase of 2°C in air

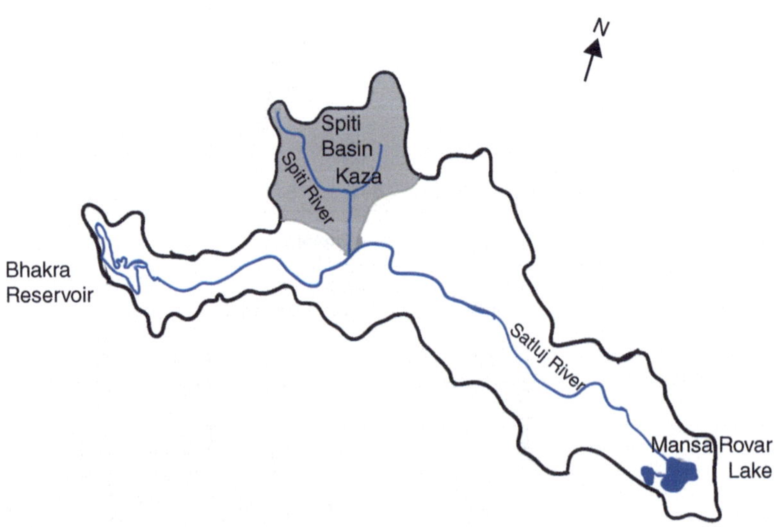

Figure 6.5 Location map of the Spiti Basin. *Source:* Singh and Kumar (1997) / with permission of Elsevier.

temperature reduced annual snow water equivalent from 1% to 7%. Changes in precipitation caused proportional changes in snow water equivalent. It is found that annual snowmelt runoff, glacier melt runoff, and total streamflow increase linearly with temperature changes (1–3°C), but the most prominent effect of an increase in temperature has been noticed on glacier melt runoff for this high-altitude basin. For example, an increase of 2°C in air temperature has enhanced annual snowmelt runoff, glacier melt runoff, and total streamflow in the range of 4–18%, 33–38%, and 6–12%, respectively. The change effect in precipitation suggests a linear increase in snowmelt runoff and total streamflow, while, in general, glacier melt runoff is inversely related to changes in precipitation. Snowmelt runoff is more sensitive than glacier melt runoff to changes in precipitation. Under a warmer climate scenario, snowmelt runoff and glacier melt runoff cause an earlier response of total streamflow and a change in flow distribution. The seasonal analysis of total streamflow indicates that an increase in air temperature produces an increase in the pre-monsoon season, followed by an increase in the monsoon season.

6.3 Frozen Soil and Hydrology

Frozen soil has three effects on the hydrological process. When soil starts freezing, soil moisture moves to the frozen front, enhancing soil moisture redistribution vertically and horizontally, and while the soil is thawing, the soil moisture is concentrated on the upper soil, where it is thawing first, resulting in surface soil swelling. Frozen soil can hold soil moisture, prevent evaporation, and keep soil wetness persistent for one to two seasons, it acts as a "water storage" (see Section 9.8, Freezing and Thawing Damage, Chapter 9). Frozen soil is also a natural water repellent, which can prevent water from infiltrating into deeper soil (Figure 6.6).

Permafrost plays a role in the water-repellent layer, and the different kinds of permafrost (continuous and discontinuous distribution) has a different water-repellent effect which has a different influence on the regional frozen soil hydrological process. The amount of subsurface ice stored in permafrost and its distribution also influence regional hydrological processes. Compared with non-frozen regions, the hydrological processes in permafrost regions are affected by the weak water permeability of the permafrost layer and have the following characteristics.

The runoff coefficient is closely related to permafrost coverage; the weak water permeability of permafrost areas makes most of the snowmelt fall into the direct runoff. Therefore, the permafrost region with high coverage usually has many hydrological characteristics, such as high runoff yield, high direct runoff coefficient, the short response time of runoff to precipitation, and a short retreat stage. With the decrease of permafrost coverage, these characteristics gradually weaken.

The annual runoff distribution is characterized by high runoff peak value and low runoff in winter. The runoff peak value in permafrost areas usually occurs in the transition period from spring to summer, when precipitation and snowmelt yield is significant; the permafrost has a strong water-isolation effect and low-infiltration rate, so there is a high runoff peak value. As the year enters summer, the active layer thaws gradually, the frozen surface decreases, the water-isolation effect weakens gradually, and the surface runoff begins to decline from the peak. When the active layer is thawed completely, the infiltration intensity increases, the water storage capacity of the basin increases, and evaporation increases. At this time, the active layer can play a role in runoff regulation to weaken the flood peak. In winter, the runoff is small because the frozen soil blocks groundwater recharge to runoff. If the regional frozen soil coverage rate is 100%, the runoff in winter may even be close to zero.

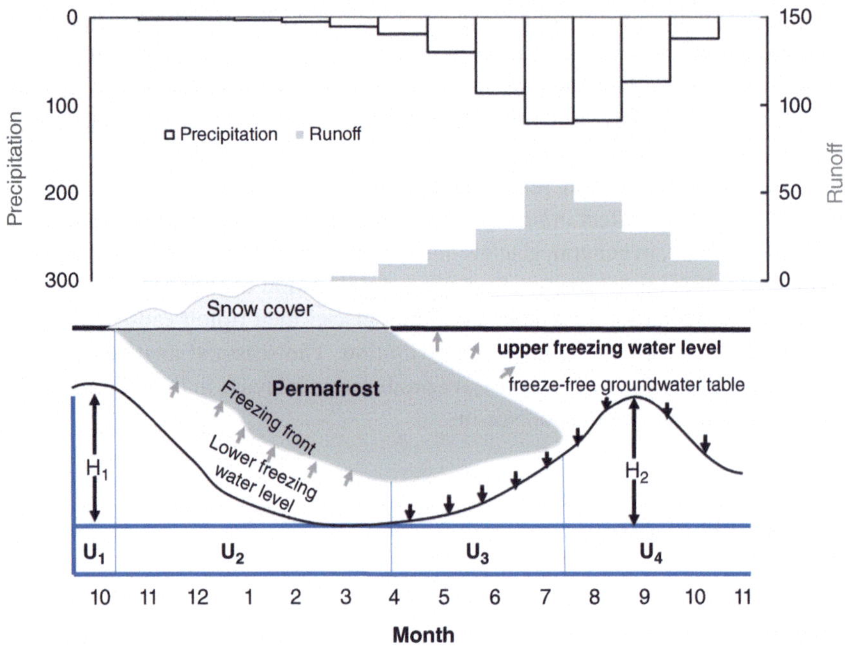

Figure 6.6 Process of groundwater supply in a cold area.

The characteristics and distribution of permafrost dramatically influence the water cycle between surface water and groundwater: groundwater level is influenced not only by the amount of snow melt and rainfall recharge, but also by the melting depth of the active layer and the recharge path of the frozen region. The change in the active layer is the main controlling factor of groundwater level change. In a large area of continuous frozen soil for many years, meteoric water, surface water and shallow water infiltrate from the local melting and bedrock fragmentation zones, and then migrate laterally and recharge to deep groundwater. Recharge and discharge conditions cross, thus, the distribution of groundwater water volume is very uneven, and the hydraulic connection with surface water is poor. In the island permafrost region, groundwater's recharge, runoff and discharge conditions are good, and the mutual conversion between groundwater and surface water is frequent and the hydraulic connection is complex, so there are many types of groundwater. In the alpine permafrost region, the recharge and discharge of groundwater are closely related to the geomorphic lithofacies zone. Generally, the alpine permafrost zone is the recharge zone. The Gobi gravel zone is the main runoff zone, and the central oasis and lacustrine zone are the discharge zone.

Permafrost has significant eco-hydrological effects: permafrost organizes the infiltration process in the active layer and retains a certain amount of water in the active layer, which is of great significance for maintaining the ecosystem in the permafrost region. In permafrost regions, the plant growth period is shorter, and the root system of vegetation usually becomes shallower and wider longitudinally. Compared with the vegetation in non-permafrost regions, its water-holding capacity is weaker. In general, the distribution and characteristics of frozen soil have a certain correlation with the vegetation types in the frozen soil area. Moss, shrub, grassland, and meadow vegetation cover correspond to different soil properties, temperature, water content and frozen soil types, so they have different eco-hydrological processes.

The hydrological process of the seasonal permafrost region is similar to that of the perennial permafrost region, because the annual average temperature of the seasonal permafrost region is higher, and the runoff generation is generally earlier than that of the perennial permafrost region. The seasonal frozen layer in the permafrost region also acts as a water-resistant layer, and its hydrological effect disappears with the thawing of the seasonal frozen layer. The hydrological characteristics of seasonal frozen soil summarized the following characteristics (Xiao et al. 2008):

1) Seasonal frozen soil increases the water storage of the soil layer. The increased water of frozen soil in different periods of the freezing–thawing cycle includes the rain and snow infiltration water in the unstable freezing period. During the stable freezing period, the amount of water is frozen at the front due to the migration of deep soil and phreatic water. Snow melt and rainfall infiltration occur during the melting period; the amount of water is released from the permafrost during thawing. As a result, the runoff coefficient of seasonal frozen soil in the freezing–thawing period is higher than that in the freezing-free period.
2) The water storage regulation of seasonal frozen soil makes the recharge time of groundwater lag: during the frozen period, the water frozen in river channels and soil will be released to recharge groundwater during the thawing period, and the water above the frozen layer can recharge groundwater in the form of gravity water after the thawing of frozen soil. Therefore, the recharge time of rainfall is obviously lagging, and only when the frozen soil is thawed can rainfall have a direct relationship with the variation of groundwater level.
3) The water in the frozen soil retained by seasonal freezing in the seasonal frozen region has a significant eco-hydrological effect: soil evaporation capacity is significantly reduced during the frozen soil existence period, and surface evaporation is almost zero. The water in the frozen soil can increase by 20–40% compared with the water before freezing, which is an indispensable groundwater resource for plants to absorb and use directly during winter and spring to maintain their survival and reproduction ability. It is the only water resource in arid and semi-arid regions, and it is the leading factor in promoting salinization and swamping and maintaining ecological balance.
4) The existence of seasonal frozen soil affects the hydraulic connection between groundwater and surface water: the water on the frozen soil accumulated in the frozen period is often imbibed or converted into a runoff by the soil in the thawing period, and the supply to the upper water is limited. For example, in the seasonal frozen regions of the Lesser Khingan Mountains in China, groundwater supply to annual runoff is usually about 12%. Due to the limited hydraulic connection between surface water and groundwater in seasonal frozen regions, they usually show different hydrochemical properties.

Some differences in hydrological processes between permafrost regions and seasonal permafrost regions are as follows. There are some remarkable periglacial landforms in permafrost regions, including frost bulges, ice cones and stone rings, which have a specific influence on the hydrological process. The hydraulic connection between surface water and groundwater in permafrost regions is usually weaker than in seasonal permafrost regions. There is less chance for groundwater in permafrost regions rising to the surface, and there is a weaker hydraulic connection between groundwater and water in permafrost regions. Similarly, water in permafrost regions can only infiltrate to the bottom of the active layer, making it difficult to recharge groundwater in permafrost regions.

Frozen soil has many distinctive effects for the soil hydrological process. Frozen soil limits moisture exchange between the atmosphere and ground, reduces infiltration ability, hydraulic conductivity, and permeability, and alters surface water and groundwater flow. The presence of a frozen soil layer, especially permafrost, can serve as a barrier to the interaction between surface water and groundwater because of the hydraulic impedance of pore ice, resulting in negligible groundwater flow in permafrost regions (French 2007).

The specific heat capacity of ice (2.1 J g^{-1} K^{-1}) almost is half of that of water (4.2 J g^{-1} K^{-1}), whereas the heat conductivity of ice at 0 °C (2.18 W m^{-1} K^{-1}) is larger than that of water (0.58 W m^{-1} K^{-1}). Thus, when water or ice changes phases as part of the freeze and thaw process, the system either releases or absorbs substantial latent heat at a rate of 334 J g^{-1}.

In a warming climate, the degradation of frozen soil is associated with an increase in air temperature, reduction of snow cover in melting season, and ecosystem transformation and land cover changes are expected to fundamentally affect the thermal and hydrological regimes of high-latitude and altitude regions where large areas of permafrost and seasonally frozen soil exist. Frozen soil degradation will change soil hydraulic conductivity, heat capacity, heat conductivity, and the releasing and absorbing of latent heat. In the Arctic, changes in permafrost conditions have led to more groundwater flow and increased base flow (subsurface runoff), especially during winter (Peterson et al. 2002). Elsewhere, as a result of the thawing of the confining permafrost, groundwater recharge and discharge rates would increase, giving rise to the deepening of groundwater flow paths and the lowering of the water table (Michel and van Everdingen 1994; St. Jacques and Sauchyn 2009; Bense et al. 2012). Permafrost degradation is also linked to shrinking wetlands, the release of soil carbon, desertification, the worsening of the ecological environment, and increased incidence of natural hazards (Hinzman et al. 2005; Hayes et al. 2014).

The largest region of permafrost and seasonally frozen soil outside the Arctic and Antarctic is the Tibetan Plateau (TP). The TP not only has climate effects but also is water resource for surrounding regions. Hydrologically, the TP is the headwater region of several large Asian rivers, including the Yellow River, the Yangtze River, the Mekong River, the Salween River, and the Brahmaputra River, which provides water for hundreds of millions of people in South and Southeast Asia. Unlike permafrost at high latitudes, permafrost on the TP is generally ice poor due to the arid climate, high evaporation, and glacial history. Ice segregation, ground heave and subsidence, and related periglacial landforms on the TP are not as common as in high-latitude permafrost regions (Yang et al. 2010). Because of these differences, changes in frozen soil on the TP could exhibit different effects on hydrology from what has been reported in the Arctic region (Hinzman et al. 2005) and elsewhere (e.g. Romanovsky et al. 2010; Frauenfeld and Zhang 2011).

On the TP, the permafrost area has been reduced from 1.50 × 106 km^2 in the 1970s to 1.05 × 106 km^2 in 2012, most likely due to global warming (Cheng and Jin 2013). The most substantial decreases in the depth of seasonally freezing ground have occurred in the central, northeastern, and northwestern portions of the plateau where the deepest propagation of seasonal frost exists. Zhao et al. (2010), Wu et al. (2012) reported that frozen soil along the Qinghai-Tibet Railway has degraded, with increases in the active layer thickness of 6.3 cm per year, and increases in temperatures at 6 m and zero-amplitude depth of 0.02 °C and 0.012 °C per year, respectively. The freeze start date was delayed by 1.7 days per decade, while the freeze end date was advanced by 4.7 days per decade during 1981–2010. Through field observations in 1997–1998, Zhang et al. (2003) reported that the freeze–thaw cycle during the monsoon season is the principal hydrological process in the active layer of the eastern TP permafrost region, and that precipitation and evaporation are the most important controls on this cycle. Variations in the active layer freeze–thaw cycle, along with vegetation cover changes, appear to be the major factors in controlling the hydrological processes in the permafrost region of the TP (Wang, Hu, and Li 2009). Given the importance of the freeze–thaw cycles that are primarily dictated by changes in soil temperature and moisture, it is necessary to investigate the long-term changes of soil temperature and moisture at different depths and examine how the changes could relate to long-term surface hydrological changes on the TP, which are essentially unknown.

Cuo et al. (2015) reported that, over the TP, soil temperature at the surface layer peaked in July and August. As soil layer depth increased, the peak of soil temperature tended to lag further behind the peak of the layer above and the magnitude of the seasonal cycle became progressively smaller. Liquid soil moisture displayed unique and different seasonal patterns between the permafrost station (PS) and the seasonally frozen soil station (SFS). PSs showed well-defined seasonal cycles in the three hydrological layers with the peak in the lower layer lagging behind that of the layer immediately above by one month. Specifically, at PSs, liquid soil moisture started to rise in April, May, and June, peaked in August, September, and October, and fell in September, October, and November, respectively, in the first, second, and third hydrological layers.

At SFSs, the seasonal cycles of liquid soil moisture were less well defined when compared to those at PSs. In the top two layers of SFSs, liquid soil moisture started to rise in February, formed a plateau lasting from April to September, and dropped in October, while the third layer did not show noticeable seasonal changes. For most of the SFSs, liquid soil moisture in the second layer was much lower than that in the first layer throughout the year, a clear contrast to most of the PSs, for which the second layer contained slightly less liquid soil moisture than the first layer. This disparity is likely related to the hydrological impedance effect of pore ice in permafrost soil. The third layer soil moisture at both PSs and SFSs was much lower than that of the first layer. At SFSs, significant increasing trends tended to occur in the early or late period of the year, while June–September was associated with decreasing or small increasing trends. The March–September soil temperature in the top layer was affected more by air temperature maxima than by air temperature minima. The ground heat conduction had more influence on deep soil temperature than air temperature maxima and minima.

In the cold months, liquid soil moisture increases with soil temperature. In May–September, the increasing soil temperature reduced soil water content. At PS deep layer, soil moisture increased with the increase of soil temperature year round. While at SFS it manifested a seasonal variability. It is because when the ice content was zero in the warm months at SFS, it was different from PS where ice content existed in the second and third layers throughout the year. The effects of precipitation were weaker than the effects of the soil temperature at PSs and similar at SFSs in general. In June–September, the positive impacts on the top layer liquid moisture by precipitation were offset by the negative impacts from soil temperature. This is likely one of the reasons that change in liquid soil moisture at SFSs tended to be small or negative in the warm months even though the northern TP has become wetter in recent decades (Cuo et al. 2013).

In other words, the increases in liquid soil moisture in the top layer in the cold months are due mainly to increases in soil temperature and enhanced soil ice melt, which is essentially the consequence of frozen soil degradation. The precipitation affects the top layer liquid moisture primarily in the warm months, while the effects of soil temperature can span the entire hydrological layers, especially at permafrost sites. For liquid soil moisture, increases in the cold months are due mainly to increases in soil temperature and enhanced soil ice melt, while changes in the warm months vary among months and stations and are the results of competing forcing between positive precipitation and negative soil temperature effects in the top layer. PSs and SFSs exhibit distinct thermal and hydrological features that are different from each other. PSs display a narrower time span of active hydrological processes (i.e. May–September) than SFSs (i.e. April–October). Both precipitation and liquid soil moisture display positive correlations with evapotranspiration (ET), and runoff during the warm months across the stations. Base flow generation is dominated by deep groundwater. ET changes are primarily associated with precipitation changes. The slight enhancement of the surface hydrological processes is indicated primarily by the statistically nonsignificant changes in evapotranspiration and runoff in most months at the study sites, and the enhancement is due mainly to the effects of precipitation changes, and subsequently to frozen soil degradation. It is expected that, besides precipitation changes, changes in liquid soil moisture also have implications for surface hydrological processes, though varying degrees. Under current conditions, permafrost soil plays a limited role in surface hydrological processes compared to seasonally frozen soil. However, as global warming continues and permafrost degrades to seasonally frozen soil, more active surface hydrological processes could be expected in the warm months in the region.

In general, surface hydrological processes were enhanced, though not in a statistically significant way. The enhancement was due mainly to increasing precipitation, with frozen soil degradation also playing some role through its effects on evapotranspiration, runoff, and base flow in cold regions.

6.4 River Ice and Lake-Ice Hydrology

River ice, lake ice, and sea ice are important components of the cryosphere, its development not only directly reflects climate change, but also has a far-reaching influence on the global water cycle. The river ice formed in the inland open water and lake ice is different. Generally, rivers and lakes in high altitudes, and middle and high latitudes may experience different degrees of ice in winter every year. The freezing and melting of river and lake ice are mainly affected by meteorological conditions, temperature, water quality, flow and human disturbance (Figure 6.7).

Ice crystals form on the surface of lakes and rivers first when the water temperature drops below freezing during icing. At the freezing point, the formation of various types of ice depends on the turbulence intensity, flow velocity, and heat loss rate. In the slow-flow region, thin ice forms on the water surface. In the low-flow velocity region, complete thin ice will form, but in the high-flow velocity region, some thin ice, intra-water ice flow, ice disk or fully mixed intra-water granular ice flow will also form.

On the banks of rivers, lakes, or reservoirs, thin and transparent shore ice is firstly formed because of the rapid loss of heat from rock and soil, small water flow, and early formation of ice crystals. The formation and development of shore ice are mainly affected by five factors: local heat exchange, shore velocity, ice density, channel morphology and water depth. At the same time as shore ice formation, if the river velocity decreases, there will be subzero supercooling water in the river. Under supercooling and mixing of the water flow, ice crystals will be generated in any part of the supercooling water, gradually forming intra-water ice of various sizes and shapes. Normally,

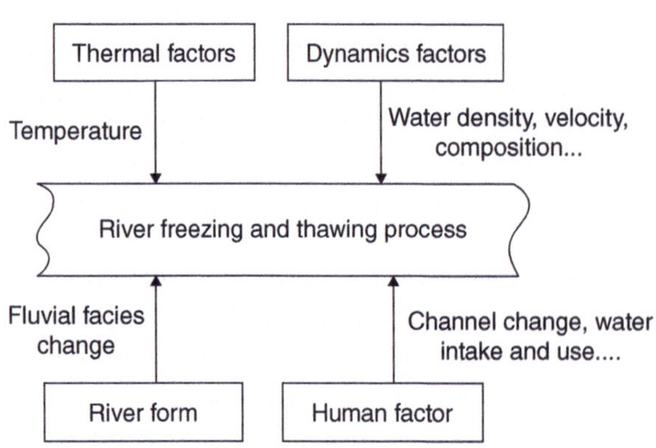

Figure 6.7 Influencing factors of the hydrological freezing–thawing process of river ice.

the amount of ice in the water declines from the water toward the river's bottom. However, when the ice volume in the water increases, it floats to the surface of the water and flows downstream with the ice crystals on the river forming floating ice. Ice at the bottom of the water is less dense than water and may rise to the surface of the water during turbulent flow, forming ice floes. The formation of ice floes depends on surface temperature and turbulence (Ding, 2017).

Study of Shen (2010) show that, in the north of China, river ice freezing is usually divided into two stages: wu seal and wen seal. The former is mainly affected by ice dynamics, and the latter is mainly affected by temperature. During the freezing period, river and lake ice freezing mainly goes through two stages: ice formation and ice thickening. Generally, lakes or reservoirs are prone to thermal action and form smooth ice sheets, while river ice sheets are mostly blocked and accumulated by floating ice generated in the glaciation period and gradually develop from downstream to upstream to form the whole ice sheet. When the ice cover covers over 80% of the total water surface of a river, lake, or reservoir, then a river, lake or reservoir enters the frozen period.

During the freezing period, the ice form depends on the ice situation and flow conditions in the upper reaches of the river, and there are two types of ice development upstream: flat sealing and vertical sealing. Among them, the flat seal ice is smooth and formed by the accumulation of ice sheets juxtaposed together; the standing seal ice is formed when the ice flows are pressed and stacked against each other due to high river velocities or strong winds, and freeze to form an ice sheet with jagged surfaces. Nevertheless, when the river flow velocity reaches a certain degree, vertically sealing at the same time, the ice will descend with the flow and seal the ice accumulation with the early generation plug. This ice jam blocking action significantly increases the flow resistance. When continuous ice is formed on the surface of rivers and lakes, the heat exchange between water and the atmosphere can only be conducted by ice, and the ice thickness will increase and change with the heat exchange.

When temperatures rise above freezing, rivers thaw and the ice begins to melt. Usually, the shore heats up faster and the shore ice melts first and breaks away from the banks. As temperatures continue to rise, the ice melts, eventually breaking under turbulence and winds, sliding and flowing again. At this time, according to the different thawing forms of river and lake ice, there can be a tranquil break and a violent break. If the change of flow is small and the effect of water flow is not strong, the ice will mainly melt in place caused by heat, with no or little ice jam or ice-dam harm, in other words, the tranquil break.

On the contrary, if the flow increases greatly, the ice sheet does not melt fully. It is mainly broken and thawed under the force of the water flow, which can easily cause ice jam or ice-dam damage, namely the violent break. The main reason for the formation of the violent break is that during the frozen period of the river channel, if the temperature difference between the upper and lower reaches is significant, when the temperature rises in spring, the snowmelt in the upper reaches melts or the river channel thaws first. The water volume increases and the water level rises rapidly, while the river channel is still sealed, and the ice water impinges on the ice of the river channel in the lower reaches. If a large amount of ice is blocked in the narrow curved channel, it is easy to form an ice dam, causing the water level to rise, forming an ice flood.

6.5 Sea-Ice Hydrology

The development of a sea ice area is comprehensive, its high albedo variation between the barrier for power to be able to and the radiation balance and the air-sea interaction. The freezing and thawing of massive sea ice will also change the water temperature and salinity of the ocean, which will affect the thermohaline circulation process of the ocean, and then cause the change of atmospheric circulation and global climate. Below, we will briefly summarize Ding (2017)'s work.

From the perspective of hydrology, 25 million km^2 of sea ice also belongs to global freshwater ice and it is a crucial component of the cryosphere. Essentially, sea ice, a product of seawater at low temperatures, is a mixture of freshwater ice crystals, salt and air bubbles. As a part of the cryosphere, sea ice is distributed at both poles and middle and high latitudes. It accounts for a certain proportion of the global ocean, is easily affected by climate, and plays an essential role in the climate system.

Sea ice is formed, developed and melted in the ocean, mainly distributed in the Antarctic and Arctic. In the northern hemisphere, the southern limit of sea ice distribution is roughly Bohai Bay, China (about northern latitude 38°). In the southern hemisphere, sea ice forms only near Antarctica and occurs from north to south latitudes 55°. Macroscopically, the evolution of sea ice can be divided into three stages: initial glacial stage, frozen stage, and final glacial stage.

Deglaciation refers to the period from the first ice date to the freeze date when sea ice is increasing. In China, the first ice day in Liaodong Bay and the northern Yellow Sea is in early November at the earliest and late November at the latest. For Bohai Bay the earliest ice day is in early December, the latest early late December, The earliest ice day in Laizhou Bay is early December and the latest mid-January.

The freezing period is from the freezing date to the thawing date. During this period of severe ice conditions, ice density is more than 70%; the period of severe ice conditions is also known as the heavy-ice age. The freezing period of Liaodong Bay in China is about two and a half months, usually from late December to March. Bohai Bay is frozen for about one and a half months, generally from early January to mid-February; Laizhou Bay is frozen for about a month, generally from mid-January to early February.

The end of the ice age is the period from thawing to the end ice date when sea ice melts as temperatures rise and sea surface temperatures (SSTs) rise. The melting period is much shorter than the growth period. The last ice date of Liaodong Bay and the northern Yellow Sea is early in mid-March at the earliest and late March at the latest. The end of Bohai Bay's ice age can begin at the end of February at the earliest end of mid-March at the latest. The thawing in Laizhou Bay's begin as early as late January, and as late as late mid-March.

6.6 Sea Ice

The Arctic sea ice can be divided into multi-year ice and one-year ice, and its sea ice extent varies greatly in season and year to year. From the perspective of the inter-decadal variation of Arctic sea ice extent, the extent of Arctic sea ice was relatively stable before the 1970s, but gradually decreased after the 1980s, and the decreasing trend accelerated in recent years, and the decreasing rate was not consistent in each sea area. Among them, the sea ice area in the East Siberian Sea has the most apparent decreasing trend, followed by the Chukchi Sea and the Beaufort Sea, while the sea ice in the Canadian Basin has the lowest decreasing rate and the sea ice thickness is also in a state of continuous thinning. In the northern hemisphere, The maximum sea ice extent in March–April is (15×10^6~16×10^6 km^2), minimum and from August to September (6×10^6~8×10^6 km^2) (Sea ice data from https://nsidc.org/).

The freezing period of Arctic sea ice is from October to March of the following year, and the melting period is from April to September of the following year. Generally, the sea ice begins to freeze in October, with a faster freezing rate that lasts until December and the average thickness is about 3 m.

From January to March of the following year, the growth rate of sea ice slowed down and reached the maximum thickness in March. The ablation began from April, accelerated from May to August, and the ablation rate was relatively slow, and the thickness was the smallest in September.

Most of the Antarctic sea ice is annual ice, and the extent of Antarctic sea ice varies not only significantly from year to year but also regionally. For nearly 30 years, the entire Antarctic sea ice increased speed by 1.3×10^4 km^2/*a*, but not a straight line up. The extent of sea ice in the southern hemisphere is maximum in September (18×10^6~19×10^6 km^2) and minimum in March (2×10^6~3×10^6 km^2) (Sea ice data from https://nsidc.org/).

Regarding spatial variation, the extent of sea ice in the Belingkojin-Amundsen Sea decreases. The Ross Sea saw the fastest increase in sea ice extent, followed by the Weddell Sea. The sea ice extent in the Indian and Pacific oceans showed a trend of small and slow increases. Regarding distribution, the annual sea ice is mainly concentrated in the Weddell Sea near the Antarctic Peninsula and the south of the Ross Sea. The sea ice in the Weddell Sea near the Antarctic Peninsula is more in the east and less in the west, while in the Ross Sea, it is the opposite. In addition, there is perennial ice on the Abbott, Cook, Shackleton, West and Fenbull ice shelves, but seasonal sea ice dominates the rest of the Antarctic continent.

In terms of the spatial distribution of sea ice, the sea ice concentration near the Antarctic Peninsula in the Weddell Sea is more in the east and less in the west, while it is the opposite in the Ross Sea. During the melting process from September to February of the following year, sea ice concentration decreases first near the eastern Weddell Sea, the Ross Ice Shelf in the Ross Sea and the Antarctic continental margin. The Weddell Sea ice is the positive feedback center of Antarctic sea ice and plays a leading role in the change of Antarctic sea ice. Interaction between north and south polar sea ice, polar sea ice of the Pacific Ocean, the Ross Sea ice at the South Pole, is the center of negative feedback; the feedback inhibition of sea ice is changing. The Ross Sea ice plays a leading role and influence on the west side of the Pacific Ocean in the Arctic area of sea ice. Arctic sea ice plays a leading role in the Atlantic ocean and affects the Antarctic sea ice in the Weddell Sea (Bian and Lin 2005).

The difference in sea ice seasonal variation and area index variation between the North and South Poles are mainly caused by the different geographical environments. The mainland surrounds the Arctic ocean. Only the Bering Strait, the Fram Strait and the Barents Sea are exchanged with an external ocean water channel; the seasonal variation of sea ice is limited within the limited area of the Arctic Ocean, so that the Arctic sea ice seasonal change is small, and as far as sea ice around the Antarctic is concerned, there is no land with the same temperature as the Arctic. Seasonal change in Antarctic sea ice is more significant than the North Pole. In addition, the Antarctic climate is influenced by the annular mode of the southern hemisphere and the Antarctic circumpolar current. As well as the vast cold storage of the Antarctic ice sheet; the temperature is kept at a low level all year round, which promotes the formation of sea ice. However, the Arctic climate is significantly affected by the Arctic Oscillation and the North Atlantic Oscillation. The glacier area is much smaller than that of the Antarctic region, and the four seasons are more distinct, resulting in higher temperature in the Arctic region, smaller sea ice area and significant seasonal and interannual variations.

6.7 Characteristics of Sea Ice Regimes in China

Ding (2017) suggested that, the Bohai Sea and the northern part of the Yellow Sea in China are the lowest latitude sea ice formation areas in the northern hemisphere. The thickness of the sea ice in the Bohai Sea and the Yellow Sea is 15–40 cm, and the ice period lasts for

90–130 days. The ice situation has significant annual and interannual variation, and the ice situation in light-ice years and heavy-ice years is also very different. In the past hundred years, 1915, 1936, 1947, 1957, 1969, and 1977 were heavy-ice years of which the spring of 1969 was the most severe.

The extent of sea ice in Liaodong Bay is large, and is affected by winter winds. The extent of sea ice continues to expand to the mouth of the bay. On the one hand, sea ice drift expands the outer edge of sea ice; on the other hand, it constantly lowers seawater's temperature, making it cool and easier to freeze. Under the influence of the counterclockwise shore current, sea ice drifts from the north coast to the east, forming a wide range of ice accumulation onn the Bayuquan. The west bank is mainly flat ice, mainly single-layer ice, but not thick.

The sea ice in Bohai Bay is mainly distributed along the 15 m isobath line influenced by circulation, and the ice area near the estuary is larger. Under the influence of winter winds, sea ice continues to drift to the west and south of Bohai Bay, forming overlapping and accumulated ice areas. Weather conditions are favorable for sea ice to freeze here, resulting in the heaviest ice years in history. In recent years, the ice situation in the Bohai Sea has been relatively light in the context of global warming.

The ice situation in Laizhou Bay is the lightest, freezing late and melting early, but it is unstable and repetitive, and the ice period is generally from early December to the end of February of the following year. Among them, the sea area near the Yellow River Delta in the west of Laizhou Bay is relatively heavy, mainly due to the low salinity of seawater, the interaction between coastal topography and wind field conducive to accumulation, and the large shoal area along the delta, while the sea area in the east is not conducive to the accumulation of sea ice, and the ice situation is relatively light. In estuaries and shoals, sea ice accumulation in extreme years can reach 3 m and fixed sea ice can reach 40 cm. After the 1980s, the sea ice in the Yellow Sea entered the stage of ice decline, but compared with the ice situation in the Bohai Sea, the ice situation change amplitude is more severe.

6.8 Precipitation Rates on the Tibetan Plateau

Water vapor is one of the vital factors in the atmosphere and is directly related to rainfall. According to the IPCC Fourth Assessment Report, in the past 100 years (1900–2005), the number of extreme precipitation events that have occurred in arid regions has rapidly increased (IPCC 2013). In general, according to the Clausius–Clapeyron equation, as temperature increases by 1 K, the water-holding capacity of the atmosphere increases by 7%. Thus, climate warming has intensified the moisture in the atmosphere (Trenberth et al. 2003). In addition, water vapor is an important greenhouse gas and plays a major role in climate systems as a dominant feedback factor with radiative effects and moisture dynamics (Trenberth et al. 2005). The water vapor in the lower troposphere is a crucial factor for precipitation, and precipitation releases latent heat, which affects the diabatic heating structure in the troposphere.

Global temperature increase has also enhanced evaporation, which has led to increased atmospheric moisture and precipitation. The simulation results of Bosilovich (2004) indicated that global moisture would increase during the twentieth century as temperature increases, as well as precipitation in the last half of the twentieth century. Trenberth (1999) estimated that for extratropical cyclones, approximately 70% of the precipitation came from moisture already in the atmosphere. The rest of the precipitation came from local evaporation, and only 20% of the precipitation came from local evaporation on a 1000-km scale.

Furthermore, contrasting analyses of various datasets showed that the Precipitable Water (PW) in atmospheric columns over the global ocean increased from 1998 to 2001. PW is significantly related to the SST (Trenberth 2005). The tropospheric PW in North America had a significantly increasing trend, but the results were not remarkable for the Eurasian continent (Ross and Elliott 1996). From 1958 to 1995, the precipitation in western China had significantly increased (Zhai et al. 2004). Water vapor pressure increased in China during 1961–2000 (Wang et al. 2011), and atmosphere-containing moisture does not equate to precipitation. The discordance of atmospheric moisture and precipitation variability implies changes in precipitation efficiency (i.e. percentage of atmospheric moisture transfer to precipitation during a storm) (Trenberth 1998).

Due to global warming, changes in spatial and temporal precipitation and PW characteristics have occurred, especially over the TP. There was a remarkably abrupt temperature change over the TP around 1980. Zhai et al. (2004) reported that the anomalies of precipitation in China turned to positive from negative around 1980. This means that precipitation has an increasing trend after 1980. Thus, due to global warming, an asymmetric change trend between the minimum and maximum temperatures occurred in the TP. The daily minimum temperature has increased faster than the daily maximum temperature; the range of daily temperatures has also increased. This increase is consistent with the rise of the near-surface temperature over the TP (Oku et al. 2006). The increased surface temperature led to the frozen ground degradation in the TP (Wang et al. 2002) and evaporation enhancement, intensified local water cycle and increased precipitation. In addition, temperature increases can give rise to glacial recessions, frozen soil (Wang et al. 2008), snow (Wang et al. 2009b), and circulation effects (Wang et al. 2009). Wang et al. (2009) also proved that the warm pool SST has impacted the increasing temperature over the TP. In contrast, Liang et al. (2006) used GPS remote sensing to retrieve the distribution of PW over TP and found three high moist centers to exist on the TP in the southeast, southwest and northwest regions. The seasonal variation of water vapor also closely relates to the 500 hPa wind field and the total atmospheric water vapor flux. The multi-year mean (1958–2001) of PW over the southeast region of the TP and its adjacent regions has exhibited a decreasing trend. The water vapor that has been transported into these regions has also suffered the same trend (Zhou et al. 2007). Cai et al. (2004) discovered that the PW over the TP and its adjacent areas varied significantly among the different regions and seasons. The summer PW over TP is the most divergent compared to other regions in China; however, the maximum of the summer PR Precipitation Rate (PR) over the TP is approximately 22%. The temporal and spatial patterns of PW and precipitation, the relationship between them and the characteristics of PR variability, are mainly due to the changing climate.

The PW is defined as the total water vapor in a column of atmosphere, that is, a column-integrated water vapor over a unit area, and is calculated as follows:

$$PW = \frac{1}{g}\int_{P_S}^{P_T} q\,dp \tag{6.1}$$

where P_T represents the pressure on top of the atmosphere (hPa); P_S represents the surface pressure (hPa); g is the acceleration of gravity ($\mathrm{m\,s^{-2}}$); and q is the specific humidity ($\mathrm{g\,kg^{-1}}$).

The PR is defined as the following:

$$PR = \frac{\mathrm{Pr}}{PW} \times 100\% \tag{6.2}$$

where P_r represents precipitation.

In past decades, the TP temperature has been rising. An abrupt change occurred around 1980. Since then, the temperature has remained continuously high (Wei et al. 2003). Precipitation levels have also experienced a remarkable change. Lin et al. (1996) revealed that, except in the northern region, precipitation over the TP increased from 1950 to 1990. Hu et al. (2003) reported a dry trend between 1951 and 2000 over the eastern edge of the TP. Ma and Hu (2005) reported stable precipitation over the main body of the TP except Qinghai and the eastern and southern regions of the TP, with the latter increasing more than any other region by approximately 10.2% in 1971–1999. Wei et al. (2003) investigated precipitation on the TP from 1961 to 2000 and found that the precipitation changed in the winter and spring in the northern TP, with the same pattern in the south and summer (i.e. precipitation made a positive departure in 1960s, but it was insufficient in the 1970s and early 1980s). During the mid-1980s and 1990s, the precipitation again made a positive departure. Across the wet seasons, the precipitation change in the northern TP was reversed in the south. The above research revealed that the spatial and temporal distributions of the precipitation change were remarkably varied across different areas of the TP; this finding is related to the complex terrain and topography. Therefore, atmospheric moisture's spatial and temporal distribution would be different.

On average, the maximum precipitation center located southeast of the TP is also one of the maximum precipitation centers in China (Dai 1990). The precipitation distribution over the TP decreased from the Brahmaputra River Valley to the northwest. The TP is influenced by southwestern monsoons, especially from March to September. Easterly winds prevail in the upper troposphere, south of 30°N of Asia. Additionally, there is a stable and vast low in the lower troposphere over the Indian peninsula, Bay of Bengal, and the TP (Ye and Gao 1979). The abundant moisture from the Bay of Bengal is transported to the TP, and the precipitation occurs with convergence and lifting. In addition, because of the obstruction of the Tanggula Mountain and the effects of the Brahmaputra River Valley, the southeastern TP is a maximum center of precipitation (Yang et al. 1987). Comparatively, the water vapor can hardly drive deep into the TP, and thus, the water vapor source and mechanism of precipitation over the inner TP are not well understood. Some studies have proposed that the precipitation over the inner TP primarily originates from local evaporation, and the PW in the atmosphere and condensation caused by lifting. A humid stream provides the water vapor from the Arabian Sea, which is transported across India and Pakistan, across Pamirs, along the southern Xinjiang Basin and into the southwestern TP (Lin et al. 1990).

The maximum specific humidity is also located in the southeastern TP ($80\,g\,kg^{-1}$). Another high moisture center is the western TP ($70\,g\,kg^{-1}$). The Qaidam Basin, north of the plateau, is a low center, with a specific humidity of only $30\,g\,kg^{-1}$. The specific humidity distribution is closely related to topography, consistent with previous studies (Liang et al. 2006). Meanwhile, the spatial distribution of humidity is affected to some degree by the southwesterly humid stream.

The PW is also related to atmospheric moisture and topography. The spatial PW distribution is more complex than the spatial distribution of humidity. The maximum centers scattered along the latitudes are related to the topography of the plateau. A maximum PW center is located along the windward slope of the Gangdise and Tanggula Mountains, which is related to the obstructed moisture along the Brahmaputra Valley. The moisture from Indian island can barely reach the Tanggula Mountain lee slope, where the PW remains inadequate. To further analyze the regional precipitation and PW variability over TP, the PW and precipitation were analyzed in four TP regions based on the precipitation and humidity distribution characteristics (Figure 6.8). Zone A is climatologically dry and cold due to the obstruction of moisture by the southern side of the plateau. Zone A is also far from the Brahmaputra Valley moisture passage. Zone B is located in the northeast region of the TP, where the precipitation is very deficient. Zone D represents the wet regions southeast of the TP, and zone C is an intermediate wet and dry zone.

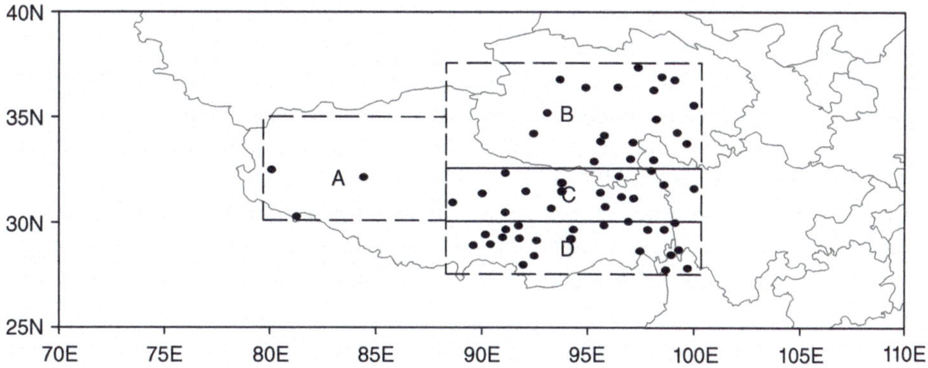

Figure 6.8 Research domains. A, B, C, and D represent four regions in the Tibetan Plateau. Black dots denote the precipitation sites in each zone.

From the interannual precipitation and PW variability in the four zones, the precipitation in zone A had an increasing trend from 1970 to 2009 (6.68 mm/10a). Although differences exist between the US National Centers for Environmental Prediction 1 and ERA-40 reanalyses, the PW presented a decreasing trend based on both reanalyses. Noticeably, moisture can barely penetrate the western TP, which is located deep into the TP. Due to global warming, remarkable temperature increases have been reported in the TP (Liu et al. 2000). Thus, in the past 40 years, the precipitation also has increased, but however, the PW has decreased. Two reasons can explain these trends: (i) the temperature rise led to a PR increase, and (ii) the intensified water cycle, which enhances evaporation, accelerated the regional moisture cycle.

The precipitation in zone B also demonstrated an increasing trend of approximately 10.8 mm/10a. However, the reanalysis data did not present an increasing PW trend; the precipitation trend was consistent with the PW. This implies that the relationship between PW and precipitation is similar to that of zone A. However, on the annual scale, the precipitation and PW changes were consistent with the data from the 1980s, but not the 1970s. This trend may be due to satellite-data assimilation after the 1980s.

In Heng-tuan Mountains, peaks as Tanggula and A'nyemaqen, the moisture that was transported by the South Asian monsoon ascended and was lifted by these mountains. Therefore, as a transition zone of wet and dry regions in the TP, the precipitation in zone C was peculiar. From the interannual precipitation and PW variability, a remarkable change occurred around 1986 in which the anomalies of the precipitation turned positive, which is consistent with the abrupt temperature change over the TP (Wei et al. 2003). However, the PW changed inversely. The consistency of the increasing precipitation with rising temperature reflects the close relationships among the variations in precipitation, PW, and temperature. Furthermore, this indicates that global warming impacts the water cycle over the TP. Further studies must confirm that these results are associated with the South Asia monsoon.

As mentioned above, the precipitation in zone D was prosperous and governed by the water vapor transported by the South Asia monsoon. In zone D, the water vapor was sufficient and largely dependent on the monsoon of southern Asia. The interannual precipitation variation was relatively low, with maximum and minimum anomaly percentages of 24.25% and −18.33%, respectively (i.e. the annual variability of precipitation was stable). The precipitation and PW changes were mainly dominated by the strength of the South Asia monsoon. Precipitation has barely decreased since 1970 (−1 mm/10a), and the PW showed a homologous decrease in this region.

These results are consistent with the fact that monsoons of South and East Asia have presented a long-term weakening trend since the mid-1970s (Li and Zeng 2005).

Except for zone B, the above discussion shows that the PW over TP has decreased since 1970. However, except in zone D, the precipitation has increased; the increased rate in zone C is especially remarkable. However, the changing precipitation and PW patterns were different in various regions. Notably, the precipitation and PW constantly change inversely.

The most direct precipitation water source is PW. The PW depends on many factors, such as moisture flux, latitude (Cai et al. 2004). Latitude is the crucial factor and affects the distribution of water vapor on a large scale. Only part of the atmospheric PW can be converted into precipitation.

The summer PR over TP is approximately 22%, the highest PR over China. Studies indicated that the summer underlain and atmosphere is a large heating source, resulting in lower atmospheric convergence around TP and subsequent lifting TP consists of a vast area and complex terrain, and the relationship between the precipitation and PW varies among different regions. In addition, the distribution and contents of water vapor are affected by many factors. In particular, the transformation of water vapor to precipitation, as well as the spatio-temporal change of PR have been attracted much attentions.

The mean PR spatial distribution results, which are based on the Eq. (6.1) (Figure 6.9), the maximum precipitation rate occurred over the slope in the southern TP. The humid warm air from the Indian Ocean lifted in the southern slope of the TP, which was the maximum center for condensed precipitation. A passage of moisture occurred from the south to the Brahmaputra and toward the north, transporting the water vapor from the Indian Ocean. The narrow valley and up-and-down terrain are not the only reason why this condensation maximum was formed, but also account for the observed maximum precipitation center. The spatial PR difference was slight due to the relatively insufficient moisture over the TP's main region, ranging from 9% to 12%. Because of the precipitous terrain, a maximum center of PR was located in the south Brahmaputra. A maximum PR center on the river basin of Qarqan, located in the northwestern TP, was connected to the maximum cloud center, where the PR reached 12%. This cloud center might have been caused by the local moisture cycle.

From the interannual PR variation, except zone B, the PR trend with ERA-40 is increasing (Figure 6.10). On average, the PR was the highest in zone D, followed by zone C, and was the

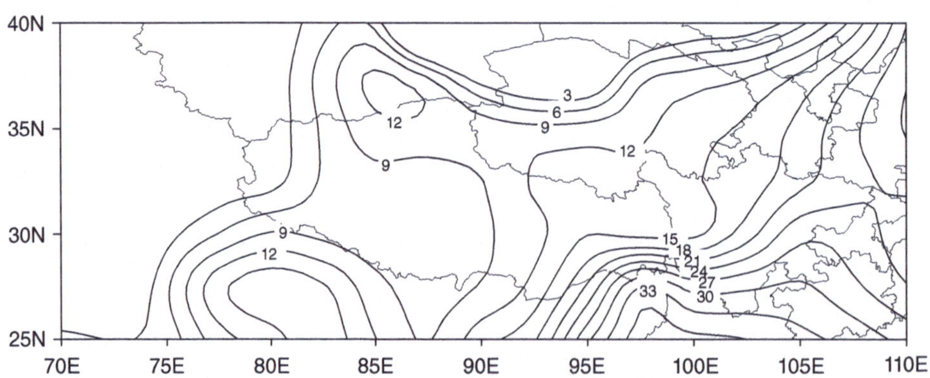

Figure 6.9 Distribution of the mean precipitation rate (1970–2009; %) over the Tibetan Plateau.

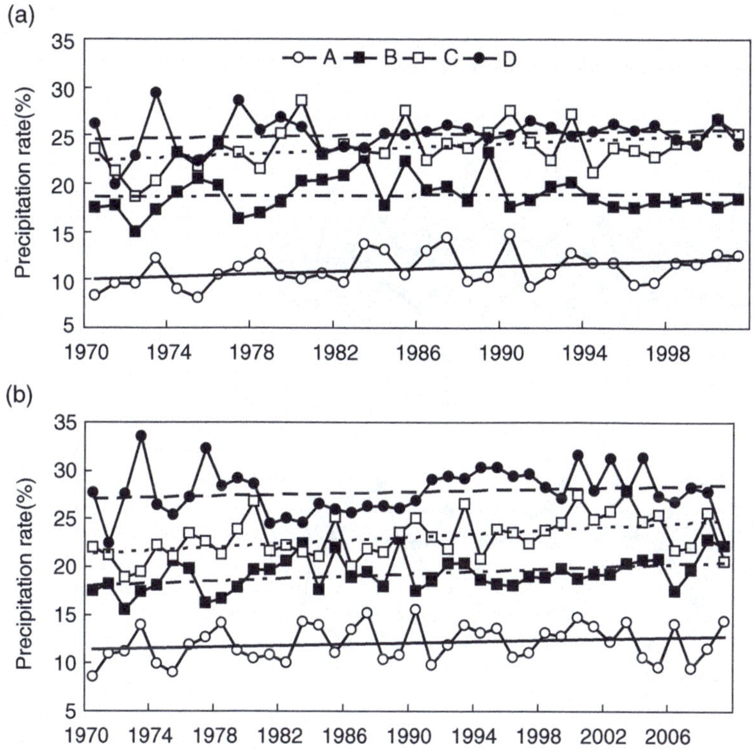

Figure 6.10 Evolution of precipitation rate in four regions (a) ERA40 reanalysis; (b) US National Centers for Environmental Prediction reanalysis; linear trends of precipitation rate in regions A, B, C, and D, are the solid line, the broken dotted line, the short broken line, the long broken line, separately.

lowest in zone A. Furthermore, the PR tendency from 1970 to 2000 was calculated, which shows that the increasing trend in zone C was the fastest (0.87%/10a to 0.91%/10a) and was the lowest in zone D.

The climate in the TP has monsoonal quality, and the precipitation also exhibits remarkable seasonal change. Therefore, the PR during the different seasons must be different. The precipitation in zone A (Figure 11a) is sparse from the annual variability of PR during the different seasons. Precipitation change in zone A has been low over the past 40 years. However, the PR decreased in winter, contrary to the increasing temperature effects. In the other seasons, PR increased rapidly in autumn, which was approximately 0.2%/a (Figure 6.11). A slight increase was observed in the summer. These variability results imply that PR change can be partly attributed to the rising temperature.

In zone B (Figure 6.11b), the PR increased in all seasons, rapidly in the winter (0.2%/a), in which the rising temperature was remarkable. However, the PR was low in the spring and summer. The PR in zone C (Figure 6.11c) has a remarkably increasing trend during the winter and spring, which also increases faster than in zones A and B during the same period. However, the PR exhibited a noticeable decreasing trend during the summer and autumn and was not very distinct in the summer. In zone D (Figure 6.11d), the maximum center of precipitation, the observed PR trend was similar to that of zone C, positive in the winter and spring and negative in the summer and autumn. The PR increase trend in winter is higher than in zone C.

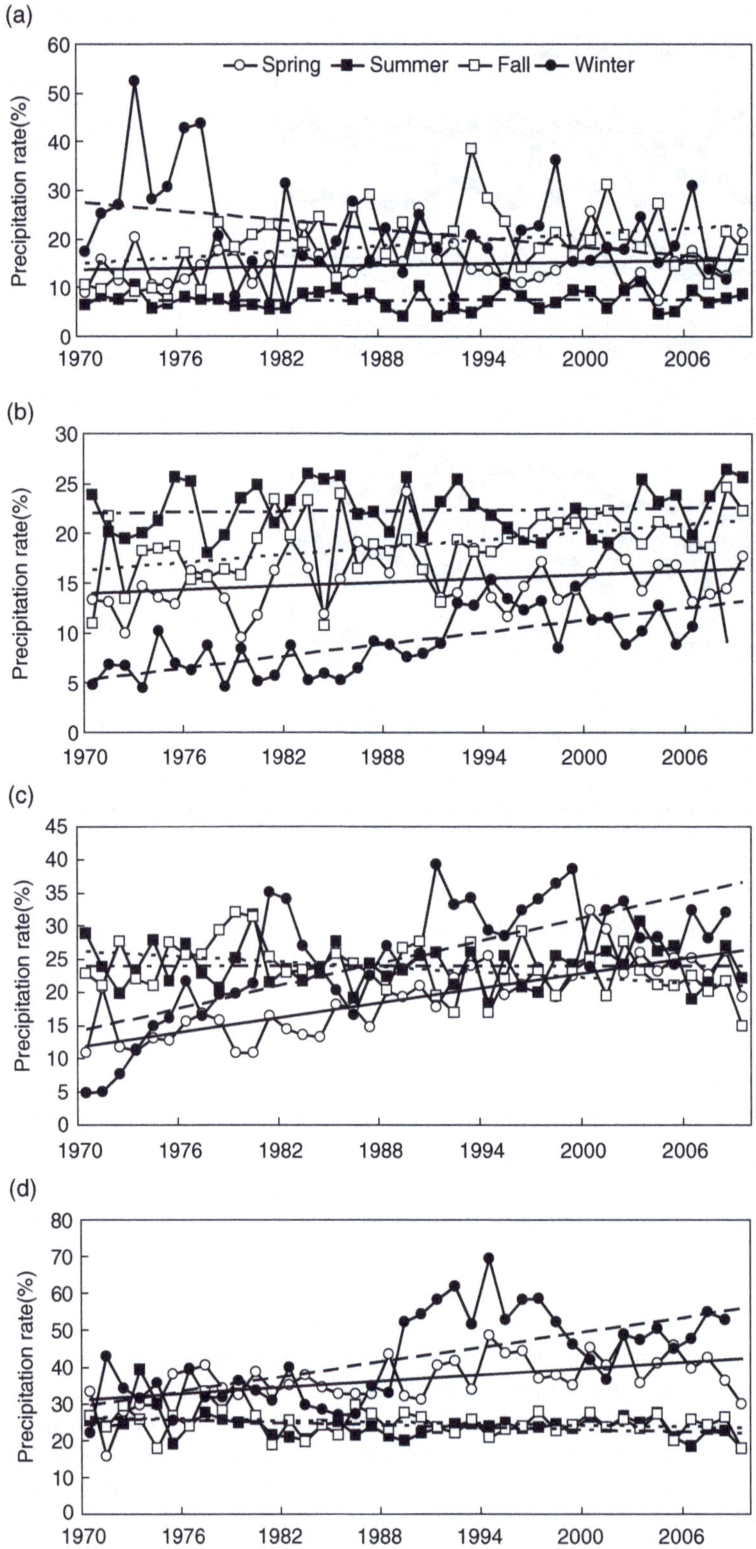

Figure 6.11 The evolution of seasonal precipitation rate (1970–2009), (a)–(d) denote the four regions separately. Linear trends of spring, summer, autumn, and winter precipitation are the solid line, the dotted broken line, the short broken line, and the long broken line separately.

From the above discussion, the PR trend is rising in the winter and spring with the increasing winter temperature and is more prominent in winter (except for zone A). The PR in winter is increasing from the west toward the east and from the north toward the south. The PR changes in the summer and autumn are small. This implies that a connection to global warming, rising winter temperatures, and PR, exists. The PR is not low in the arid regions, zones A and B. However, the seasonal variability is small.

As discussed above, the precipitation over the TP has a significant monsoonal character. Monthly PR in the four zones was calculated to analyze the PR annual change feature. PR in zones B, C, and D had prominent monsoon characteristics, and zone A can be divided into wet and dry seasons (Figure 6.12). Three patterns of changing PR were identified. The annual PR range was most significant in zone A, with a PR maximum in November, which was lower in other months. However, November was not the period of the maximum precipitation; June should have been. The PR was observed to associate more closely with the rising winter temperatures. The second pattern was seen in zone B, where the precipitation and PR changes were consistent. Pattern 3 was located in zone C. Although the precipitation had a monsoonal character in zone C, the annual PR variability was not high, and the PR maximum occurred during the winter, in which rainfall was low. In addition, the PR minimum occurred in the spring when the rainfall was also insufficient. In humid zone D, the PR maximum occurred during the winter half-year, which might have been attributed to the adequate summer water vapor, which produced much rainfall. The atmosphere was easily saturated in most cases. Meanwhile, a large amount of latent heat was released, which prevented the further formation of precipitation. A similar phenomenon occurred in zones C and D, in which the PR exhibited obvious interannual variability.

The above analyses conclude that a precipitation mechanism in different climate zones was observed. In some regions, such as zones A and C in winter, the PR was higher, although the PW was inadequate. The PR was low in some regions that possessed sufficient PW and more precipitation (e.g. zone D). The precipitation mechanism differences were not only exhibited in spatial distributions, but also temporal distributions. The PR was different during the four seasons. With rising winter temperatures, the PR in the winter half-year demonstrated an increasing trend, although the PR did not increase in all regions.

Generally, if the external forcing climate systems do not change on the decade scale, then the PR should be stable. However, with rising temperatures, especially in winter, the precipitation, PW and PR possess a different variability on a different temporal scale. In conclusion, since 1970, the precipitation over the TP demonstrated an increasing trend except for its southeast region. Meanwhile, a slightly decreasing trend in southeastern TP was observed, which may be attributed to the South Asia monsoon degradation. Since 1970, a decreasing trend in PW over the TP, remarkably in the southeast of TP, was observed; the mean rate was approximately 0.078 mm/10a. The PR increased significantly with rising temperatures in the TP within the last 40 years, and zone C suffered the most rapid increase of approximately 0.87% per decade. On average, the PR in the humid southern region of the TP was higher than that of the dry northern area. The PR exhibited significant interannual and annual variability, which was higher in the winter half-year than in the summer half-year. Finally, the PR in the humid zones was higher than in the dry zone, especially during the winter. However, the PR maximum over the western TP was observed during the wet seasons. The PR in the humid region was observed in the dry seasons, and higher PR occurred during the sufficient rainfall season in the arid region.

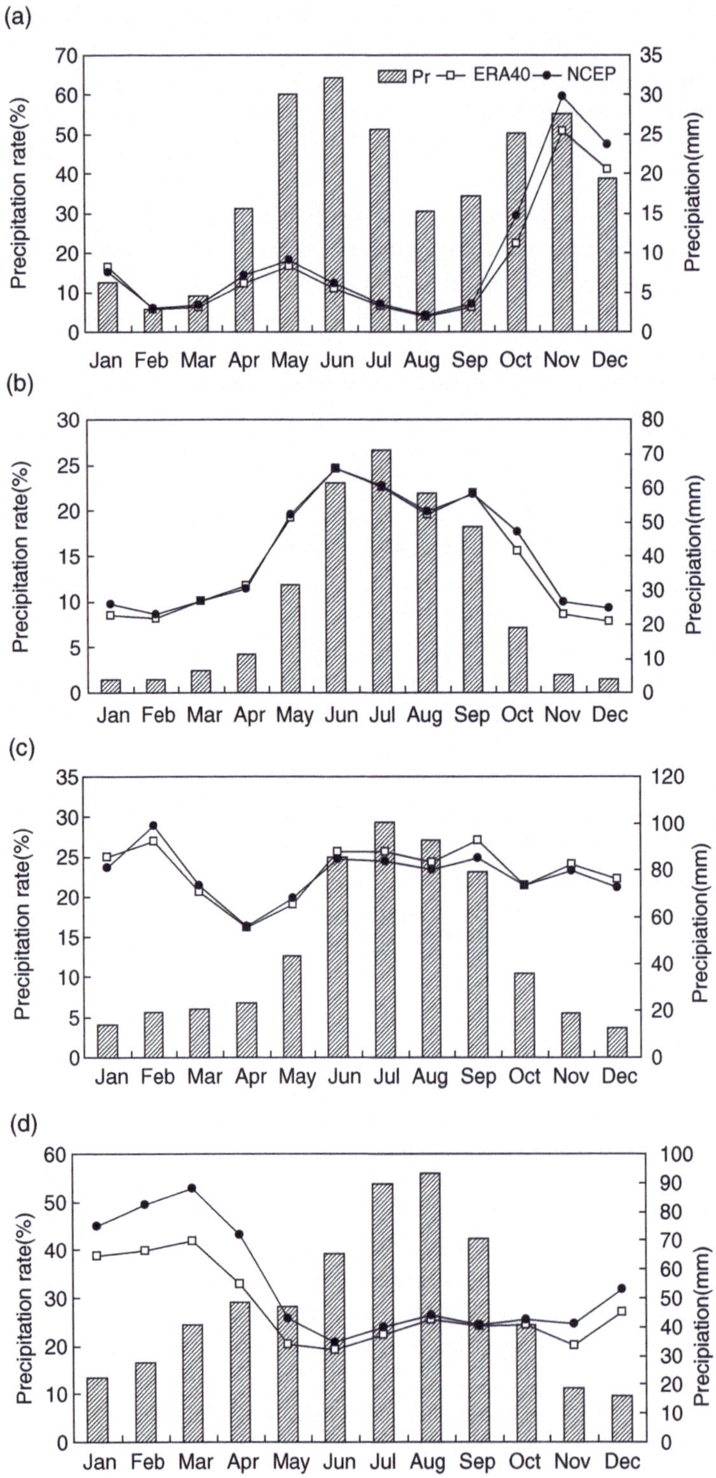

Figure 6.12 Monthly variability of precipitation rate with mean from 1970 to 2009, (a), (b), (c), (d) donate four regions A, B, C, and D separately.

6.9 Precipitation Recycling on the Tibetan Plateau

Precipitation recycling, a primary aspect of the hydrological cycle, is essential in balancing the water and energy cycles. Usually, precipitation over a land region is derived from two sources: (i) water vapor evaporated within the region, and (ii) water vapor evaporated outside of the region and later transported into it (Brubaker et al. 1993). The first source indicates the importance of interactions between hydrology and regional climate (Brubaker et al. 1993;). Many studies have defined the precipitation recycling ratio (PRR) as the ratio of precipitation evaporated within the region versus the total precipitation (Budyko 1974; Brubaker et al. 1993). Using this definition, many models that calculate the PRR have been derived (Budyko 1974; Brubaker et al. 1993; Dominguez et al. 2006).

The diabatic heating over TP caused by the high elevation and complex surface processes plays a powerful role in the East Asian monsoon system (Wang and Cui 2011). Certain unique surface characteristics in the TP (e.g. frozen ground, snow cover, and glaciers) may have a particular influence on regional climate change in other parts of East Asia. The soil moisture generated by spring thawing of frozen ground in the TP is closely related to the wet season's start in the TP (Wang and Shang 2007), and the soil moisture over the TP relates to the onset of the southern Asian monsoon (Shang and Wang 2006). In terms of global warming, studies have reported that the temperature of the TP increases more quickly than in other regions at similar latitudes. Under the global warming, most TPs precipitation conversion rate has increased markedly (Wang and Cui 2011). With temperature increases, the water cycle over the plateau (even the entire East Asian region) will be affected by the thawing of frozen ground, snow melting, and glacier change. These changes will further influence diabatic heating processes between the surface of the TP and the atmosphere above, and these changes will impact the circulation and the monsoon in East Asia. Moreover, the southern Asian monsoon has retreated in the past decade (Wang 2001; Li and Zeng 2001). That means the moisture transport northward from the Indian Ocean will weaken. However, as the southern TP is affected by the South Asia monsoon, how and what changes in PPR will occur in the southern TP also require study. The precipitation recycling over the TP is a multi-scale process.

The moisture recycling process can be described as follow. The area of the study region is A (in a one-dimensional model, the area is regressed into a length scale, L, as in Budyko (1974)). F_{in} represents the moisture transported into a land region, and F_{out} represents the moisture transported out of a land region. E is the evaporation, and P is the total precipitation. In previous studies, precipitation was considered to be from two sources: (i) moisture transported into the region, or (ii) moisture evaporated within the region (Budyko 1974; Brubaker et al. 1993). Therefore, P_m is defined as the fraction of precipitation from evaporation, and P_a represents the fraction from moisture flux. The average precipitation in a region is composed of two parts, specifically, the advective component and the local evaporative component, and it is defined as the following:

$$P = P_a + P_m \tag{6.3}$$

As in Brubaker (1993), the vertical flux of P_a, P_m, and E are treated as constants equal to their average values; therefore, the evaporated and advected moisture within the region increases linearly. The horizontal flux of advected moisture (Q_a) through the region (of area A) is taken as the arithmetic mean of the incoming and outgoing advective moisture:

$$Q_a = \frac{F_{in} + \left(F_{in} - P_a A\right)}{2} = F_{in} - \frac{P_a A}{2} \tag{6.4}$$

Similarly, the horizontal flux of locally evaporated moisture (Q_m) is defined as the following:

$$Q_m = \frac{0 + (E - P_m)A}{2} = \frac{(E - P_m)A}{2} \tag{6.5}$$

Considering that the atmosphere is well mixed, the ratio of P_a versus P_m will equal that of Q_a versus Q_m:

$$\frac{P_a}{P_m} = \frac{Q_a}{Q_m} = \frac{F_{in} - \frac{P_a A}{2}}{\frac{(E - P_m)A}{2}} \tag{6.6}$$

The PRR is defined as $\beta = \frac{P_m}{P}$. Thus, the PRR can be derived from Eqs. (6.3)–(6.6) as the following:

$$\beta = \frac{EA}{EA + 2F_{in}} \tag{6.7}$$

Because of the incredibly complex terrain of the TP, there are remarkable differences between the route and extent of moisture transport in different regions of the TP (Wang and Cui 2011). Regional differences also occur in terms of precipitation. Considering the various climate features in different regions of the complex terrain of the TP, the TP was divided into four separately investigated parts. The precipitation over the western TP in summer is mainly derived from evaporation (i.e. it is regionally recycled). The moisture of precipitation in winter is mainly adverted from outside of the region. In the four regions, evaporation and the PRR have similar trends during the year. When using climatologically averaged variables and constant terrain, the efficiency at which evaporation becomes precipitation is nearly constant. Therefore, the PRR is critically decided by the amount of evaporation determined using the equation:

$$E \approx P + \nabla \bullet \vec{Q} \tag{6.8}$$

where E, P and $\nabla \bullet \vec{Q}$ are the monthly evaporation, precipitation, and divergence of moisture flux, respectively.

From the above analysis, one can conclude that the PRRs are complicated over the TP because of the complicated terrains and water vapor sources in the area. The evaporation mainly decides the value of the PRR.

As global warming increases, the TP warms, as well. This phenomenon may be one of the most critical factors affecting the TPs PRR trends. The hydrological cycle over the TP is unique compared to large river basins. The TP is rich in snow, frozen soil and glaciers that provide considerable stores of water and energy. Once the climate warms, those water stores will release massive amounts of latent heat that may affect the circulation, the hydrological cycle, and perhaps even the climate of East Asia.

However, a ratio value of 0.62 occurs in July in the southeastern TP. In summer over region C, the ratio is about 0.3. However, in late spring and early summer (May–July in Kang et al. (2004)), the mean value in region D is about 0.4, which is almost twice that of the Yangtze River Basin (with a mean value of less than 0.25). The lower reaches of the Yangtze and Yellow Rivers are located in eastern China, where the East Asia monsoon prevails each summer. Mass amounts of moisture are carried to land from the western Pacific and northern Indian Ocean. Precipitation here is mainly derived from advected moisture arriving from outside eastern China. Thus, the amount of local recycled moisture is small, resulting in a low recycling ratio during summer. In contrast, the TP is located deep inland and is highly elevated; therefore, moisture from the ocean hardly reaches the main body of the TP. Though the Indian monsoon can bring massive moisture to the TP via the Brahmaputra Valley (Yang et al. 1987; Lin and Wu 1990), the steep southern slope and the corner of the Brahmaputra Valley obstruct the moisture from further penetrating the plateau (Gao et al. 1985). Therefore, moisture deep in the TP is mainly locally recycled. It is notable that though moisture in summer can hardly reach the west of the TP and precipitation is mainly dependent on regional circulation, the precipitation in region D is mainly derived from advected moisture. The recycling ratio in region A should be higher than in region D, which seems contrary to the result of this work. This can be explained by the difference in evaporation between regions A and D. Evaporation in region D is significantly larger than in region A during the summer. Fin (i.e. the advected moisture) is higher in region D than in region A. Region D is the highest of the four regions, which implies a lower PRR. However, evaporation in region D is higher than the corresponding value of Fin, and as a result, the recycling ratio in region D is the highest. Due to abundant vegetation, the reason for high evaporation in region D may be due to its low albedo (Wei and Li 2003). Low albedo implies more short-wave radiation absorption, leading to more land heating and greater evaporation.

For region A, PRR decreased from 1979 to 2008. The most significant decreasing trend occurs in the autumn, at −0.038 per decade. In summer, the rate is −0.024 per decade. Annually, it decreases by −0.025 per decade. The surface air temperature has also been increasing in recent years (Wu et al. 2005; Oku and Ishikawa 2006), which should result in intensified local precipitation recycling. However, northwestern China (including the TP) is suffering a climatic shift from warm-dry to warm-humid (Shi et al. 2008; Wu et al. 2005; Mao et al. 2008; Tang et al. 2011). The moistening indicates that more water vapor is transported into these areas. That transport may balance the increased evaporation caused by the temperature increase and may even cause a decrease in the recycling ratio. For region B, precipitation and the recycling ratio are both increasing. The trend in the annual recycling ratio is 0.031 per decade, and the highest increasing trend occurs in the summer with approximately 0.078 per decade. The evapotranspiration at the Rikaze site (in western region D) is decreasing, and this finding is in good agreement with the decreasing PRR trend in region D. Region C is a transient region between wet and dry. The changing character in region C is similar to that in region D.

Annual variations of the PRR in the four regions have remarkably seasonal characters. The maximum PRR occurs in July and August, and the minimum occurs in January and December. The highest PRR occurs in July for region D at approximately 0.62. Peak values in regions C and D are generally in July and August, and the East Asian summer monsoon may cause the phenomena.

In conclusion, the interannual and seasonal mean variation of the PRR for region A is decreasing. The variation is increasing in Regions B, C, and D. The largest decreasing trend in region A occurs in autumn. The largest increasing trends in regions B, C, and D occur in summer. The decreasing trend for region A is a decrease of approximately 0.025 per decade. The largest increasing trend occurs in region B and is approximately 0.031 per decade. Annual variation in the four regions has a significantly seasonal characteristics. The annual maxima occurs in summer, and the minima occurs in winter.

7

Ecology and Vegetation in Cold Regions

Plant growth and distribution depend on hydrothermal conditions. Plant growth is constrained by thermal conditions in cold region. Thus, plant categories and their spatial distribution are largely dependent on temperature distribution.

Global warming is affecting a broad spectrum of physical, ecological, and human/cultural systems that may be irreversible on century time scales and have the potential to cause rapid changes in Earth's systems. The response of the ecosystem succession in cold regions to climate change is globally conspicuous. Systematic ecology and vegetation distribution are described more thoroughly in relevant monographs, and ecology system, as part of the climate system, are essential to the formation and evolution of the climate system. This chapter is an introduction of the status of the ecology and vegetation and their response to climate change in a few special cold regions, the Tibetan Plateau (TP), middle and high latitudes, and the Arctic region.

7.1 Overview

As climate production, the categories and distribution of ecology and vegetation depend on precipitation and temperature distribution. Ecological species and distribution over cold regions are largely dominated by temperature, and are different from other climate zones.

Table 7.1 shows some primary ecology species and coverage. The basic ecological landscape in cold regions is dry steppe, desert. The ecology in mid-high latitudes is dominated by forests, while the polar region is dominated by tundra. The distribution of such ecological species is a production of climate, which in turn determines the relation between land surface and atmosphere in these areas, thus constituting different external forcing of atmosphere that affect climate.

The amount and distribution in ecological species over cold regions also affects the exchange of carbon between land and atmosphere, and this issue is becoming more and more important due to global warming.

7.2 Ecology and Vegetation Characteristics on the Tibetan Plateau

7.2.1 Spatial Distribution of Ecology and Vegetation

Chen and Kaung (2019) summarized the spatial distribution of ecology and vegetation of TP. Their results are introduced below. In terms of climatic zone, the TP spans the humid area of the middle subtropical zone, the arid area of the middle temperate zone, the humid area of the north

Climatology in Cold Regions, First Edition. Chenghai Wang.

Table 7.1 Ecosystem cover area (Unit: 10^2 km^2).

Regions Ecosystem types		TP	Mid-high	Polar
Grassland	High covered grass	1983.60	84 222.8	/
	Medium covered grassland	4700.93		
	Low covered grassland	5935.82		
Desert	Gobi	2036.43	38 480	668 294.3
	Sandy land	542.87		
	Saline alkali land	329.21		
	Other unused lands	374.64		
Forest	Forest land	1469.94	91 283.72	/
	Shrub	108		
	Open woodland	550.48		
	Other forest lands	38.79		
Water body and wetland	Rivers and canals	81.72	/	/
	Lake	476.16		
	Reservoir pit and pond	13.30		
	Permanent glacier and snow	391.96		
	Beach	352.60		
	Swamp	355.12		
Farmland	Paddy field	20.49	27 929.56	/
	Dry land	214.65		
Settlement	Urban land	7.99	/	62 000
	Rural residential area	10.18		
	Other construction land	10.00		
Other	Bare land	741.42	/	/
	Bare rock and rocky ground	4065.47		

TP, Tibetan Plateau. *Source:* Chen and Kaung (2019), Liu, Sun et al. (2013), Julien et al. (2006), and Ward and Paul (2001).

subtropical zone, the arid area of the warm temperate zone, the semi-arid area of the plateau sub-cold zone, the semi-humid area of the plateau sub-cold zone, the arid area of the plateau sub-cold zone, the semi-arid area of the plateau temperate zone, the arid area of the plateau temperate zone, and the humid and semi-humid area of the plateau temperate zone.

The TP contains complex ecosystems. The grassland ecosystem is the largest ecosystem in the region, with an area of $12\,620.36\times10^2$ km^2, accounting for 48.89% of the total area of the whole region, most of which are low-coverage grasslands, accounting for 47.03% of the total area of the whole grassland ecosystem. Due to natural environmental factors, the second largest area of the TP is other ecosystems, with an area of 4806.89×10^2 km^2, accounting for 18.62% of the ecosystem in the whole region, of which the main land type is bare rock gravel land, accounting for 84.58% of the whole other ecosystem. The next largest area is desert ecosystem, with an area of 3283.15×10^2 km^2, accounting for 12.72% of the ecosystem of the whole region, of which the main land type is the Gobi Desert, accounting for 62.03% of the whole desert ecosystem. Important forest areas in China are distributed in the TP, and China's rare primitive forest areas are also located, with a forest ecosystem area of $3167.21 \times 10^2 km^2$, accounting for 12.27% of the ecosystem of the whole region, of which the main land type is forest land, accounting for 46.41% of the whole forest ecosystem; the area of water body and wetland ecosystem is $1670.86 \times 10^2 km^2$, accounting for 6.47% of the ecosystem of the whole region, of which lakes, permanent glaciers, swamps and beaches account for 28.50%, 23.46%, 21.25%, and 21.10% of the whole water body and wetland ecosystem respectively. Due to the high altitude, thin air, and harsh climate, in addition to river valleys, Alpine animal husbandry can only be developed on the vast plateau, and the area of farmland ecosystem in the region is $235.15 \times 10^2 km^2$, accounting for only 0.91% of the ecosystem in the region, of which dry land accounts for 91.28% of the farmland ecosystem. At the same time, there are few places suitable for human habitation and survival, and the area of settlement ecosystem in the region is $28.07 \times 10^2 km^2$, accounting for only 0.11% of the ecosystem in the whole region, of which urban land, rural residential areas, and other construction land account for 28.10%, 36.28%, and 35.62% of the whole settlement ecosystem respectively.

Chen and Kaung (2019) reported that, the farmland ecosystem over the TP includes paddy fields and dry land, with an area of 20.49×10^2 and $214.65 \times 10^2 km^2$ hectares, respectively. 80.90% of farmland ecosystems are distributed in the TP's temperate zone and 10.28% in the middle subtropical zone. Among them, 64.52% of the paddy fields are distributed in the TP's sub-cold zone and 87.20% of dry lands are distributed in the TP's temperate zone. The secondary classification of forest ecosystem in the TP is: $1469.94 \times 10^2 km^2$ forest land, $108 \times 10^2 km^2$ shrub, $550.48 \times 10^2 km^2$ open woodland, and $38.79 \times 10^2 km^2$ other forest lands. From the overall distribution, 67.50% of the forest ecosystem is distributed in the TP's temperate zone; 20.78% is distributed in the middle subtropical zone; 11.62% is distributed in the sub-cold zone of the TP. The secondary classification of grassland ecosystem in the TP is: high cover grass $1983.60 \times 10^2 km^2$, medium covered grassland $4700.93 \times 10^2 km^2$, low cover grassland $5935.82 \times 10^2 km^2$. In terms of overall distribution, 50.34% of grassland ecosystem is distributed in the TP's sub-cold zone; 47.92% is distributed in the TP's temperate zone. The secondary classification of water body and wetland ecosystems in the TP is as follows: rivers and canals $81.72 \times 10^2 km^2$, lakes $476.16 \times 10^2 km^2$, reservoirs and ponds $13.30 \times 10^2 km^2$, permanent glaciers and snow $391.96 \times 10^2 km^2$, beaches $352.60 \times 10^2 km^2$, swamps $355.12 \times 10^2 km^2$. In terms of distribution, 62.49% of the water and wetland ecosystems are distributed in the TP's sub-cold zone, 36.80% in the TP's temperate zone, and a small amount in the northern subtropical zone, middle subtropical zone, warm temperate zone and middle temperate zone. The TP is vast and sparsely populated, and 83.09% of the settlement ecosystems are mainly distributed in the TP's temperate zone and 9.69% of them in the TP's sub-cold zone, including $7.89 \times 102 km^2$ of urban land, $10.18 \times 10^2 km^2$ rural residential area, and $10 \times 10^2 km^2$ other construction land. It is mainly distributed in Xigaze City, Lhasa City, Yushu area, Haidong area, Hainan and Gannan area. The secondary classification of desert ecosystems in the TP is as follows: is $542.87 \times 10^2 km^2$ sandy land, $2036.43 \times 10^2 km^2$ the Gobi Desert, $329.21 \times 10^2 km^2$ saline alkali land, $374.64 \times 10^2 km^2$ other unused land, of which 50.08% are distributed in the TP's sub-cold zone and 49.90% in the TP's temperate zone. Other secondary classifications of ecosystem are as follows: $741.42 \times 10^2 km^2$ bare land, $4065.47 \times 10^2 km^2$ bare rock and rocky ground, of which 51.08% is distributed in the TP's temperate zone, 48.10% in the TP's sub-cold zone, and 5.72% in the mid-temperate zone.

The distribution of vegetation corresponds with the terrain height. Xu et al. (2009) suggested that, the TP can be divided into three latitude bands: north region (35° ~ 40°N), middle region (30° ~ 35°N), and south region (south of 30°N). In north region, there are three major mountain systems, the Qilians, Altuns, and Kunluns. The distribution of vertical zone spectrums of vegetation on the north slopes of each range is as follows: mountain grassland or mountain desert grassland except for the baseband at the east end of the Lianshans; the west end of the Kunluns is alpine grassland; the baseband of the mountain vegetation vertical zone on the north slope of the whole TP is mountain desert; the upper limit height of the desert zone rises from east to west, and the spectral width of the zone is widest in the Altuns. Taking this as the boundary, the spectral width of the desert belt in the east increases gradually from east to west. The width of the west spectrum decreases gradually. The spectral width of the desert zone in the eastern part of Qilians is 250 m, 400–500 m in the middle section, 630–1000 m in the west section, Mount Arkin is 1200 m, and Qiakeke Lake in the Kunluns 300 m. The upper limit of distribution is 2000, 2300, 2500, 2850, and 3500 m above sea level, respectively. The belt spectrum is mainly desert grassland, mountain grassland, and alpine grassland. There is a mountain forest belt or forest grassland belt in the east. The forest belt is distributed between 2500 and 3300 m. There are alpine meadows and shrub meadow belt development. With the elevation of the mountain base, the height of the sub-ice and snow zone increases gradually from east to west. Overall, the vegetation belt spectrum of the Kunluns, the western Qilians, and the Altuns is simple, being composed of four vegetation belts: desert grassland, mountain grassland, alpine grassland, and alpine meadow. The belt spectrum of the middle and eastern Qilians is complex forests, grasslands, shrubs, and meadows (Zeng and Li, 1990; Zeng, 1997).

For south region, vegetation is mainly distributed in the central area of TP at 30–35° N, including Gangdises, Tanggulas, Nianqing Tanggulas, Bayankalas, Hengduans and other mountain ranges. The lower limit of the sub-ice and snow zone increases gradually from east to west. The mountain forest belt is mainly distributed in the mountains east of Bomi and the forest zone is the dominant zone. The upper limit of the forest zone increases gradually from east to west. The upper limit height of the forest belt in the east is 3400–3700 m, 4300–4400 m in the western TP. There is no alpine grassland vegetation zone in the vertical zone spectrum of mountain vegetation. It is mainly composed of broad-leaved forests, mixed forests, coniferous forests, shrub grasslands, alpine meadows, alpine deserts, shrub meadows, and other vegetation zones. There is no forest zone in the vertical zone spectrum of mountain vegetation in the plateau area west of Bomi. There, it is mainly shrubs, alpine meadows, alpine grasslands, and a combination of alpine desert zones (Xu et al. 2009).

The vertical spectrum distribution of mountain vegetation in the south of 30° N is characterized by the combination of shrub, alpine grassland, desert and alpine meadow. The mountain vertical spectrum is distributed in mountains with a base elevation of more than 4000 m, such as the west of Alisaga (83°E) and Mount Everest. The vertical band spectrum of mountain vegetation with a forest zone is distributed in the mountains with a base elevation of under 4000 m. Moreover, the forest is the dominant zone, and there are many types of forest, such as rainforest, seasonal rainforest, evergreen broad-leaved forest, coniferous forest and mixed forest. The upper limit of the forest zone can reach about 4200 m. The vegetation spectrum is mainly the combination of forest zone and mountain shrub, shrub meadow, and alpine meadow zone. The vertical band spectrum of mountain vegetation containing shrub grassland and alpine grassland is distributed in the mountain area west of 92°E (Xu et al. 2009).

7.2.2 Vegetation Variations Respond to Climate Change

As an important factor to evaluate vegetation growth status, the normalized difference vegetation index (NDVI) cannot only reflect the vegetation growth status and distribution pattern on a global or regional scale, but also indicate the change trend in a long-time series. It is closely related to important vegetation characteristic parameters such as vegetation productivity and leaf area index (LAI). Therefore, NDVI is used to evaluate the vegetation coverage. From 1981 to 2005, the annual change rate of vegetation coverage on the TP reached 8.69% (Liu et al. 2013). The multi-year averaged coverage of coniferous forest and broad-leaved forest was the highest, but the vegetation coverage of both coniferous forest and broad-leaved forest showed a downward trend, and the broad-leaved forest decreased the fastest. The coverage of other vegetation types increased in varying degrees, and the coverage of farmland and alpine meadow increased the fastest. The change of vegetation coverage has significant regional differences. The rising area is located in the middle, northwest, and southwest of the plateau, accounting for 70.5% of the whole region. The reduction area is located in the southeast, northeast and west of the TP, accounting for 27.5% of the whole region. The areas with unchanged vegetation coverage are scattered, accounting for 2.0% of the whole area. The surface vegetation improvement area of the TP is mainly located in the low-coverage area, and the degraded area is mainly located in the high-coverage area, and the area of the vegetation improvement area is much larger than that of the degraded area. The overall change trend of interannual variation of vegetation coverage and precipitation shows a certain degree of lag. In some years, precipitation increases, but vegetation coverage decreases. The interannual variation of vegetation coverage and the interannual variation of temperature are similar, and their changes are synchronous. On the one hand, it shows that vegetation cover change is sensitive to climate change. On the other hand, it also shows that precipitation and temperature are not the absolute limiting factors of vegetation cover change. Human activities often dominate the regional differentiation characteristics of vegetation cover change.

Liu et al. (2013) investigated the precipitation of broad-leaved forest, alpine desert and temperate grassland present a decreasing trend during 1981–2005. The precipitation of other vegetation is increasing, and the interannual variability of coniferous forest is the largest. From different vegetation types, the interannual change rate of temperature in coniferous forest is negative, and other vegetation types increase temperature in varying degrees. The change of vegetation coverage is positively correlated with the change of precipitation. The vegetation types in positive correlation areas are mainly alpine meadow, alpine grassland, and temperate grassland, and the change of precipitation has a great impact on vegetation coverage. The negative correlation regions are mainly located in the northwest, east, southeast and southwest, accounting for 33.14% of the whole region. The vegetation in the negative correlation area is mainly alpine desert, coniferous forest, broad-leaved forest, temperate meadow, and shrub. The interannual change rate of precipitation is small and has marginal impact on vegetation coverage. Uncorrelated regions are scattered, accounting for 2.99% of the whole region. The maximum value of the positive correlation coefficient between vegetation coverage and precipitation is 0.8. The maximum absolute value of negative correlation coefficient is 0.73. The change of vegetation coverage is positively correlated with the change of temperature as well. The region is located in the northwest, middle east, and southwest of the TP, accounting for 63.52% of the whole region. In the positive correlation areas, the altitude is high and the temperature is low. The vegetation is mainly alpine meadow, alpine grassland, and alpine desert. The increase of temperature can promote the photosynthesis of plants and prolong the growth period of vegetation, which is conducive to the growth of plants. The negative correlation region is located in the northeast, southeast, and west of the plateau, accounting for 33.25% of the whole region. The negative correlation area is the main gathering area of towns, population and industry on the TP. The vegetation is mainly coniferous forest, broad-leaved forest, temperate grassland, temperate meadow, and shrub. The change of vegetation cover is mainly affected by human activities, and the effect of warming on the change of vegetation cover is not obvious. Irrelevant areas are scattered, accounting for 3.23% of the whole region. The maximum value of the positive correlation coefficient between vegetation cover and temperature is 0.83, and the maximum absolute value of the negative correlation coefficient is 0.74.

Liu et al. (2013) also indicated that, the changes of coverage of coniferous forest and broad-leaved forest were negatively correlated with the changes of precipitation and temperature. The change of alpine desert coverage is positively correlated with the change of temperature, but negatively correlated with the change of precipitation. The changes of coverage of other vegetation types were positively correlated with the changes of precipitation and temperature. The correlation coefficient between temperate grassland coverage change and precipitation change is the largest, and the correlation coefficient between alpine desert coverage change and temperature change is the largest.

Han et al. (2019)'s results show that, from 2000 to 2018, the change trend of NDVI in parts of Sichuan and Yunnan province in China, the middle east and south of Qinghai province, the southeast and north of Tibet showed an obvious increasing trend, indicating that the vegetation situation in these areas has improved in the past 19 years. In Xinjiang and Gansu, southeast and northeast Qinghai, southeast Tibet and remote areas in the west, NDVI showed a decreasing trend. Especially in the south of Hotan, most of Nyingchi, parts of the Qilians and most of the Golog Tibetan Autonomous Prefecture, the trend of decline is obvious, that is, the vegetation degradation is more serious in these areas. From 2000 to 2018, the total area of grassland vegetation degradation was 60.3 km^2, and the total area of restoration and improvement was 69.13 km^2. Therefore, the area where vegetation growth has been improved in the whole TP is larger than the area where vegetation has been degraded. Among them, the area of vegetation improvement accounts for 27.35% of the total area of the TP, and the degraded area accounts for 23.75%. Therefore, although the vegetation on the TP has deteriorated locally, it is still in a state of recovery as a whole. In most relatively arid areas, vegetation growth is positively correlated with precipitation. In humid areas, it is negative correlation. NDVI has a good relationship with temperature and precipitation on the TP. The correlation coefficient between NDVI and air temperature is 0.909. Although the correlation between NDVI and precipitation is smaller than that of temperature, the correlation coefficient can also reach 0.793. Water plays a very important role in plant nutrient transport, structure maintenance, and various physiological and biochemical processes. Therefore, there is an obvious positive correlation between vegetation growth and precipitation. The net photosynthetic rate of plant leaves increases with the increase of temperature. Too high or too low temperatures will affect the physiological and biochemical processes of plants. From the linear relationship between them, it can be seen that when the temperature increases by 1°C, NDVI increases by 0.128. For every 100 mm increase in precipitation, NDVI will increase by 0.172 (Han et al. 2019).

Results of Chen et al. (2020) showed that, the net primary productivity (NPP) of vegetation refers to the amount of organic matter accumulated by green plants through photosynthesis in unit area and unit time. It is the core index indicating the health status of land surface ecosystem. From 2000 to 2015, the annual average NPP of vegetation on the TP generally decreased from southeast to northwest. In terms of interannual variation, vegetation NPP has fluctuated and increased in the past 16 years, including a steep rise in 2005 and a high fluctuation from 2005 to 2015. The vegetation NPP increase area (change rate > 10%) on the TP is mainly concentrated in the source area of the three rivers, the north of Hengduan Mountain area, the middle and lower reaches of the Yarlung Zangbo River and the middle east of the Naqu region, while the vegetation NPP decrease area (change rate < −10%) is mainly distributed in the upper reaches of the Yarlung Zangbo River and the Ali plateau. In the past 16 years, the center of gravity of vegetation NPP on the TP has generally moved to the southwest, indicating that the increment and growth rate of vegetation NPP in the southwest is greater than that in the northeast. There are significant regional differences in the correlation between vegetation NPP and climate factors on the TP. The regions with significant correlation between vegetation NPP and precipitation are mainly located in the middle of the TP, the southeast of the TP, and the middle and lower reaches of the Yarlung Zangbo River Basin, while the regions with significant correlation between vegetation NPP and temperature are mainly located in southern TP, the north of Hengduan Mountain area, the middle and north of the TP. There are significant temporal and spatial differences in the relative effects of climate change and human activities on the change of vegetation NPP in the TP (Chen et al. 2020).

7.3 Ecology and Vegetation Characteristics in Middle and High Latitudes

It is complex to describe the spatial distribution of ecology and vegetation in the middle and high latitudes in the northern hemisphere. Therefore, this section mainly introduces the variation of ecology and vegetation in the middle and high latitudes in the northern hemisphere. Many studies suggested the increasing trend in the middle and high latitudes in the northern hemisphere is due to the increased temperature (Myneni et al. 1997; Liu et al. 2010; Yang et al. 2016).

Variations in the amplitude and timing of the seasonal cycle of atmospheric CO_2 have shown an association with surface air temperature, which is consistent with the hypothesis that warmer temperatures have promoted plant growth during summer and/or plant respiration during winter in the northern high latitudes. The photosynthetic activity of terrestrial vegetation according to satellite retrievals increased from 1981 to 1991 in a manner that is suggestive of an increase in plant growth associated with a lengthening of the active growing season (Myneni et al. 1997). The regions exhibiting the greatest increase lie between 45 and 70°N, where remarkable warming has occurred in the spring due to an early disappearance of snow. Both the satellite data and the CO_2 record indicated that the global carbon cycle has responded to interannual fluctuations in surface air temperature, though small at the global scale, but are significant at the regional scale (Figure 7.1).

The hot-spots of vegetation in the middle and high latitudes in the northern hemisphere can be divided by several regions, such as northern China, northern Asia, Europe, eastern America and central America.

7.3.1 Northern China

In northern China, vegetation coverage has been improved in the past 40 years (1971–2011), with forests and farmland increasing most rapidly (Liu et al. 2013). The greening area was larger than the degraded area in the northern China, and the vegetation of forest and farmland recovered most obviously, while the degraded area of bush, grassland and desert was larger than others. In spring, the vegetation is mainly decreasing, but in summer, autumn and winter it is mostly increasing. There was a close relationship between vegetation and precipitation in northern China under the warming climate. Vegetation change in northern China was more significantly correlated with precipitation than with temperature, indicating that the sensitivity of vegetation change to precipitation was higher during these years. Besides climatic factors, human activities were also important factors affecting the vegetation change.

In northern China, the relationship between precipitation and temperature can be divided into four categories such as warm-dry, warm-wet, cold-dry, and cold-wet. The occurrence of cold-dry and cold-wet relationship becomes later from west to east, and the duration of cold-dry relationship has obvious interannual variation feature. On an interannual time scale, spring and the growing season becomes longer from east to west. In other seasons, the duration of the cold-wet relationship decreases with the decrease of latitude. Except for winter, the duration of cold-wet relationship

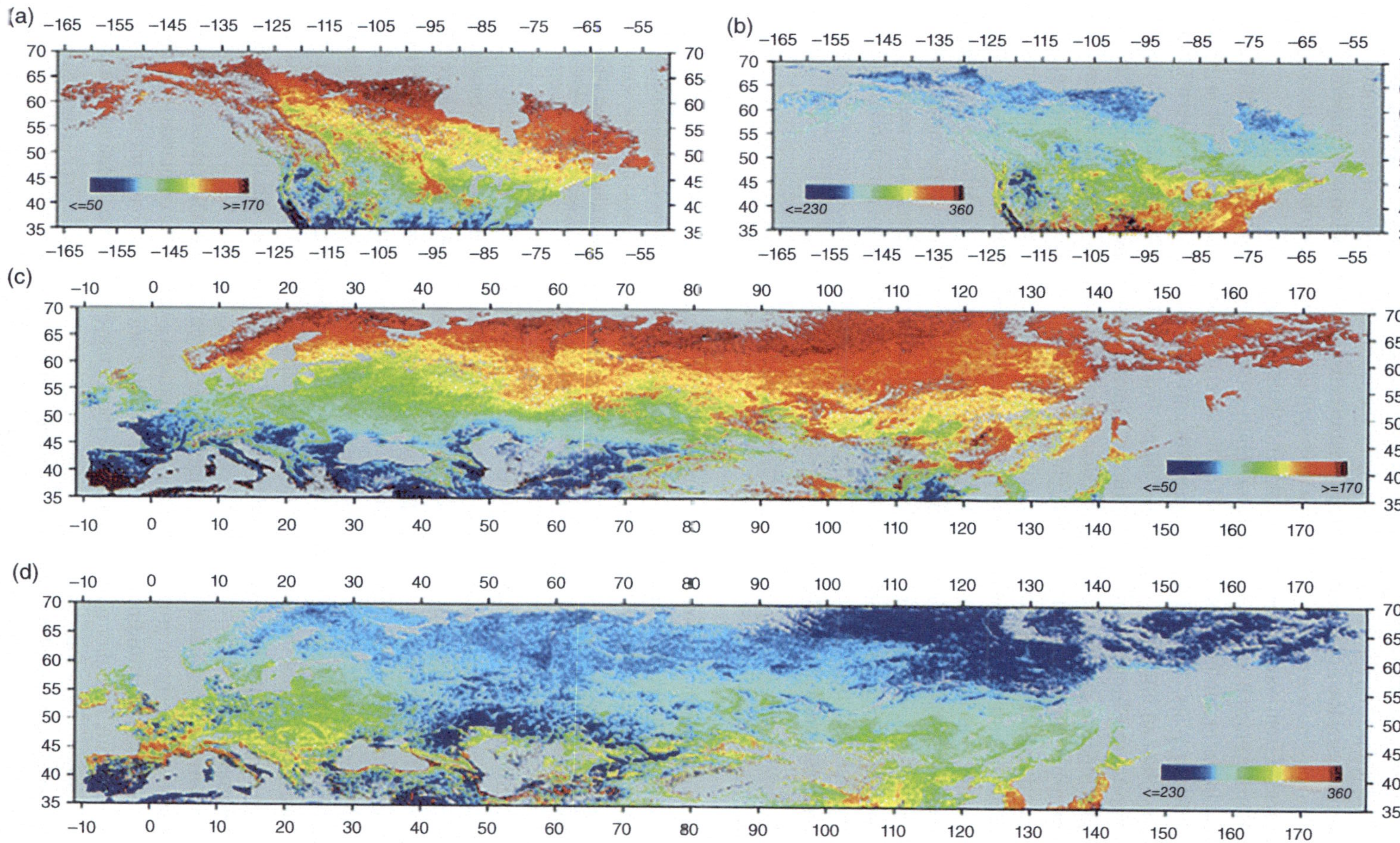

Figure 7.1 Spatial patterns of vegetation phenology. Only the first cycle is displayed for pixels with multiple modes. The color legend indicates the day of year while the gray indicates the areas either without vegetation or without good quality of temporal data. The horizontal axis represents the longitude degree and the vertical axis represents latitude degree. (a) Greenup onset in North America, (b) dormancy onset in North America, (c) greenup onset in Europe and Asia, and (d) dormancy onset in Europe and Asia. *Source:* Adapted from Zhang et al., 2004.

becomes longer from north to south. In the 1950s, the relationship between precipitation and temperature is manifested mainly by the cold-dry category. Since the 1970s, the relationship between precipitation and temperature is manifested mainly by the warm-dry category, and continued into the 1990s. After entering the 21st century, the relationship between precipitation and temperature is manifested mainly by the warm-dry type, especially in spring. The spatial distribution of the vegetation index decreased from southeast to northwest in northern China.

The vegetation coverage in northeastern China greatly changed in different season, with higher in summer and lower in winter (Ding et al. 2021). The vegetation cover of northern China greatly changed in different years. The change of vegetation cover is marginal in northwestern China. Human activities had an impact on vegetation change. From 1982 to 2015, the changed areas of vegetation in northern China are only 3.21%. In the growing season, the vegetation in the sub-region of cold temperate coniferous forest and deciduous broad-leaved mixed forest showed a negative relationship with meteorological drought; meanwhile, vegetation in the sub-region of warm temperate deciduous broad-leaved forest, temperate grassland, and temperate desert showed positive relationship with meteorological drought. There was a positive correlation between vegetation and precipitation in northern China in spring and summer but negative correlation in autumn. There was a positive correlation between vegetation and reference crop evapotranspiration in different seasons, with higher correlation in spring and summer. There was a negative correlation between forest of high coverage and drought and a weaker negative correlation between grassland, cropland, and drought. The region of high-coverage vegetation and various vegetation types had strong resistance to drought.

7.3.2 Europe

There is an increasing trend of vegetation over the whole Europe except for southern Europe. Land surface mean temperature is stable or slightly lower (up to −1.5 K) over the whole continent, except for southern areas for which the increase is up to 2.5 K. The arid and semi-arid areas of southern Europe become more arid. The rest of Europe presents an increase in its wooded land proportion, while seasonal amplitude in northern Europe has decreased (Julien et al. 2006).

The summer mean temperature has the strongest impact on vegetation productivity in Europe (Movchan 2019). The increase of summer temperatures caused the strongest statistically significant response of the vegetation in Europe. The spatial heterogeneity of such response across a latitudinal gradient and varying sensitivity of different types of the vegetation to climate change have been confirmed. The increased summer temperatures negatively influenced the productivity of croplands in western, eastern and southern Europe. The effect has been absent in northern Europe. In addition, the impact intensity shifts toward lower latitudes. Even in the middle latitudes, the productivity of croplands decreases under high temperatures without proper adaptation. The absence of the dependence for the higher latitudes can be explained by the fact that the increase of summer temperatures did not reach critical values and did not pass over the threshold in northern regions over Europe.

Variations of forest ecosystems have a lower threshold to extremely high temperatures. Forest is significantly more responsive to growing summer temperatures compared to croplands. Moreover, this reaction varies in the latitudinal direction. The negative effect is predominant in lower and middle latitudes, while in higher northern latitudes, the increase of summer mean temperatures stimulated the growth of the forest biomass.

Tree growth responds to a climates across a diverse and complex landscapes within different latitudes. Like forests, the productivity of savannas was also reduced with the summer temperatures in the southern and middle latitudes but enhanced in northern regions. The mean summer temperatures reached a threshold in southern Europe and their subsequent growth will lead to reducing vegetation productivity. At the same time in northern regions, the threshold has not been reached; therefore, summer temperatures increasing will stimulate the growth of vegetation.

Nowadays, dynamics of vegetation productivity in Europe is not of concern. The NPP changes were observed for less than 10% of the total vegetation area and generally had a poor positive trend. Nevertheless, despite the absence of negative evidence, analysis of relations between the vegetation productivity and temperature regime clearly shows that most European vegetation operates at its temperature optimum and further temperature increases will lead to inhibition of plant growth (the exception is northern Europe). Summer temperatures have the strongest impact on the vegetation productivity dynamic within all concerned temperature parameters. In western Europe, this negative impact is related to the maritime climate in Eastern Europe.

Although different countries have adopted different means of addressing the challenges of forest vegetation management in Europe in the 21st century, some common themes are still evident (Mccarthy et al. 2011). In all countries, there is a consensus that vegetation management is a critical silvicultural operation to achieve forest establishment, regeneration, growth, and production. It appears that herbicides are still in use to some degree, although at a lower intensity in many of the northern countries compared to other regions. The most common alternatives to herbicides are mechanical methods to cut vegetation and achieve soil cultivation; overstorey canopy manipulation to control vegetation by light availability; and in some instances, the use of mulches or biological control. Any reductions in herbicide use do not seem to have been driven solely by participation in forest certification schemes. Other factors, such as national initiatives or the availability of additional resources to implement more expensive nonchemical approaches, may be equally important. The development of more cost-effective and practical guidance for managers across Europe on nonchemical control methods can best be brought about by future collaborative research into more sustainable and holistic methods of managing forest vegetation, through the identification of silvicultural approaches to reduce or eliminate pesticide use and through gaining a better understanding of the fundamental mechanisms and impacts of competition.

7.3.3 North America

In North America, plant growth over extensive land areas was closely coupled to El Niño–Southern Oscillation (ENSO) effects on regional climate (Potter et al. 2008). Areas strongly tied to recent (2002–2003) ENSO climate effects were located mainly in northwestern Canada, interior Alaska, the northern Rocky Mountains of the United States, and throughout northern Mexico.

Vegetation cover anomalies early in the growing season have a statistically measurable effect on precipitation and surface temperature later in summer over the North American grasslands (Wang et al. 2006). In particular, higher means and/or decreasing trends of vegetation cover anomalies tend to be followed by lower rainfall but higher temperatures during July through September. Initially enhanced vegetation may deplete soil moisture faster than normal and thereby induce drier and warmer climate anomalies via the strong soil moisture–precipitation coupling over the North American grasslands. Consistent with this soil moisture–precipitation feedback mechanism, interactions between temperature and precipitation anomalies in this region indicated that moister and cooler conditions are also related to increases in precipitation during the preceding months. Because vegetation responds to soil moisture variations, interactions between vegetation and precipitation generate oscillations in NDVI anomalies at growing season time scales, which are identified in the temporal and the spectral characteristics of the precipitation–NDVI system. The precipitation–NDVI systems are as follows: (i) long-term interactions between the two anomalies tend to enhance one another, (ii) short-term interactions (less than two months) tend to damp one another, and (iii) intermediary period interactions (four to eight months) are oscillatory. The vegetation may influence summertime climate variability via the land–atmosphere hydrological cycles over these semi-arid grasslands Freden et al. (1974).

The North American monsoon leads to a large increase in summer rainfall and a seasonal change in vegetation in the southwestern United States and northwestern Mexico (Forzieri et al. 2011). Seasonal biomass productivity, as captured by the time-integrated NDVI (TINDVI), is an excellent means to synthesize vegetation dynamics. High TINDVI occurs for ecosystems with a short period of intense greening tuned to the North American monsoon or with a prolonged period of moderate greenness continuing after the monsoon. These cases represent different plant strategies (deciduous versus evergreen) that can be adjusted along spatial gradients to cope with seasonal water availability. Long-term trends in TINDVI may also indicate changing conditions favoring ecosystems that intensively use North American monsoon rainfall for rapid productivity, as opposed to delayed and moderate greening. A persistence of these trends could potentially result in the spatial reorganization of ecosystems in the NAM region.

Vegetation is an important factor in global climatic variability and plays a key role in the complex interactions between the land surface and the atmosphere in North America (Hong et al. 2007). Focus is usually on the spatial and temporal variability of vegetation and its relationship with land–atmosphere interactions. The vegetation water content (VegWC), the LAI, the NDVI, the land surface temperature (Ts) are important variables. There are three regions that have climatically different characteristics: the North America Monsoon System (NAMS) region, the Southern Great Plains (SGP) region, and the Little River Watershed in Tifton, Georgia. The introduction of the normalized vegetation water content (NVegWC) derived as the ratio of VegWC and LAI corresponding to the amount of water in individual leaves has been estimated and this yields significant correlation with NDVI and surface temperature (Ts). The analysis of the NVegWC–NDVI relationship in the three regions listed above displays a negative exponential relation, and the Ts–NDVI relationship is inversely proportional. The correlation between these variables is higher in arid areas such as the NAMS region, and becomes less correlated in the more humid and more vegetated regions such as the area of eastern Georgia. The regional distribution of vegetation reflects the relationship between the biological vegetation characteristics of water and the growing environment (Hong et al. 2007).

7.4 Ecology and Vegetation Characteristics in the Arctic Region

Arctic plants have a compressed growing season: they initiate growth rapidly in spring, and flower and set seed much sooner than plants that grow in warmer conditions. Their peak also occurs at a much lower temperature than plants from farther south. Compact cushions of vegetation keep the plants close to the warm soil and shield the tender central growing shoot (Figure 7.2). The height of Arctic plants is also governed by snow depth. Plants that protrude above the snow are subject to strong winds, blowing snow, and being eaten by caribou, and are common in the Arctic. These plants have the ability to stop growth at any time and resume it promptly when conditions improve. They can even survive being covered by snow and ice for over a year.

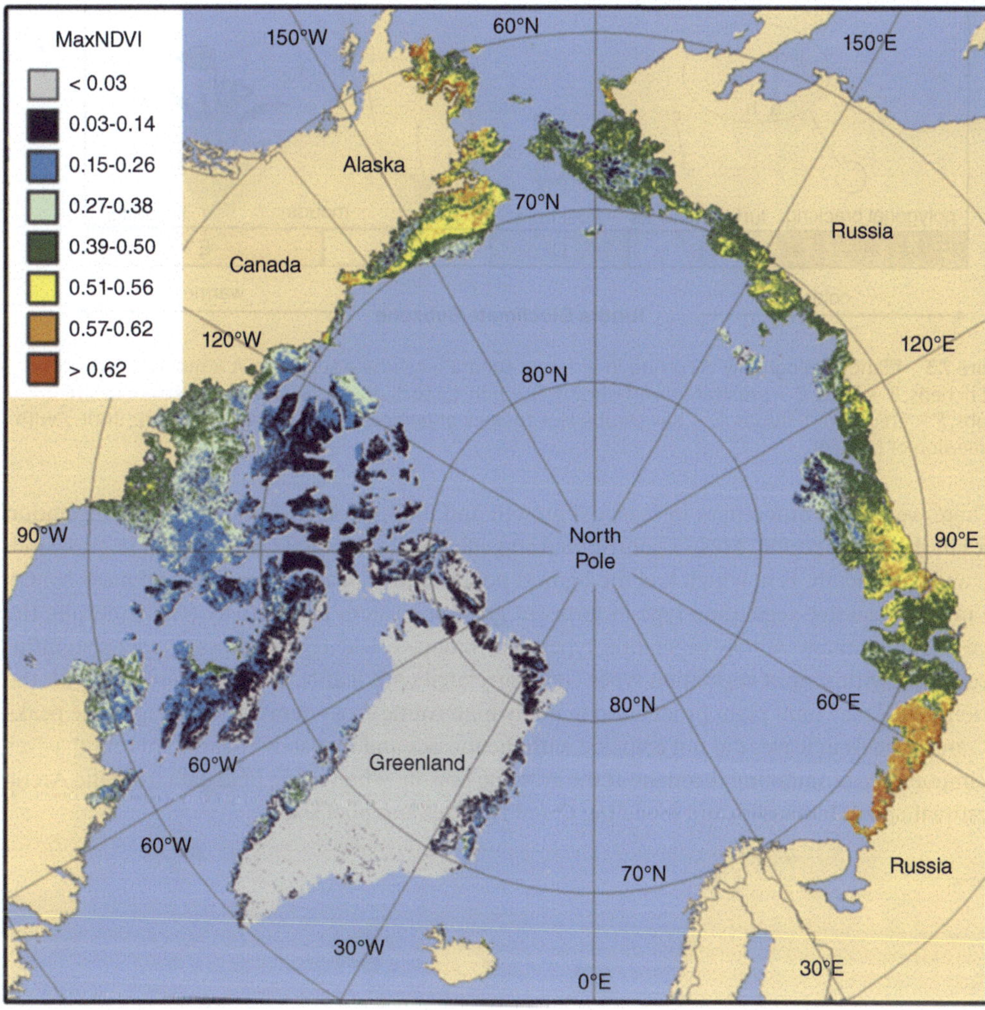

Figure 7.2 Circumpolar maximum normalized difference vegetation index (NDVI) of Arctic tundra. Areas of high NDVI (oranges and reds) correspond to regions of dense shrub cover. Yellow and green areas generally have peaty soil surface horizons and are continuously vegetated with mosses, sedges, and dwarf shrub areas approximately corresponding to areas of incomplete vegetation cover. Low NDVI (blues) corresponds to areas with sparse, very-low-growing plant cover on mineral soils (Bhatt et al. 2010).

The dominant trend of normalized NDVI was a decrease from south to north, correlating with bioclimate subzones and vegetation units in the Arctic region (Raynolds et al. 2006). NDVI also decreased at higher elevations and with higher substrate pH value. In the coldest bioclimate subzone, increased elevation was not correlated with decreased NDVI. In the warmest tundra bioclimate subzone, especially in Alaska, NDVI did not decrease with the first several hundred meters of elevation. NDVI in this subzone varied more by region than by elevation or substrate chemistry, and was lowest in recently glaciated areas such as the Canadian Shield. Phytomass (above-ground plant biomass) was calculated from NDVI using a relationship derived from ground clip-harvest data. Phytomass for the tundra bioclimate subzone was estimated at 2.5×10^{12} kg, with most of this in the warmest subzone, at the lowest elevations, and on acidic substrates (Figure 7.3).

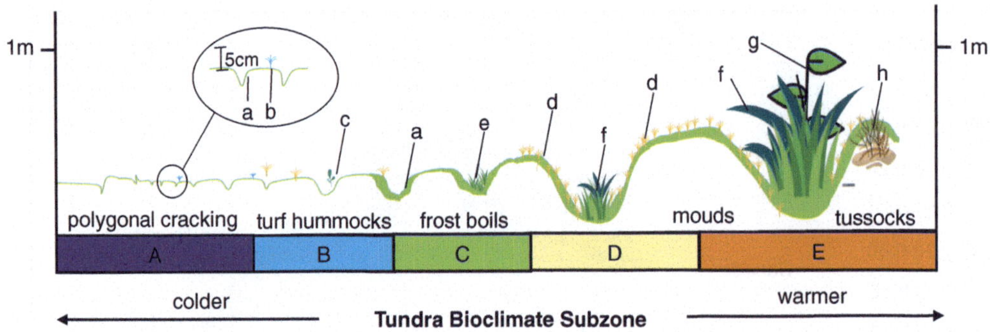

Figure 7.3 Plant physiognomy occurring in different tundra bioclimate subzones. A – mosses, liverworts and lichens, B – forbs, C – prostrate dwarf-shrubs, D – non-tussock graminoids, ehemiprostrate dwarf-shrubs, F – erect dwarf-shrubs, G – low shrubs, H – tussock graminoids. *Source:* Raynolds et al., 2006 / with permission of Elsevier.

There were large differences in seasonal growth and production within and among vegetation types in the Arctic region. Mosses contributed considerably to the total C and N pool in grassland, fen, and Salix heath. Fen, which had the highest pool of leaf N, leaf chlorophyll, and moss N, was the most productive vegetation type in terms of gross ecosystem productivity (GEP), despite the lowest total biomass. Across vegetation types, leaf biomass, leaf N, and moss N pool size influenced GEP. Within most vegetation types GEP correlated with leaf N, in correspondence with the notion that N may limit plant production in many high Arctic ecosystems. The timing of the peaks in C and N pools in leaves did not coincide with the mosses and in woody tissues. This emphasizes the importance of sampling throughout the growing season, when using field data from the Arctic to estimate plant biomasses and modeling C and N fluxes and pool sizes.

8

The Carbon and Nitrogen Cycle in Cold Regions

Cold regions store abundant of carbon (C) and nitrogen (N). For example, frozen ground can seal large amounts of organic carbon microorganisms for a thousands years, and with rising temperatures, the carbon stored in the frozen ground will be released. Thus, cold regions are generally regarded as sources of carbon. The release of carbon (CH_4, CO_2, etc.) is mainly carried out via heterotrophic respiration effects of microorganism. Its restoration is undertaken by photosynthesis in vegetations.

The carbon and nitrogen cycles in cold regions plays a robust role in climate change. Carbon releasing from frozen soil attracts much attention. This chapter introduces some characteristics of the carbon and nitrogen cycles in cold regions and will concentrate on three typical cold regions such as the Tibetan Plateau (TP), high-latitudes, and polar regions.

8.1 Carbon, Nitrogen, and Greenhouse Gases on the Tibetan Plateau

The TP has an annual mean temperature below 0 °C, with low annual precipitation, strong solar radiation, shallow soils, and a very short growing season (Liu et al. 2018). The high altitude and low temperature also reduce soil microbial activity and slow down the decomposition of organic matter such as apo plankton and plant and animal residues, which further promote the accumulation of soil carbon on the TP (Fang et al. 1996).

The transformations between soil organic carbon (SOC) and soil inorganic carbon (SIC) fundamentally depend on the circulation of CO_2. There usually has an unbalanced transformation mechanism among atmospheric CO_2, SOC, and SIC (Pan 1999, Figure 8.1). CO_2 in the atmosphere could be assimilated by plants as organic matter through photosynthesis. Soil organic matter (SOM) from the debris of animals, plants, and soil microorganisms could be mineralized directly for carbonates, stored in the soil as the inorganic C formed. In addition, CO_2 gas released from the decomposition of organic matter could generate HCO_3^- when it reacts with the soil moisture (SM), which is transformed into bicarbonate and turns into carbonate ultimately by means of a leaching process (Li et al. 2016). Consequently, the carbon lost from organic matter is sequestrated as inorganic form, Li, et al., (2016) proposed this process as Figure 8.1.

The organic carbon stocked in the surface layer (0–20 cm) of grassland soils on the TP is about 10 Pg C (Wang, Zeng, et al. 2019a,b). The SOC stock (0–200 cm) is about 16.5 Pg C and the SIC stock (0–100 cm) is about 14 Pg C (Wang et al. 2019). The soil carbon stocked on the TP gradually decreases from southeast to northwest, and there is no consensus on the role of change factors on the soil carbon stock on the TP (Wang et al. 2019).

Climatology in Cold Regions, First Edition. Chenghai Wang.

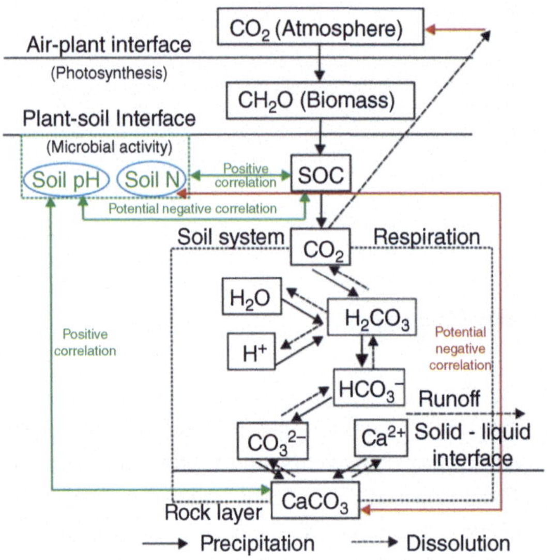

Figure 8.1 The model and mainly influencing factors for carbon transfer in the earth surface ecosystems. The black part is quoted from Pan (1999); the green part has been demonstrated in other studies; and the red part shows new results (Li et al. 2016). SOC, soil organic carbon.

Liu et al. (2012), reported the spatial distribution of SOC showed a significant relationship with soil properties and vegetation composition as well as climate conditions. The significant relationships between SOC density and belowground biomass (BGB) and SM can be observed. BGB and SM may be the dominant factors influencing SOC density in the topsoil. The averaged densities of SOC at a depth of 1 m were about 7.72 kg C m^{-2}. SOC densities were concentrated in the topsoil (0–20 cm) and fell exponentially as soil depth increased. SOC density has a negative relationship with temperature and a positive relationship with precipitation diminishing with soil depth. SOC was concentrated in the topsoil, and SOC density correlates well with BGB. SOC was predominantly influenced by SM, and to a much lower extent by temperature and precipitation (Liu et al. 2012).

The land types on the TP are mainly grasslands and wetlands. Gross primary production (GPP), ecosystem respiration (Re), and net ecosystem production (NEP) were markedly different among the vegetation zones, with lower Re and GPP in deeper water. For example, Hirota et al. (2006) measured ecosystem CO_2 fluxes in different vegetation zones during the growing season of 2002. Their results showed that diurnal variation of CO_2 flux was highly correlated with variation in photosynthetic intensity and soil temperature. The relationship between CO_2 flux and these environmental variables varied among the vegetation zones. Seasonal CO_2 fluxes, including GPP, Re, and NEP, were strongly correlated with aboveground biomass, which was in turn determined by water depth. In the early growing season, temperature sensitivity (Q_{10}) for Re varied from 6.0 to 8.9, which depends on vegetation zone. Q_{10} decreased in the late growing season. Estimated NEP for the whole deep-water wetland over the growing season was 24 g C m^{-2}. Water depth is the major environmental control of seasonal variation in CO_2 flux, whereas photosynthetic photon flux density (PPFD) controls diurnal dynamics.

Huang et al. (2022) estimated the gross primary productivity (GPP) and net ecosystem exchange (NEE) in Qinghai grasslands. The GPP of the Qinghai grasslands fluctuated from 1979 to 2018, with an annual mean value of 118.78 g C m^{-2}. The NEE of the Qinghai grasslands fluctuated from 1979 to 2018, with a mean value of −5.16 g C m^{-2}. After 2000, GPP increased, and NEE decreased in a fluctuating manner. There were clear regional differences in GPP and NEE. GPP was low in most areas of Qinghai, and GPP was high in eastern and southern Qinghai. The southern, southeastern, and northeastern Qinghai were mainly carbon sinks, and the northwestern Qinghai and the region between the southeastern and northeastern Qinghai were mainly carbon sources.

Grazing generally decreased GPP and increased NEE in the Qinghai grasslands from 1979 to 2018. There was spatial heterogeneity in the effect of grazing on GPP and NEE. Under grazing, GPP and NEE were significantly decreased mainly in eastern Qinghai, and GPP and NEE were significantly increased mainly in southern and eastern Qinghai. NEE was most affected by grazing in eastern Qinghai.

Zhang et al.(2009) summarized a schematic diagram of annual C partitioning in diverse C pools in the K. humilis meadow plant–soil system. Figure 8.2 shows that, when separating Re into above-ground and total soil respiration, they have individual contributions of 0.42 and 0.58, respectively (Zhang et al. 2009). These values matched the contributions of 0.40 and 0.60 despite the fact that both newly fixed carbon and stored carbon were involved in the measured respiration. Adopting the relative contribution of root and soil microbial respiration to total soil respiration in the same article, roots lost only 2.9% of annual fixed carbon through autotrophic respiration, while it transferred seven times more carbon to the soil by rhizodeposition, which is similar to results in the grassland at Sourhope (Leake et al. 2006), but much higher than the fraction of 7–12% reported in warmer climate regions for a *Lolium perenne* and *Trifolium repens* grassland (Kaštovská and Šantrùèková 2007), *Lolium perenne* plants (Butler et al. 2004), *Bromus erectus* plants (Warembourg and Estelrich 2000), and *Zea mays* plants (Haller and Stolp 1985; Jones and Darrah 1993). In annual assimilated carbon, 2.476 t ha^{-1} remained above ground in shoot, 2.868 t ha^{-1} in root and

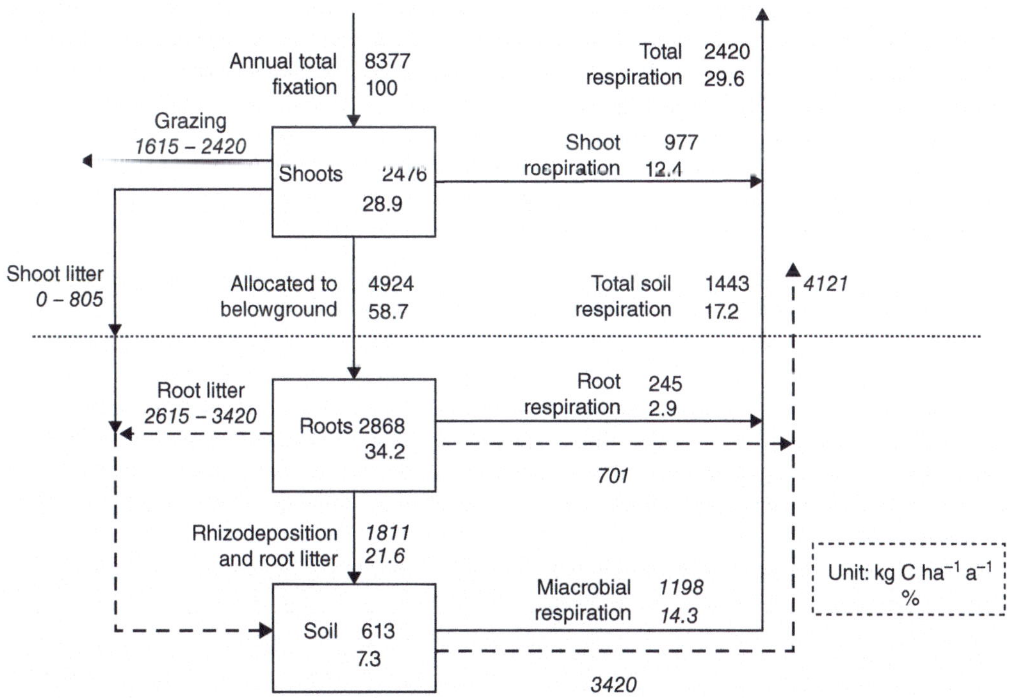

Figure 8.2 Schematic diagram of annual carbon partitioning in diverse carbon pools in the *Kalanchoe humilis* meadow plant–soil system representing the annual carbon fluxes. The absolute estimates (upper numbers in Kg C ha^{-1}) and their percentage (bottom numbers) of the total annual carbon fixation are indicated. Solid lines indicate carbon fixed in current year and broken lines indicate previously stored carbon. Numbers in italics are cited or recalculated from Cao et al. (2008), Du et al. (2007), and Zhang et al. (2009). The value of 2615–3420 kg C ha^{-1} a^{-1} for root litter is the minimum value required to balance total soil loss through belowground respiration. The amount of shoot carbon removed or returned to the soil is determined by grazing intensity (Wu et al. 2010).

0.613 t ha^{-1} in soil. Carbon storage was reported to be 11.1 and 78.4 t ha^{-1} in roots and soil (0–20 cm) in this grassland (Zhao 2009). Therefore, the carbon turnover rate in root and soil was about 25.8% and 0.78% per year, corresponding to mean residual time of 3.9 and 128 years, assuming the humification coefficient for the root-derived carbon is 10% (Kuzyakov et al. 2001) and soil-carbon stock approximately constant. The measured maximum aboveground biomass of this vegetation is reported to be 550 g DW m^{-2} (Du et al. 2007). A previous investigation using a closed chamber method indicated that total soil respiration in this ecosystem lost 5564 kg C ha^{-1} annually (Cao et al. 2008). It appears that the direct carbon flow into belowground carbon pools through living plant carbon translocation was insufficient to balance the respiratory carbon loss. Under high grazing intensity and with little return of aboveground litter to soil, the plant–soil system may become a weak source of atmospheric CO_2. Without litter input, the belowground carbon pool would have a net loss of 643 kg C ha^{-1} a^{-1}. Under a light grazing regime (e.g. three sheep unit hal), above ground litter return (94 g DW m^{-2}) transformed the system into a weak carbon sink, about 162 kg C ha^{-1} a^{-1}. About 701 kg C ha^{-1} a^{-1} was lost through root respiration of stored carbon.

A study by Wang et al. (2002) showed that, the organic carbon content can reach up to 33.52 Pg of carbon stored in the 0–0.75 m depth of soil subtending various types of grasslands on the TP. Carbon emissions from the grassland soils were estimated based on the two major modes of emission: (i) natural soil respiration, and (ii) shifts in net C flux to/from soil due to land-use changes and their potential influence on organic matter decomposition. Annual soil respiration-driven CO_2 emissions from the grassland soils of the plateau reached 1.17 Pg C $year^{-1}$, accounting for 26.4% of China's total soil respiration and 1.73% of global soil respiration. Because the grassland area accounts for 1.02% of the global terrestrial land and 16.9% of the total terrestrial land in China, this CO_2 emission rate is significantly higher than the country's mean annual rate (approx. 4.2 Pg C $year^{-1}$) and even higher than the global mean rate (approx. 68 Pg C $year^{-1}$). In the past 30 years, approximately 3.02 Pg C have been emitted from the grassland soils of the TP due to land-use changes and grassland degradation. The total CO_2 emissions rate from the grassland soils of the TP reached 1.27 Pg C $year^{-1}$.

The dynamics of aboveground carbon and soil surface SOC at the soil depth of 0–20 cm for typical grasslands shows different patterns of temporal variation in different ecosystems. The extent of temporal variation increases with the increase of SOC in the ecosystem; SOC increases first and decreases quickly; Spatially, SOC density obtained for the equilibrium condition declines gradually from the southeast to the northwest on the plateau and showed a high heterogeneity in the eastern plateau; SOC density in the alpine grasslands shows remarkable response to climate change during the 42 years, and the net carbon exchange rate between the alpine grassland ecosystems and the atmosphere increases from 1990 to 2000 as compared with that before 1990 (Zhang et al. 2007).

The deep organic soils stored in wetlands have been accumulating carbon for 4000–5000 years. However, temperature increases due to climate change and drainage of wetlands may provide conditions to reverse this trend, leading to overall carbon loss. Alpine wetland meadow could function as a carbon sink due to its high soil organic content and low decomposition. In total, annual GPP were estimated at 575.7, 682.9, and 630.97 g C m^{-2} in 2004, 2005, and 2006, respectively. Meanwhile, the Reco were equal to 676.8, 726.4, 808.2 g C m^{-2}, and thus the NEE were 101.1, 44.0, and 173.2 g C m^{-2}; in addition, the alpine wetland meadow was a moderate source of CO_2 and the observed carbon dioxide fluxes in the alpine wetland meadow were higher than other alpine meadow such as the *Kobresia humilis* and shrubland meadows (Zhao et al., 2010).

Based on the linear regressions and the soil pits classification, Wu et al. (2017), summarized a conceptual model which included these environmental factors and contributions of the factors to the SOC content as Figure 8.3. In this diagram, the geomorphology determines the drainage conditions, and geomorphology and drainage conditions affect the vegetation. The vegetation interacts with permafrost through mulching effects, and permafrost presence can supply water for plant growth. Parent materials are related to soil particle size through pedogenesis. For the sites with permafrost, the depth of active layer explains 37.9% and the vegetation cover accounts for 17.4% of the SOC variations.

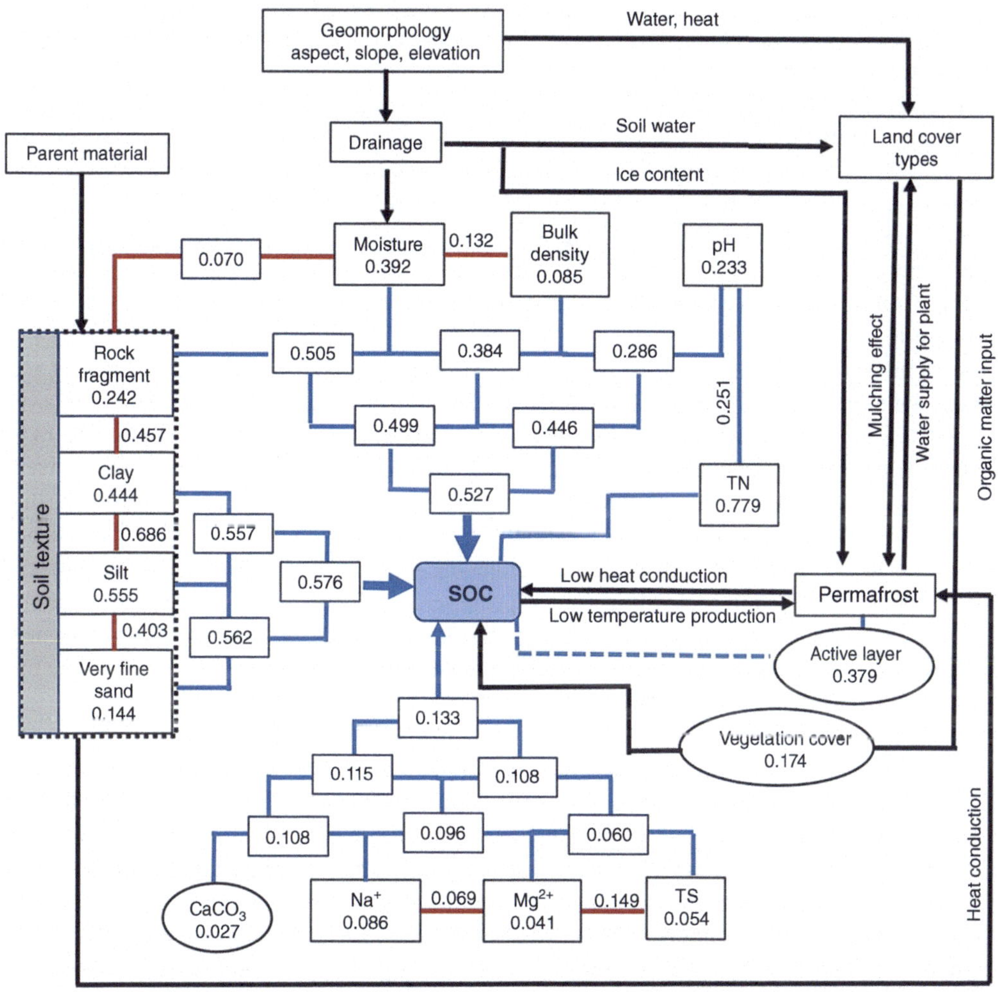

Figure 8.3 A conceptual model of the relationship among the environmental conditions and the contribution of variables to the soil organic carbon (SOC) stocks for the upper 200 cm of soils. The values show the proportion of the explained variance when using the factors as predictors for the others. The gray area highlights the SOC; the black line with the arrow shows the effects of one factor on the other; the blue line shows the contributions of variables to the SOC variations (also the pH to the TN); the red line shows the explained variances of the linear relationship between the variable; the dashed blue line shows the contribution of ALT to the SOC for the 18 sites; the dashed black line indicates the soil particle size; and the black ovals show part of effects of vegetation and permafrost. The $CaCO_3$ did not significantly correlate with SOC for upper 200 cm stocks, but it significantly correlated with SOC stocks for 0–50 cm and 50–100 cm layers. *Source:* Adapted from Wu et al. 2017.

Fan's (2006) full work is described below, the CO_2 fluxes on the TP ecosystem varied seasonally, with the maximum in summer and the minimum in winter. From October of the first year to April of the second year, the ecosystem is a weak source of atmospheric carbon; from May to September, it is a carbon sink, especially from June to August, and averaged over the whole year, it is an atmospheric carbon sink. The temporal and spatial variation of the fluxes is also obvious, with the largest fluxes in the southwest and southeast borders of the TP and in the northeast in summer, and the smallest in the hinterland and northwest of the TP in winter. The change in net CO_2 uptake in response to increasing temperature shows that, the net CO_2 fluxes increase by leaps and bounds with increasing temperature, which has important climatological implications. Ecosystems on the TP have the ability to regulate the greenhouse effect, and the warmer the climate, the stronger this regulation is likely to be.

Kato et al. (2004) investigated the CO_2 exchange between the atmosphere and an alpine meadow ecosystem on TP, the maximum rates of CO_2 uptake and release derived from the diurnal course of CO_2 flux (FCO_2) were -10.8 and $4.4\,\mu mol\,m^{-2}\,s^{-1}$, respectively, indicating a relatively high net carbon sequestration potential as compared to subalpine coniferous forest at similar elevation and latitude. The daily CO_2 uptake during the measurement period indicated that the alpine ecosystem might behave as a sink of atmospheric CO_2 during the growing season if the carbon lost due to grazing is not significant. The daytime CO_2 uptake was linearly correlated with the daily PPFD in each month.

The methane (CH_4) cycle on the TP is sensitive to climate change and subsequent freezing and thawing dynamics. The TP permafrost region was a CH_4 sink of $-0.86 \pm 0.23\,g\,CH_4\text{-}C\,m^{-2}\,yr^{-1}$ over 2012–2016, a rate higher than that of many other permafrost areas, such as the Arctic tundra in northern Greenland, Alaska, and western Siberia. Soil temperature and soil water content were dominant factors controlling CH_4 fluxes; however, their correlations changed with soil depths due to freezing and thawing dynamics (Yun et al., 2018). This region was a net CH_4 sink in the fall, but a net source in spring, despite both seasons experiencing similar topsoil thawing and freezing process. The opposite CH_4 source–sink function in spring versus fall was likely to have been caused by the respective seasons' specialized freezing and thawing processes, which modified the vertical distribution of soil layers that are highly mixed in the fall, but not in spring. Furthermore, the traditional definition of four seasons failed to capture the pattern of the annual CH_4 cycle.

The results of Jin's (2015) show that, for the net exchanges of CH_4 and CO_2 over the TP for the period 1979–2100, the region currently acts as a net CH_4 source with $0.95\,Tg\,CH_4\,y^{-1}$ emissions and $0.19\,Tg\,CH_4\,y^{-1}$ soil uptake, and a photosynthesis C sink of $14.1\,Tg\,C\,y^{-1}$. By accounting for the net CH_4 emission and the net CO_2 sequestration since 1979, the region was found to be initially a warming source until the 2010s with a positive instantaneous radiative forcing peak in the 1990s. In response to future climate change projected by multiple global climate models (GCMs) under four representative concentration pathway (RCP) scenarios, the regional source of CH_4 to the atmosphere will increase by 15–77% at the end of 21st century. NEP will continually increase from the near neutral state to around $40\,Tg\,C\,y^{-1}$ under all RCPs except RCP8.5. Spatially, CH_4 emission or uptake will be noticeably enhanced under all RCPs over most of the TP, while statistically significant NEP changes over a large scale will only appear under RCP4.5 and RCP4.6 scenarios. The cumulative greenhouse gas fluxes since 1979 will exert a slight warming effect on the climate system until the 2030s, and will switch to a cooling effect thereafter. Overall, the total radiative forcing at the end of the 21st century is 0.25–$0.35\,W\,m^{-2}$, depending on the RCP scenario.

There is a significant northwest-southeast trend of CH_4 concentration in the troposphere of the TP, and there are four fixed high-value centers in the north and south of the TP, which are located in Ali, Nagqu, Shannan, and Yushu, respectively. Feng et al. (2017) considers that, in the middle and upper troposphere, CH_4 concentrations are basically distributed in a valley-like pattern with the lowest values in the lower altitude and the highest values in the north and south sides of the TP. At the top of the troposphere, the low value zone is the dividing line, and the CH_4 concentration is obviously high in the southern TP and low in the northern TP. The CH_4 concentration on the TP tends to rise slowly with time, with an average rate of 0.0018 ppm a^{-1}, the fastest rise occurs in summer while the slowest occurs in the fall. The CH_4 on the TP has obvious single-peak seasonal variation, with highs in summer and fall and lows in winter and spring, which is different from the double-peak characteristic in winter and summer in the eastern region. The seasonal variation is more obvious as the altitude rises.

There was high spatial variability for soil organic carbon density (SOCD), soil total nitrogen density (STND), and soil total phosphorous density (STPD) in the drained alpine wetland. Spherical models best described the structure of the semi-variograms for SOCD and STPD, and an exponential model for STND, with the range parameter of less than 4 m (Bai et al., 2010). Similar spatial distribution with lower or higher patches of carbon, nitrogen, and phosphorous storage were observed; SOCD, STND, and STPD were significantly negatively correlated with soil moisture, and significantly positively correlated with bulk density. However, no significant correlations were observed between SOCD, STND, and STPD and soil pH values (Bai et al., 2010). Wetland drainage might lead to higher carbon, nitrogen, and phosphorus densities in top 0.30 m soils due to peat compaction; thus, it is necessary to incorporate water table fluctuations or the whole depth of peat layers to estimating precisely carbon, nitrogen, and phosphorus storage (Bai et al., 2010).

Zhang et al. (2020) investigated that, inland waters are large sources of CH_4 to the atmosphere. However, considerable uncertainty exists in estimating the emissions of this potential greenhouse gas from global streams and rivers due, in part, to a lack of direct measurements in the high-altitude cryosphere and poor accounting for ebullition. CH_4 ebullition rates decrease exponentially whereas diffusion declines linearly with increasing stream order. Nevertheless, the mean ebullition rate (11.9 mmol CH_4 m^{-2} d^{-1}) from these streams and rivers, which have large organic stocks in surrounding permafrost, abundant cold-tolerant methanogens, shallow water depths, and experience low air pressure, were six times larger than the global mean value and reached a maximum of 374.4 mmol CH_4 m^{-2} d^{-1}. Upscaled total emissions from sampled third- to seventh-order waterways of the east TP are estimated to be 0.20 Tg CH_4 yr^{-1}, 79% of which was attributed to ebullition. These CH_4 emissions are approximately 20% of CO_2 emissions (2.70 Tg CO_2 yr^{-1}) in terms of carbon release and two times greater in terms of CO_2-equivalent emissions. When upscaled to first- to seventh-order waterways, emissions were 0.37–1.23 Tg CH_4 yr^{-1}.

Strong CH_4 emission at the rate of 26.2 ± 1.2 and 7.8 ± 1.1 µg CH_4 m^{-2} h^{-1} was observed for a grass community in a *Kobresia humilis* meadow and a *Potentilla fruticose* meadow, respectively. A shrub community in the *Potentilla* meadow consumed atmospheric methane at the rate of 5.8 ± 1.3 µg CH_4 m^{-2} h^{-1} on a regional basis; plants from alpine meadows contribute at least 0.13 Tg CH_4 yr^{-1} over the TP (Cao et al., 2008).

Zhou et al. (2017) analyzed that, the CH_4 fluxes in grasslands are sensitive to the changes in precipitation and soil nitrogen concentrations. The aboveground biomass and litter fall were the most important factors promoting and limiting the CH_4 uptake rate of alpine meadows, respectively. The CH_4 concentration decreases with increasing altitude on the TP. The seasonal variation of CH_4 content in the upper troposphere is obvious, and its averaged concentration is high in July–September, followed by June and October, and lowest for the rest of the year. The analysis of the spatial distribution of CH_4 over the TP showed that the concentration of CH_4 in the northern TP was higher than that in the southern TP. During the summer monsoon, with the strong convective transport over the TP and the blockage of the South Asian high pressure above the TP, the CH_4 concentration in the upper troposphere increased significantly and accumulated continuously, and the maximum value appeared from the end of August to the beginning of September. Based on the analysis of the monsoon index, it was found that the strong convective transport under the influence of the summer monsoon was one of the main reasons for the formation of high values of upper tropospheric CH_4 on the TP, and the maximum values of upper tropospheric CH_4 concentrations appeared about half to one month later than the peak of the monsoon index, and the high values of CH_4 concentrations decreased rapidly with the retreat of the summer wind.

The $\delta^{13}C$ values derived from 198 samples range from −28.6 to −15.1‰ versus PDB (Pee Dee belemnite) and exhibit a clean relation to different vegetation communities from forest (−25.9 ± 1.2‰) to shrub (−24.7 ± 1.4‰), steppe (−23.1 ± 1.3‰), alpine meadow (−23.6 ± 0.7‰), alpine desert steppe (−21.3 ± 1.6‰), and alpine desert (−18.9 ± 2.5‰) (Lu et al., 2004). The observed variability in $\delta^{13}C$ can be attributed to the mean annual precipitation and temperature, which control the distribution of vegetation types on the TP, and cause the change in carbon isotope composition of modern soils at any given altitude. Mean annual precipitation and temperature are the major factors affecting the $\delta^{13}C$ values of surface soils (Lu et al., 2004).

8.2 Carbon, Nitrogen, and Greenhouse Gases at High Latitudes

Tundra, woodland, and forest ecosystems of the northern high latitudes (north of 50°N) contain about one-quarter of the global terrestrial organic carbon, much of which is stored in the soil. Soil carbon is sensitive to climate changes, and has the potential for accelerating soil microbial processes, thus leading to increased emissions of CO_2 and CH_4 to the atmosphere in a warmer and wetter future (Zhuang et al., 2006). Estimates of the annual net CO_2 balance for the northern high latitudes during the 1990s indicated that, these regions function as a sink of several hundred teragrams of carbon per year. For CH_4, there is general agreement that the northern high latitudes are currently a source of this potent greenhouse gas, though its magnitude is poorly defined. How the combined CO_2 and CH_4 budget for the region will change over the 21st century depends on the complex responses of northern high-latitude ecosystems to climate change, atmospheric CO_2 concentration, and disturbance regimes, especially for permafrost thawing and fires. Terrestrial ecosystems of the northern high latitudes (above 50°N) exchange large amounts of CO_2 and CH_4 with the atmosphere each year. Northern high-latitude regions are a net source of carbon for the atmosphere at 276 Tg C yr^{-1}, and the source will increase by up to 473 Tg C yr^{-1} by the end of the 21st century compared with the current emissions (Zhuang et al., 2006).

Land areas above 50°N (about 23% of the vegetated global land area) are currently accumulating about 0.4 Pg C y^{-1} (about 30% of the estimated global terrestrial sink) and this sink could persist until the end of 21st century (Zhuang et al., 2006). This was due mainly to an increase in forest productivity and biomass in response to increasing atmospheric CO_2, temperature and nitrogen deposition, and includes an estimate of the effect of boreal forest fire, which was estimated to diminish the sink approximately by the amount of carbon emitted to the atmosphere during fires (Zhuang et al., 2006). White et al. (2001) concluded that, regional mean results indicated that, nitrogen deposition contributed about 18% to the sink by the 2080s. As expected, climate change (temperature, precipitation, solar radiation, and saturation pressure deficit) and nitrogen deposition produced a carbon source without increasing atmospheric CO_2. Forest areas expanding both south and north, will halve the current tundra area 2100. This expansion will contribute about 30% to the sink by the 2090s. Tundra areas which were not invaded by forest will fluctuate from sink to source. High -atitude carbon sink exists at present and, even assuming few effects of nitrogen deposition, no forest expansion and continued boreal forest fires, the sink is likely to persist at its current level for a century.

Soil carbon and nutrient stores are large in many high-latitude ecosystems. The distribution of the SOC in the permafrost organic carbon pool in the historical period (1986–2005), which indicates the SOC density, is more than 60 kg m^{-2} in western Siberia and south of Hudson Bay in Canada. The SOC density is relatively small at the edge of the permafrost, where it is lower than 10 kg m^{-2}. The difference in the SOC in permafrost regions may be related to different soil types. Because the SOC is primarily distributed in the shallower soil layers, the release of carbon where the active layer thickness (ALT) is above 3.5 m; therefore, the stored carbon in the shallow permafrost estimated during the historical period is approximately 667.99 Gt.

Grosse et al. (2011) summarise as follows, the northern high-latitude SOC pool is a dynamic stock affected by variations in inputs (organic litter quality and quantity), the mode and timing of carbon stabilization (permafrost aggradation, cryoturbation, peat accumulation, sedimentation) and carbon destabilization (microbial decomposition, combustion), and carbon export (via dissolved and particulate phases, inorganic and organic state, gas fluxes). Individual processes of these carbon pool dynamics can act on different timescales, complicating an overall assessment. As a result, the net increases or decreases of the SOC pool are inexorably tied to both the processes and form of carbon stabilization. Rates of burial, cryoturbation, inundation, and permafrost aggradation become important factors determining SOM quality (form and relative lability or recalcitrance) and together with disturbance history can dictate whether organic carbon in a soil is most vulnerable to combustion, leaching, or microbial decomposition if subjected to thawing and subaerial exposure. Moreover, interactions among soil processes and the physical and biological systems occur at different times with different responses and durations. For example, one possible local scenario might be: slow increase in active layer depth → slow landscape drying → slow increase in fire probability → fast fire occurrence → fast thermokarst → slow carbon accumulation and fast CH_4 emissions under anaerobic conditions. Meanwhile other landscapes less vulnerable to fire owing to poor soil drainage may respond only to changes in temperature (more plant production and/or decomposition). Thus, heterogeneity of the landscape is further complicated by the timing and duration of various physical and biogeochemical feedbacks.

Grosse et al. (2011) summarized the disturbance processes impacting northern high-latitude SOC stocks and feedback as Figure 8.4. Disturbance or the perturbation of a normal state or regime, of the northern high-latitude soil C cycle is an important natural mechanism rapidly facilitating such changes. Important research issues concern the degree to which disturbances affect SOC storage or loss in northern high-latitude soils, and how past disturbance regimes determine current SOM quality. Disturbance of the northern high-latitude SOC pool as an event or process forced onto this system results in a significant redistribution of C among major reservoirs. Natural disturbances alter key ecosystem factors in ways that affect the dynamics of the northern high-latitude SOC pool (Figure 8.4). Disturbances can be categorized by their time-horizon of influence, the complexity of involved processes, the size of area affected, and the total impact on a SOC pool or the C cycle. Furthermore, press and pulse disturbances can be distinguished. *Presses* are impacts driven by decadal to century-scale changes of factors controlling a system. *Pulses* are one-time or episodic short-term events that are often extreme in nature and impact a system very rapidly. Press disturbances with impact on northern high-latitude SOC include, for example, paludification, aridification, top-down permafrost thaw, and changes in pedological processes and vegetation structure (Figure 8.4). Pulse disturbances include wildfires, thermokarst formation, thermokarst lake drainage, flooding, and rapid erosion of soils. By crossing environmental thresholds, press

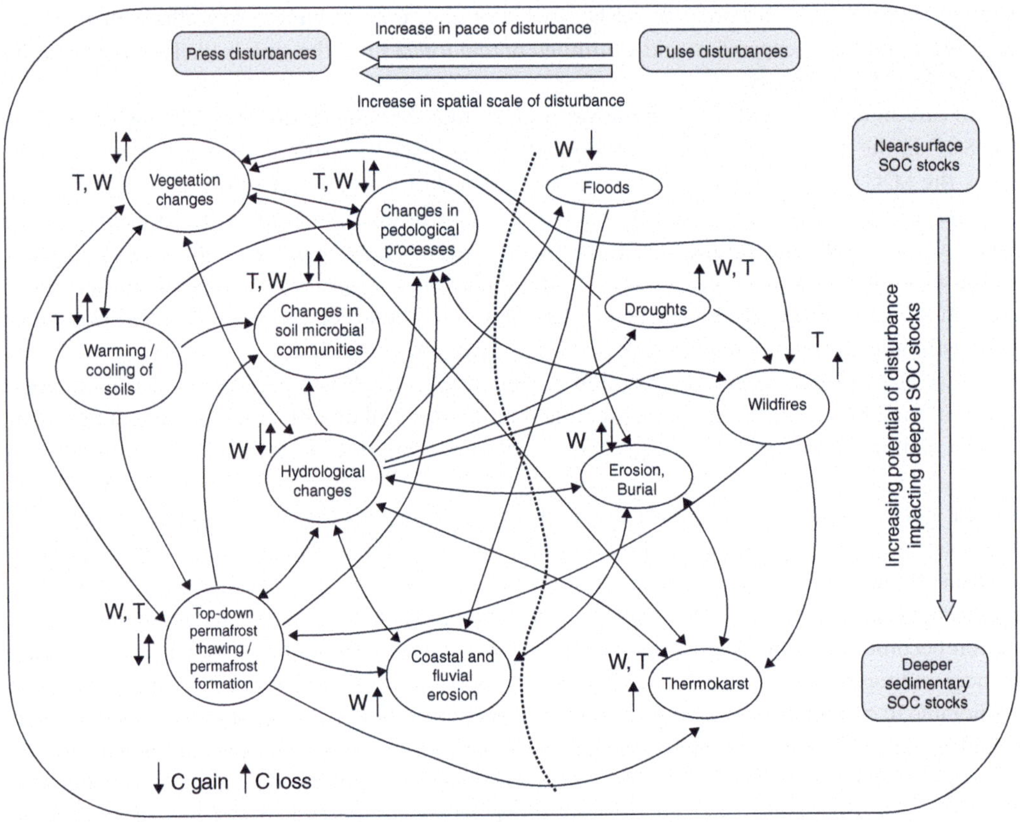

Figure 8.4 Key disturbance processes impacting northern high-latitude soil organic compound (SOC) stocks and feedback between them. Press disturbances are slower (decades to centuries) but affect larger areas, whereas pulse disturbances can be fairly rapid (seasons to decades) but are often local with sometimes widespread occurrence. Disturbances have a different potential to impact SOC stocks at greater depths. Feedback between disturbances include for example the impacts of wildfires on vegetation changes, and subsequent soil warming and top-down permafrost thawing. Disturbances affect SOC stocks mainly via changes related to water (W), or temperature (T), or both. For example, wildfires affect SOC in the upper soil organic layers by combustion (temperature), or thermokarst affects SOC stocks by warming soils (temperature) and melting ground ice (water). Small black arrows indicate whether SOC is lost or sequestered due to certain disturbances; some disturbance types can result in SOC losses or gains depending on environmental preconditions. *Source:* Grosse et al. (2011) / John Wiley & Sons.

disturbances could trigger pulse disturbances. For example, widespread top-down permafrost thawing could trigger local rapid thermokarst or increased soil erosion, or changes in vegetation structure could result in increased risk of wildfires. There are also feedbacks between different pulse disturbance processes. For example, wildfires can trigger local thermokarst in some permafrost types or floods soil erosion. Though natural disturbances are an important component in shaping the northern high-latitude SOC pool, changes in their frequency and intensity may have important impacts on this carbon pool. It is therefore essential to understand how global change will alter disturbance frequencies and intensities in the northern high-latitude region.

A study by Robinson (2002) indicated that, at high latitudes, decomposition rates and soil nitrogen availability are pivotal in determining ecosystem responses to climate change. The controls on nitrogen availability are less well characterized than those on decomposition. Because of the concomitant small rises in soil temperature together with interactions between the controls, the increase of mean summer temperatures at high latitudes may not necessarily cause significant increase of decomposition and nitrogen mineralization.

Large amounts of carbon are buried in the permafrost under the ALT. As the temperature increases, the ALT deepens and the carbon stored in the permafrost will become increasingly vulnerable to decomposition and subsequent release into the climate system. The assumed height between the top and bottom of the permafrost represents the height of the carbon pools, and the SOC density is used to estimate the amount of organic carbon stored in the permafrost. The SOC is the carbon stored in SOM, which enters the soil via the decomposition of plant and animal residues, root exudates, living and dead microorganisms, and soil biota. SOM contains approximately 58% SOC: therefore, SOM can be converted to SOC by multiplying by 0.58.

SOC changes when the global mean surface temperature increases by 1.5°C and 2°C relative to that in the historical period under different RCP scenarios (Figure 8.5). A negative ATL represents regions where the SOC densities decrease, which also implies that a large amount of SOC is released into the atmosphere. The most significant reduction in the SOC is approximately 8–32 kg m^{-2} in southern Canada. When the global mean surface temperature increases by 1.5°C, the SOC density in southern Canada decreases by 8–20 kg m^{-2}. When the global mean surface temperature increases by 2°C, the reduction of the SOC density in southern Canada can reach up to 32 kg m^{-2}. To quantify the amount of SOC released from the permafrost, the ALT decreased as the height of the carbon pools decreased and used the SOC density to estimate the amount of SOC released from the permafrost. When the global mean surface temperature increases by 1.5°C, the permafrost region releases 102.66 Gt, 96.22 Gt, and 97.18 Gt

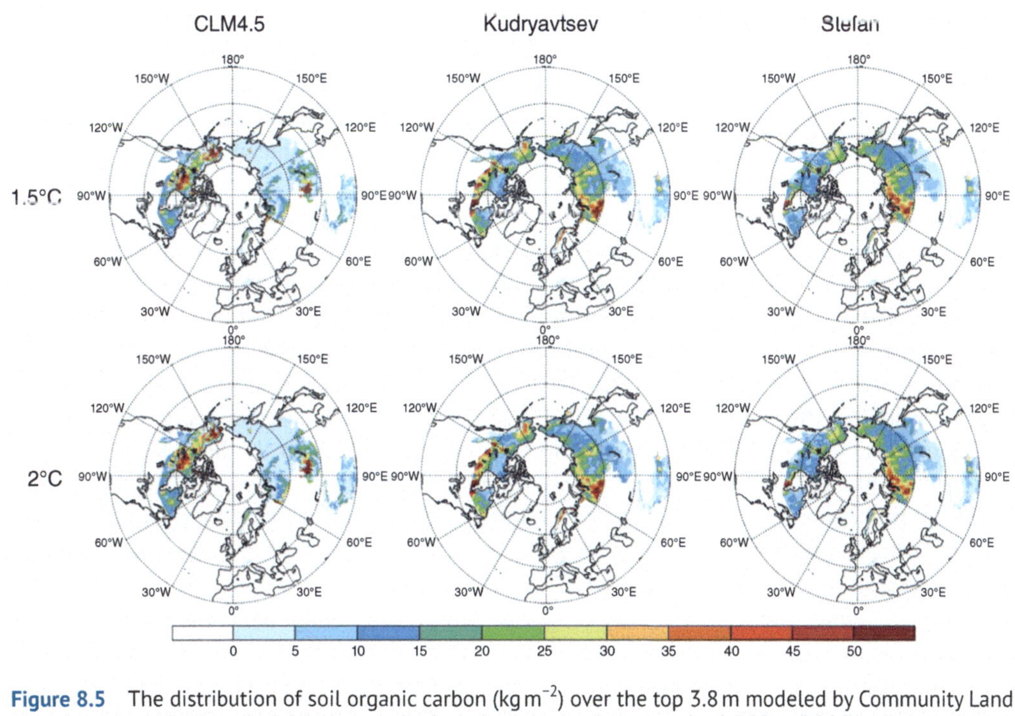

Figure 8.5 The distribution of soil organic carbon (kg m^{-2}) over the top 3.8 m modeled by Community Land Model version 4.5 when the global mean surface temperature increases by 1.5 °C and 2 °C under the three representative concentration pathways.

of carbon under RCP2.6, RCP4.5, and RCP8.5, respectively. When the global mean surface temperature increases by 2°C, the permafrost region releases 116.92 Gt and 114.99 Gt of carbon under RCP4.5 and RCP8.5, respectively. The SOC release amount accounts for nearly one-sixth of the carbon stock in the permafrost organic carbon pool in the historical period.

Robinson (2002) reviewed the control of decomposition and soil N supply at high latitudes. Net mineralization represents the balance between the key processes of gross mineralization and adsorption of ammonium onto soil and immobilization in decomposer biomass. For example, 6–7% of total soil nitrogen was present in microbial biomass in a sub-Arctic podzol and 2.2–19.6% of total soil N was present in fungal biomass in various horizons in a mature *Pinus sylvestris* stand. In most Arctic environments, the predominant form of mineral nitrogen is ammonium, but in high Arctic soils of neutral/basic pH, nitrogen may be present mainly as nitrate. Simple organic forms of nitrogen, such as amino acids, have been shown to be important in tundra nitrogen cycles.

Nitrogen mineralization is strongly regulated in the Arctic by low soil temperatures, high or low soil moisture contents and the recalcitrant nature of organic matter. Although net nitrogen mineralization is lower in Arctic soils (0.1–0.6 $g\,m^{-2}\,year^{-1}$) than soils of temperate regions (6–7 $g\,m^{-2}\,year^{-1}$), nitrogen release varies widely among the different types of Arctic ecosystems. In the foothills of Alaska's North Slope, fivefold differences were found in nitrogen mineralization among dry, moist, and wet ecosystems using *in situ* soil incubations. These differences were mainly due to the quality of the SOM and microclimate. Despite a flush of inorganic nitrogen on soil thaw, nitrogen mineralization in Arctic soils during the winter may be a more important component of annual nitrogen mineralization rates than recognized hitherto, since Clein and Schimel (1995) have shown that nitrogen was mineralized in tundra and taiga soils kept in the laboratory at −2 and −5°C. Nitrification occurs in some Arctic ecosystems, especially with a high soil pH, although NH4+ is the dominant form of inorganic nitrogen in cold, wet tundra soils. Increases in soil temperature have been shown to stimulate strongly nitrogen availability and mineralization. However, that there was no difference in the amount of nitrogen mineralized between 3 and 9°C in moist tundra or wet sedge tundra soils, but mineralization was increased by up to ten times in the range 9–15°C. Analogous results were evident when no greater net nitrogen mineralization was found at a high Arctic site in Spitsbergen, when mean soil temperatures at 5 cm depth were increased by 0.7°C from 6.1 to 6.8°C, although the rise in soil temperature was small. By contrast, in a fellfield and treeline ecosystem at Abisko, when mean soil temperatures were increased from 7.8°C (fellfield) and 9.1°C (treeline heath) to give soil temperatures of c. 10°C at 3–5 cm depth, there was little effect on nitrogen mineralization. Nitrification is found to be more sensitive to increases in temperature than ammonification (Paul and Clark 1989), for example, by results from the CLIMEX soil-heating experiment in a small, forested catchment in southern Norway. Nitrification in the 0–10 cm soil layer increased from approximately 0.2 $g\,m^{-2}\,year^{-1}$ in the control to 2 $g\,m^{-2}\,year^{-1}$ in response to increases in soil temperature of +3°C (summer) to +5°C (winter) by heating cables at 1 cm depth in the litter layer (Verburg et al. 1999). Although bulk soil moisture content was not significantly increased, greater net seasonal total nitrogen mineralization was found in treatments where a 58% experimental increase in summer precipitation was made at the polar semi-desert study site. The topographic soil moisture gradient, rather than differences in plant species, was a fundamental control of gross N mineralization in alpine tundra. The interaction of temperature with moisture on rates of N mineralization in Arctic soils has been identified as a research priority. The role of organic matter quality in controlling rates of nitrogen mineralization in six tundra soils was more important than differences in soil temperature in the field. The chemistry of SOM was important, but secondary to soil temperature as a control on nitrogen cycling. These groups were saprotrophs capable of using organic nitrogen from humus, and saprotrophs using nitrogen from dead wood or litter. In conclusion, the net nitrogen mineralization rate is low in Arctic soils compared with that in temperate ecosystems, but different types of tundra soil exhibit a range of rates. Nitrogen mineralization is seasonal, although little is known about rates during winter, and nitrate and organic nitrogen could be as important in plant nutrition as ammonium at some sites. Soil temperature, soil moisture content, and the chemistry of SOM are also likely controls on nitrogen availability, as they are on carbon mineralization, but their effects on nitrogen mineralization are currently poorly understood. The importance of individual or groups of decomposer fungal species in nitrogen mineralization is little studied in Arctic soils, as is nitrogen supply in natural substrata on fungal abundance and activity.

Phelps et al. (1998) suggest that, ice cores removed from shallow ice-covered tundra lakes near Barrow, Alaska, and taiga lakes near Anchorage, Alaska exhibit increasing concentrations of methane with depth. Methane concentrations in the ice cores increased from 0 μm in the top 15 cm to a maximum of 23 μm in the lowest 15 cm of tundra lake ice and to a maximum of 147 μm in taiga lake ice. Methane concentrations in the water beneath the ice reflect a similar pattern, with values near 5 μm early in the ice-covered season, increasing up to 42 μm in the tundra lakes, and up to 730 μm in the taiga lakes. Methane levels increase in the water beneath the ice during the course of the winter due to decreasing water volume, exclusion from growing ice, and continued methane production in thawed sediments. Since the ice layer prohibits gas exchange with the atmosphere, the methane is not oxidized, as it would be during the summer months, allowing the winter accumulation and storage of methane in the ice and lake waters. Efflux measurements, taken with floating chambers on the taiga lakes, indicated a large pulse of methane released during the period of ice melt and spring turnover. The efflux from one lake ranged from 2.07 g CH_4 m^{-2} in 1995 to 1.49 g CH_4 m^{-2} in 1996 for the 10-day period immediately after ice melt. Estimation of methane efflux using a boundary layer diffusion model and surface water concentrations during the entire ice-free period in 1996 predicted an efflux of 1.79 g CH_4 m^{-2} during the same 10-day period, compared with 2.28 g CH_4 m^{-2} for the remainder of the summer season. This observation suggests that almost as much methane efflux can occur during a brief period immediately after ice melt as occurs during the remainder of the ice-free season. Since measurements of methane efflux from high-latitude-lakes are generally made after this breakup period, the overall contribution to atmospheric methane from high-latitude lakes may be twice that of current estimates.

Euskirchen et al. (2006) consider that, in terrestrial high-latitude regions, observations indicated recent changes in snow cover, permafrost, and soil freeze–thaw transitions due to climate change. These changes may result in temporal shifts in the growing season and the associated rates of terrestrial productivity. Changes in productivity will influence the ability of these ecosystems to sequester atmospheric CO_2. The stability of snow cover and permafrost decreases from 1960 to 2100. Decreases in snow cover agree well with satellite observations from the National Oceanic and Atmospheric Administration collected between 1972 and 2000. There has been trend toward an earlier thaw date of frozen soils and the onset of the growing season in the spring by approximately two to four days from 1988 to 2000. Between 1988 and 2000, satellite records yielded a slightly stronger trend in thawing and the onset of the growing season, averaging between five and eight days earlier. Both the simulations and satellite records indicated that, trends on the day of freezing in the autumn are weaker, such that overall increases in growing-season length are due primarily to an earlier thaw. Although regions with the longest snow cover duration displayed the greatest increase in growing-season length, these regions maintained smaller increases in productivity and heterotrophic respiration than those regions with shorter duration of snow cover and less of an increase in growing-season length. Concurrent with increases in growing-season length, a reduction in soil carbon and increases in vegetation carbon, with greatest losses of soil carbon occurring in those areas with more vegetation were found, and suggested that this trend could be reversed in the future.

Grosse et al. (2011) summarized as follows, disturbances caused by climate warming in Arctic, sub-Arctic, and boreal environments can result in a significant redistribution of carbon among major reservoirs with potential global impacts. The current northern high-latitude SOC pools can be divided into: (i) near-surface soils where SOC is affected by seasonal freeze–thaw processes and changes in moisture status, and (ii) deeper permafrost and peatland strata down to several tens of meters depth where SOC is usually not affected by short-term changes. Press disturbances of relatively slow but persistent nature such as top-down thawing of permafrost, and changes in hydrology, microbiological communities, pedological processes, and vegetation types, as well as pulse disturbances of relatively rapid and local nature such as wildfires and thermokarst, could substantially impact SOC stocks. Ongoing climate warming in the high-latitude region of North American could result in crossing environmental thresholds, thereby accelerating press disturbances and increasingly triggering pulse disturbances and eventually affecting the carbon source/sink net character of northern high-latitude soils.

8.3 Carbon, Nitrogen, and Greenhouse Gases in the Polar Regions

The recent warming in high latitudes is affecting a broad spectrum of physical, ecological, and human cultural systems in polar regions. Some of these changes may be irreversible on century time scales, and have the potential to cause rapid changes in the earth system. Given the large stores of carbon in northern high-latitude regions, the response of the carbon cycle of the Arctic to climate change is a major issue of global concern.

Walsh et al. (1989) suggested that, surface air temperatures of the Arctic increased by 1.2–1.5 °C from 1880 to 1980, in contrast to a global warming of only 0.4–0.5 °C; since 1980, six of the warmest years in the past century have been observed. Polar enhancement of a temperature rise, induced possibly by anthropogenic release of "greenhouse" gases, CO_2, N_2O, CH_4, and freons, to the atmosphere, is attributed to altered ice/snow albedo at sea level, namely melting of sea ice. A 5% decline of sea-ice extent in the Arctic and Antarctic from 1979 to 1987 may have resulted in increased light availability within previously ice-covered polar regions. If such a short-term trend were to continue, it might lead to a negative biogeochemical feedback, that is enhanced extraction of atmospheric CO_2 during marine photosynthesis. As a consequence of deep vertical mixing in the Antarctic Ocean, however, primary production during the austral summer may have actually declined in response to a reduction in extent of meltwater regions, where stratified water columns allow carbon fixation ten times that of open water. In contrast, within shallow adjacent seas of the Arctic Ocean, where shelf regions are 10-fold larger than those of the Antarctic, the positive global consequences of greenhouse warming at polar latitudes will probably be felt first. Specifically, the Pacific-influenced regions of the Chukchi and East Siberian Seas, where sufficient nutrients and shallow depths prevail, now have annual primary productions of $>200\,g\,C\,m^{-2}\,yr^{-1}$, 10 times that of other high Arctic shelves, and may supply 50% of the carbon respiration demands within the halocline of the deep Canadian and Eurasian basins via brine-mediated runoff. Continued melting of ice in the Arctic could increase by an order of magnitude the present CO_2 sink of $\sim 0.1 \times 10^9\,t\,C\,yr^{-1}$.

The Arctic has experienced the most rapid change in climate of anywhere on Earth, and these changes are certain to drive changes in the carbon budget of the Arctic as vegetation changes, soils warm, fires become more frequent, and wetlands evolve as permafrost thaws. In the context of the global carbon cycle, the Arctic plays an important role in the dynamics of global CO_2 and CH_4. The Arctic has been a sink for atmospheric CO_2 in recent decades, between 0 and 0.8 $Pg\,C\,yr^{-1}$, representing 0–25% of the global net land/ocean flux in the 1990s. The Arctic is an important source of atmospheric CH_4 (32–112 $Tg\,CH_4\,yr^{-1}$), mainly because of the large wetlands throughout the region.

In terrestrial environments, carbon from biotic activity is predominantly stored in two large reservoirs, SOM (soil carbon) and living biomass (vegetation carbon). The role of northern high-latitude terrestrial ecosystems in the global carbon cycle has long been of interest because of the large stocks of soil organic carbon in the region. Soils in high-latitude ecosystems are generally considered to have been storing large quantities of carbon in unglaciated regions prior to the last glacial maximum and also in areas that have since been deglaciated since the last glacial maximum. In general, this accumulation is considered to have been promoted by cold and wet soils that inhibit decomposition of dead plant tissue that enters the SOM pool. There is a considerable range in the estimates of soil carbon stored in high-latitude regions because of several issues associated with soil depth.

McGuire et al. (2009) reviewed the carbon cycle changes to climate change in the Arctic region, and summarized that the Arctic has been a sink for atmospheric CO_2 of between 0 and 0.8 $Pg\,C\,yr^{-1}$ in recent decades, which occupies up to 25% of the global net land/ocean flux during the 1990s. The Arctic is a substantial source of CH_4 to the atmosphere (32–112 $Tg\,CH_4\,yr^{-1}$). Sensitivity of the carbon cycle of the Arctic during the remainder of the twenty-first century is highly uncertain. For instance, Schuur et al. (2008) estimated that non-peatland vegetation types of northern high-latitude regions contain 750 Pg carbon in the upper 3 m, while Gorham (1990, 1991) estimated that carbon stocks in northern peatlands range from 200 to 450 Pg, depending on the depth of peatlands. In addition, there are large amounts of carbon that are accumulated in now-frozen soils of unglaciated regions prior to the last glacial maximum (Zimov, Davydov et al. 2006, Zimov, Schuuret al. 2006).

In comparison with SOM storage, carbon stored in vegetation of global terrestrial ecosystems is much less. In the Arctic region, vegetation carbon storage of boreal forests is quite different, ranging from 2 to 70 Pg C, with the uncertainty that it could be larger than 10 Pg carbon because of uncertainties in estimating BGB, primarily coarse and fine roots (Li et al. 2003).

Apart from the European sector of the Arctic region, the Arctic region was on average a modest CO_2 sink that is less than 0.5 $Pg\,C\,yr^{-1}$. The substantial interannual variations of surface–atmosphere CO_2 fluxes are caused primarily by climate fluctuations and associated variation in wildfires on land CO_2 exchange (Bousquet et al. 2000; Baker et al. 2006). McGuire (2000) summarized the uncertainty estimate for the flux from the entire Arctic-related factors, including (i) the sub-grid scale mixing processes in numerical models, (ii) a sparse observational network that has limited temporal and spatial representative, (iii) differences in measurement techniques and standards across monitoring networks are commonly not readily comparable. Study by Stephens et al. (2007) suggested an overestimation on the order of 30% of the magnitude of the northern extratropical CO_2 sink as inferred by the models.

Tundra boreal forest ecosystems are prominent in the Arctic region. Their carbon recycle is unique. Results of Schulze et al. (1999), Aurela et al. (2001), and Hobbie et al. (2002) showed that, a high degree of spatial and temporal variability exists in Arctic tundra and boreal forest ecosystems, making it difficult to ascertain whether these ecosystems as a whole are currently acting as longer-term sinks or sources for atmospheric CO_2. The interannual and across-site variability of CO_2 exchange in Arctic tundra ecosystems is driven primarily by growing-season dynamics and moisture conditions. Growing-season rates of CO_2 uptake by tundra ecosystems are closely related to the timing of snow melt, especially in the earlier snowmelt season (Groendahl et al. 2007).

Nordstroem et al. (2001) suggested that pulse emissions during the thaw period are important to the annual carbon budget. Overall, tundra regions in the Arctic are sources of carbon to the atmosphere under dry and mesic conditions and are carbon sinks under wet conditions. Results of Sitch et al. (2007) showed that tundra is greening in the Arctic, suggesting an increase in photosynthetic activity and net primary production. Results of Myneni et al. (1997) showed that carbon sequestration in tundar regions was likely due to increased production resulting from enhanced CO_2 uptake, increased nutrient availability and longer growing seasons. Goetz et al. (2005) suggested that the decrease of carbon is a result of drought stress, nutrient limitation, and increased disturbance. The estimation results for tundra and boreal forest ecosystems of Canada, Alaska, Northern Europe, and Russia showed that terrestrial ecosystems across the Arctic acted as a sink of 0–600 Tg C yr^{-1} near the end of the twentieth century (Balshi et al. 2007).

Inland waters are another major landscape feature of the Arctic. Lakes and rivers play an important and prominent role in the global carbon budget as conduits of terrestrial organic matter to the ocean, as sources of both terrestrially and aquatically produced CO_2 to the atmosphere, and as sinks of carbon stored in sediments and sequestered in aquatic production. There is considerable variability in carbon cycling processes within the inland waters of the pan-Arctic that contributes to the uncertainty of our lake and river CO_2 efflux estimates. For example, Kling et al. (1991) and Cole et al. (1994) argued that although aquatic primary production can contribute to varying degrees, the carbon balance of freshwater ecosystems is dominated by inputs from land. Kling et al. (1991) estimated that CO_2 evasion from Arctic lakes and rivers is approximately 20 Tg C yr^{-1} for tundra in the Arctic. Estimation results of Cole et al. (2007) showed that, about half of the C input from terrestrial ecosystems is discharged to the ocean, with about 40% lost to CO_2 efflux from the surface and 10% stored in sediments of inland waters. The per-area flux of CO_2 to the atmosphere is generally higher from smaller lakes than from larger lakes (Christensen et al. 2007). For instance, Jonsson and Karlsson (2003), and Eugster et al. (2003) observed that substantial losses of terrestrial carbon exist in small and medium sized lakes in both the sub-Arctic and Arctic environments. Therefore, carbon cycling in smaller lakes in the Arctic regions is intimately coupled to terrestrial processes. In comparison to tundra lakes, carbon recycling in boreal forest lakes is generally higher (Algesten et al. 2003).

Although the Arctic Ocean is smaller compared to the global oceans, the per-area primary production and air–sea fluxes of CO_2 are many times the global average, particularly in inflowing shelf seas like the Barents Sea and Bering-Chukchi Seas. The Pacific waters flowing through the Bering Strait follow varied routes within the Arctic Ocean in deep-water formation, which transports dissolved organic and inorganic carbon from the surface ocean to the intermediate and deep waters, which is globally important in sequestering anthropogenic CO_2. At present, the Arctic Ocean is undergoing rapid changes, including decreasing sea-ice coverage (Stroeve et al. 2007), which have great potential to alter processes important to carbon exchange with the atmosphere. Both cooling and primary production promote a flux of CO_2 from the atmosphere to the surface ocean. However, as much of the central Arctic Ocean surface is covered by sea ice, at least during the winter season, gas exchange is hampered even if it does not necessarily provide a completely impermeable barrier. One consequence of the seasonal ice cover is that the surface waters of the central Arctic Ocean are under-saturated with respect to CO_2, in other words, the loss of CO_2 from surface waters through fixation and associated flux of organic carbon to deeper water exceeds the net physical flux of CO_2 from air to sea.

Surface methane exchange in the Arctic region is prominent. Recent inversion studies indicated that the Arctic is a source of CH_4 to the atmosphere of between 15 and 50 Tg CH_4 yr^{-1} (Bergamaschi et al. 2007; Chen and Prinn 2006). The interannual variability of CH_4 in the Arctic is associated with climate variability (Bousquet et al. 2006). Results of Walter et al. (2001a, 2001b) indicated that CH_4 emissions in high-latitude regions occupy one-quarter to one-third of global emissions. Results of Zhuang et al. (2004) showed the annual net emission rate at the end of the century for the region north of 45°N at 51 Tg CH_4 yr^{-1}. Lake systems in sub-Arctic and Arctic regions are surrounded by permafrost soils that are rich in carbon accumulated during the Pleistocene, and significant CH_4 emissions from boreal, sub-Arctic, and Arctic lakes is observed by Bastviken et al. (2004). Walter et al. (2007) estimated that lakes in Siberia, Alaska emit 15–35 Tg CH_4 yr^{-1}.

Substantial stocks of carbon are stored in the Arctic. McGuire et al. (2009) summarized a general picture of arctic carbon cycle as in Figure 8.6. Results of Jobbagy and Jackson (2000) showed that soil organic carbon is stored mainly in peatlands and deep permafrost soils in Siberia. Due to wet and cold physical conditions that are not conducive to the decomposition of SOM, most of carbon storage has accumulated. Besides, the Arctic also stores 10–20% of the world's vegetation carbon. There are large stocks of dissolved inorganic carbon in the Arctic Ocean. Also, beneath the ocean floor and beneath both subterranean and submerged permafrost of the Arctic, there may be substantial stocks of CH_4 stored as a gas phase. The Arctic plays an important role in the global dynamics of both CO_2 and CH_4. Top-down atmospheric analyses indicate that the Arctic is a sink for atmospheric CO_2 of between 0 and 0.8 Pg C yr^{-1} (Figure 8.6), which is between 0% and 25% of the net land/ocean flux of 3.2 Pg C yr^{-1} estimated for the 1990s by the IPCC's Fourth Assessment Report (Surabi et al. 2007). The growth of trees in the boreal forest is responsible for most of the

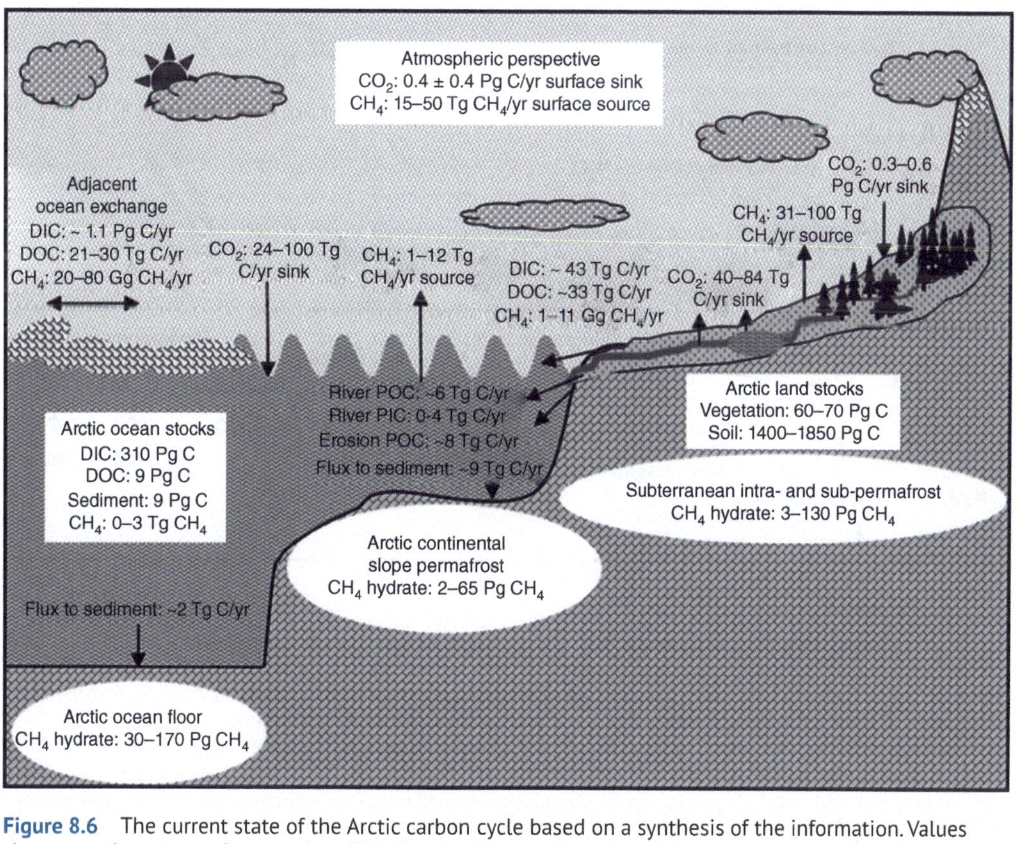

Figure 8.6 The current state of the Arctic carbon cycle based on a synthesis of the information. Values shown are the ranges of uncertainty. DIC, dissolved inorganic carbon; DOC, dissolved organic carbon. *Source:* McGuire et al. (2009) / John Wiley & Sons.

sink activity in the Arctic. Surabi et al. (2007) argued that Arctic is a source of CH_4 to the atmosphere of between 15 and 50 Tg CH_4 yr^{-1}. Results from Sarmiento and Gruber (2006) indicated that, from the global perspective, the transfer of carbon from land to ocean systems in the Arctic roughly reflects the 11% of global river discharge of water from land to ocean represented by the pan-Arctic drainage basin. Approximately 80 Tg C yr^{-1} are transferred from land to ocean via rivers (Figure 8.6), which is approximately 10% of the estimated 0.8 Pg C yr^{-1} transferred from land to ocean via rivers globally; in addition, coastal and wind erosion are responsible for the transfer of approximately another 8 Tg C yr^{-1}, with coastal erosion responsible for about 75% of the erosional transfer; the annual transfer of carbon to sea floor sediments of the Arctic is approximately 11 Tg C yr^{-1}, which is about 5% of the estimated 0.2 Pg C yr^{-1} transferred to ocean floor sediments throughout the globe.

In arctic region, bryophytes and dwarf shrubs preferentially occupy these lower slope positions where there is greater moisture availability (Figure 8.7, Stewart et al. 2004), where lichens are excluded by competitive displacement, and by their inability to tolerate prolonged hydration episodes. Both the proximity of permafrost to the soil surface, and moisture retention by bryophytes, help to create a moist environment in these low-lying areas, which is likely to be crucial in maintaining high rates of N_2-fixation. In addition, nutrient transport at the microtopographical scale from the higher mounded areas may increase dissolved organic and inorganic nutrients, such as

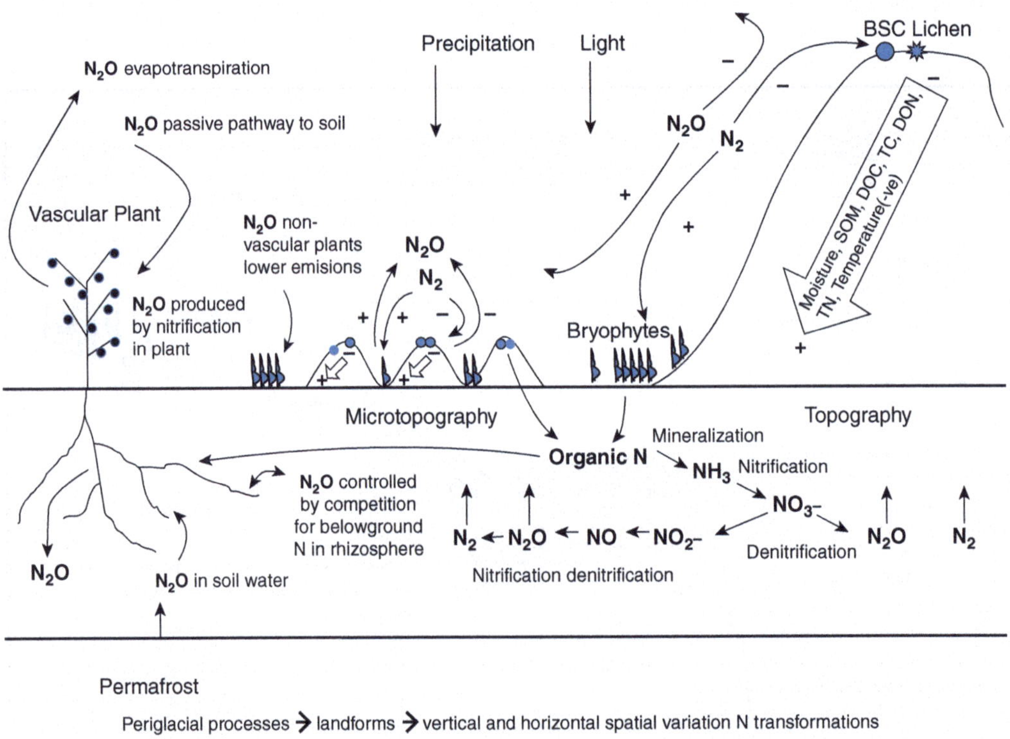

Figure 8.7 Key factors controlling gaseous nitrogen (N) input and output pathways in Arctic ecosystems. Topography and microtopography are associated with abiotic gradients of higher soil moisture, soil organic matter (SOM), dissolved organic carbon (DOC), total carbon (TC), dissolved organic nitrogen (DON), and total nitrogen (TN) and lower soil temperature downslope. These gradients drive higher N_2-fixation rates (N_2) and N_2O emissions (N_2O) in lower-lying areas. The influence of non-vascular plants and the above and belowground systems of vascular plants on N_2O exchange are noted (Stewart et al. 2004).

carbon and phosphorus in hollows, which in turn may promote N_2-fixation in these hollows. Furthermore, this effect may be just as important at the landscape scale since there is great potential for movement of dust and surface runoff from higher to lower elevation ecosystems.

8.4 Outlook

Based on the discussions above, carbon and nitrogen and their transport to atmosphere in different cold regions are summarized in Table 8.1. The soil-carbon pool of the TP is an important component of the global ecosystem carbon cycle, and it is important to elucidate the spatial and temporal dynamics and patterns of soil-carbon stocks in the region. Soil-carbon stocks on the TP exhibit certain spatial distribution patterns due to inherent factors such as soil properties, geographic factors, climatic factors, and biological factors, while climate change and human activities cause interannual fluctuations in soil-carbon stocks through different degrees of variability, especially the loss of soil carbon due to ecosystem degradation, which is also an important issue to be addressed. On the whole, the soil-carbon stock estimation technology on the TP has not yet reached a mature stage, and the estimation results are highly uncertain. To improve the accuracy of

Table 8.1 Carbon and nitrogen and its transport to atmosphere in cold regions.

	Total amount			Sink(+) or source(−) for atmospheric	
	C	CH_4	N	C	CH_4
Tibet Plateau	10 Pg C (0–20 cm SOC; grasslands; Wang et al. 2019) 16.5 Pg C (0–200 cm SOC; tundra; Wang et al. 2019) 14 Pg C (0–100 cm SIC; tundra; Wang et al. 2019) 33.52 Pg C (0–0.75 m SIC; grasslands; Wang et al. 2002) 7.72 kg C m^{-2} (1 m SIC; Liu et al. 2012)		0.93 kg N m^{-2} (1 m TN; Liu et al. 2012)	14.1 Tg C yr^{-1} (Jin et al. 2015)	0.95 Tg CH_4 yr^{-1} (Jin et al. 2015) 0.86 ± 0.23 g CH_4-C m^{-2} yr^{-1} (permafrost region; Yun et al. 2018)
High latitude	608 Pg C (tundra, boreal, and sub-Arctic peatlands; Robinson 2002)		0.4 Pg C yr^{-1}; (White et al. 2000) −276 Tg C yr^{-1} (net; Zhuang et al. 2006)		−36 Tg C yr^{-1} (net; Zhuang et al. 2006)
Polar region	1400–1850 Pg C (McGuire et al. 2009)	35–365 Pg C (McGuire et al. 2009)	0 and 0.8 Pg C yr^{-1} (McGuire et al. 2009)		32 and 112 Tg CH_4 yr^{-1} (McGuire et al. 2009)

Note: soil organic carbon (SOC), soil inorganic carbon (SIC) stock.

soil-carbon stock estimation on the TP, the role of soil in the global ecosystem carbon cycle under climate change conditions, such as: (i) establishing standards and criteria for soil-carbon stock estimation; (ii) enrichment of the soil-carbon stock field measurement database; (iii) optimization of estimation methods, especially for taking into account the unique natural environmental characteristics, human activities and the results of the measured data on the TP.

A number of factors control carbon turnover in high-latitude soils, but because they are unique to northern systems, they are mostly ignored by biogeochemical models used to predict the response of these systems to global change. Those factors include: (i) many northern systems are dominated by mosses, whose extremely slow decomposition is not predicted by commonly used indices of litter quality; (ii) cold temperature, permafrost, waterlogging, and substrate quality interact to stabilize SOM, but the relative importance of these factors, and how they respond to climate change, is unknown; (iii) recent evidence suggests that biological activity occurring in winter can contribute significantly to annual soil-carbon fluxes; (iv) processes at the landscape scale, such as fire, permafrost dynamics, and drainage, control regional carbon fluxes, complicating the extrapolation of site-level measurements to regional scales.

9

Climate Disasters in Cold Regions

Disasters in cold regions are frequent and most of them are induced by climate change. Disasters in cold regions include snow- and ice-melt floods, glacial-lake outbursts and flows of glacial debris and are related to snow and ice activity. Snow- and ice-melt floods mainly include glacial meltwater, snow meltwater, and floods formed by mixing the two with rainfall, which are closely related to climate change and mostly occur in spring and summer. The most common glacial-lake outbursts in alpine areas are moraine blocked-lake outbursts and glacial-blocked outbursts, which are notable for their sudden and large runoff modulus, making them very difficult to predict. Glacial-debris flow is also common in alpine areas and are generally characterized by sudden, violent, short-lived and devastating outbreaks.

This chapter will introduce disasters in cold regions in detail, including their definitions, classification, characteristics, and causes.

9.1 Overview

The temporal and spatial scales of weather and climate in cold regions are different, and the disaster-pregnancy environments are different as well. Therefore, the occurrence and impact of different types of disasters in cold regions have specific temporal and spatial scales. Some disasters may be localized, while others have the characteristics of ditch, watershed, regional or even larger scales. Among them, snow/ice avalanches are instantaneous and localized, and mostly occur in mountainous areas. Glacier-debris flows, glacial-lake outbursts, and glacier-surging disasters occur at an hourly scale, with the spatiality of a gully scale. Frost disasters and iceberg disasters occur at a daily scale, but spatially at a regional scale. For blizzard, low-temperature rain, and snow-freezing disasters, their occurrence is at the daily or monthly scale, with the spatiality of a larger regional scale. The temporal scale of coastal freeze–thaw erosion disasters is seasonal, inter-annual, or inter-decadal, with the spatiality of a circum-Arctic scale. Sea-level rise caused by snow/ice melt in cold regions, has a temporal scale at inter-decadal or an even longer time scale, and spatial scale at global scale. The disaster types in cold regions are different, and there are differences in the disaster environment and scope, the scope and type of disaster-bearing bodies, and the exposure factors. With different spatial distribution, the degree of harm and the way of damage are also different. With the deepening of the understanding of the mechanism of disasters in cold regions and the innovation of disaster prevention and mitigation technology, the disaster-affected areas and disaster-bearing bodies in cold regions will also change accordingly. Disaster bodies and

Climatology in Cold Regions, First Edition. Chenghai Wang.

different types of disasters have obvious differences in the degree of disaster damage and post-disaster recovery.

Disasters in cold regions around the world are widely distributed in space and have various types, and most of them occur in low-latitude, high-altitude mountains and high-latitude regions, and have a significant impact on the human economy and society.

9.2 Snowmelt Flood

Snowmelt floods are formed by melting snow, referred to as snow floods. They generally occur in the spring and summer in the middle and high latitudes and high mountain areas, and can be divided into two types: plain and mountainous. Floods caused by heavy snowfall are commonly accompanied by a rapid recovery in temperature, which accelerates the melting of the snowpack. Unlike normal floods, snowmelt floods are accompanied by large amounts of ice and melt ice, which can be extremely destructive in all directions. In some mid-latitudes and high latitudes, where winters are long and severe and the snowpack is deep, the following spring and summer, when temperatures rise above 0°C, the snowpack melts to form floods, and snowmelt floods are closely related to the amount of accumulated snowpack, with significant differences in spatial and temporal distribution (Abulimitijiang-Abulikimu et al., 2015). The main factors affecting the size and process of snow ablation floods are the area of the snowpack, snow depth, snow density water holding capacity and snow surface freezing depth, the heat of snow melt, the topography, orientation, climate and land use of the snowpack site, and these factors influence each other. The phenomenon and formation of snowmelt flood is shown in Figures 9.1 and 9.2.

Snowmelt floods are the result of the combined influence of snow, heat conditions, topography and other factors, and therefore have different characteristics from rainfall floods, Vormoor et al. (2016) summarized these characteristics mainly in the following aspects: (i) the changes of flood and temperature process are basically the same, but in the early stage of warming, the temperature rises more quickly, the river flow does not change much and is slowly rising. When the

Figure 9.1 Snowmelt flooding, from National Oceanic and Atmospheric Administration. Snowfall during autumn and winter supplies most of the water to most mountainous regions. Topography, winter weather patterns, and a host of other factors can combine to produce snow packs that can have an excess of 50 in. of water equivalent each year. *Source:* Lincoln County (WY) Emergency Management / NOAA.

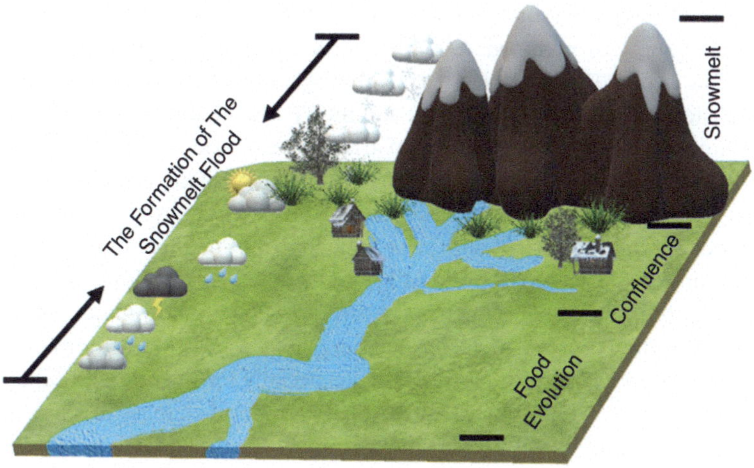

Figure 9.2 The formation of the snowmelt flood.

temperature reaches a certain level, the flood process rises more than the temperature. The flooding process is steeper than the temperature process, and the whole flooding process lags behind the temperature process, with the flood peak lagging behind the temperature peak; (ii) there is a clear daily variation in the flooding process, with a peak and a trough. Flood peaks usually occur in the late afternoon and valleys in the evening, and the timing of peaks and valleys varies from river to river due to the proximity of each hydrological station to the snowpack; (iii) although floods occur at the start of spring, the inter-annual variation in temperature at the start of spring is also significant due to the extreme instability of spring temperatures and the wide variation in the rate of warming from year to year.

Unlike rainfall, which reaches the soil almost immediately, snow stores the water for a period until it melts, delaying the arrival of water in the soil for days, weeks, or even months. Once it does reach the soil, the water either soaks into the ground or runs off. If more water runs off than soaks in, snowmelt flooding occurs.

Snowmelt flooding typically occurs every year in the northern United States, with most snowmelt events being minor and localized. Eight of the most significant floods were related to snowmelt in the 20th century. The effect of snowmelt on potential flooding, mainly during the spring, causes concern for many people around the world. Besides flooding, the rapid snowmelt can trigger landslides and debris flows. In alpine regions like Switzerland, snowmelt is a major component of runoff. In combination with specific weather conditions, such as excessive rainfall falling on melting snow, it may even be a major cause of floods. From record of the United States Geological Survey, in Switzerland, snowmelt forecasting is being used as a flood-warning tool to predict snowmelt runoff and potential flooding. In some parts of the world, such as in the Pacific Northwest of the United States, annual springtime flood events occur when rain falls on existing snow packs, known as a "rain-on-snow event." Runoff during rain-on-snow events has been associated with mass-wasting of hill slopes, damage to riparian (areas alongside streams) zones, downstream flooding and associated damage, and loss of life. The amount of forest cover can have an influence on the magnitude of rain-on-snow events.

The Tibetan Plateau (TP) is a typical and frequent snowmelt flood region. Many studies have been focused on northern TP, and the arid and semi-arid regions on the northeastern TP (the Hexi Corridor), where the flood caused by snow melting not only provides irrigation for agriculture, but also brings about disasters due to the paroxysmal characteristics. China covers a vast area and a variety of climate regions, and its snow-covered regions are widely distributed geographically with evident spatial differences (Tan et al. 2016). The Northern Xinjiang Uygur Autonomous Region (hereafter, Xinjiang), Northeastern China–Inner Mongolia, and the TP are the three main regions with stable snow cover in China (Ke et al. 2016). As a typical arid and semi-arid region of Central Asia with a significant lack of freshwater resources, the surface runoff in Xinjiang is mainly supplied by snowmelt (Chen et al. 2020). In northeast China, snow plays an important role as a natural reservoir in winter and a source of water in spring, with snowmelt contributing more than half of the runoff during the main crop sowing season (April and May; Qi et al. 2020). Melting snow is also an important hydrological process on the TP, which is the source region of many major Asian rivers, and is considered as being the Asian water tower (Immerzeel et al. 2010). Furthermore, snowmelt is an important water source in other parts of China, especially for the arid and semi-arid areas in north China (Zhang et al. 2015; Wu et al. 2021). Spatiotemporal changes in China's snow properties (e.g. snow-cover extent, snow-cover phenology, snow depth, snow density and snow water equivalent) have drawn wide attention in recent decades (Chen et al. 2006; Dai and Wigley 2000; Tan et al. 2016; Yang et al. 2020).

TP is one of the China's three main stable snow-cover regions. According to the above studies, from 1951 to 2017, the snowmelt increased significantly on the TP. The annual mean snowmelt runoff ratios are generally more than 10% in almost all third-level basins in Western China and the annual snowmelt runoff ratios decreased in most third-level basins in China. Under the representative concentration pathway version 2.6 (RCP2.6), RCP4.5, and RCP8.5 scenarios, the projected snowmelt in China in the near future (2011–2040), mid-future (2041–2070), and distant future (2071–2099) may decrease by 10.4% (15.8%; 13.9%), 12.0% (17.9%; 21.1%) and 11.7% (24.8%; 36.5%) compared to the reference period (1981–2010), respectively. Most of the projected annual mean snowmelt runoff ratios in third-level basins in different future periods are lower than those in the reference period. Low-temperature regions can tolerate more warming, and the snowmelt change in these regions is mainly influenced by precipitation; however, the snowmelt change in warm regions is more sensitive to rising temperatures. The spatial variability in snowmelt changes may lead to regional differences in the impact of snowmelt on water supply.

Continuous rising temperatures will bring forward the snow melting time. Observed data from the Spanish Pyrenees for 1950–1999 shows reduced winter precipitation leading to reduced discharge in both winter and spring due to less snow accumulation and melt (López-Moreno and Garcia-Ruiz 2004), while an observed streamflow record for 151 catchments in the Nordic countries (Norway, Sweden, Finland, Denmark, and Iceland) shows earlier snowmelt peaks in all except Iceland (Wilson et al. 2010). For the western United States, observed trends also show earlier snowmelt (Clow 2010; Hamlet et al. 2007).

Spring snowmelt floods are formed by the melting of the winter snowpack, and snowmelt floods generally occur between April and June. Peak flood flows occur from May to June. For rivers located near the same latitude, snowmelt floods occur earlier in the plains than in the mountains. In the Altai Mountains of China, snowmelt floods generally occur from April to May and end as late as June.

Snow and ice-melt floods are formed by glacial meltwater and snow meltwater as the main source of recharge. Those with glacial meltwater as the main source are glacial melt floods, those with snow meltwater as the main source are snowmelt floods, and those formed by heavy rainfall mixed with glacial and snow meltwater are mixed rainfall and glacial melt floods and mixed rainfall and snowmelt floods, respectively. Rivers recharged by glacial melt water or snowmelt water alone are rare in alpine areas, and are generally recharged by a mixture of glacial meltwater, snowmelt water, and rainwater. Snowmelt floods are seasonal floods and are closely related to climate change. Every year, when the temperature rises above 0°C, the ice and snow melts into liquid water. The stronger the solar radiation, the larger the glacier area and the thicker of the pre-snowpack, all making the melt more intense. The water melted by glaciers and snow forms part of the surface runoff that recharges the rivers directly, and part of it recharges the rivers as shallow groundwater through infiltration, forming spring and summer floods.

9.3 Freezing Rain

Freezing rain is rain maintained at temperatures below freezing by the ambient air mass that causes freezing on contact with surfaces. Unlike a mixture of rain and snow or ice pellets, freezing rain is made entirely of liquid droplets. The raindrops become supercooled while passing through a subfreezing layer of air hundreds of meters above the ground and then freeze upon impact with any surface they encounter, including the ground, trees, electrical wires, aircraft, and automobiles, forming so-called glaze ice that can accumulate to a thickness of several centimeters and cover all exposed surfaces. The phenomenon of freezing rain is shown in Figure 9.3.

Freezing rain is hazardous for travelers because the ice can create slick spots on roads, causing motorists to lose control of their automobiles without timely warning. Bridges, overpasses and elevated roadways are especially susceptible to icing as they are surrounded on all sides by the cold air and freeze more quickly. In addition, ice caused by freezing rain can rapidly add weight to tree branches and power lines, causing them to snap or break. In addition to these fallen branches causing damage to whatever they land on, power outages may also occur.

Usually, precipitation first falls from a cloud as snow (when it is cold enough high up where the cloud is). If it falls through warmer air before reaching the ground, it can melt and turn to rain droplets. On rare occasions, if it then falls through cold air again just before hitting the ground, the droplets can become "supercooled" and this means that they are still falling in a liquid form, even though their temperature has fallen below zero. When this "supercooled" droplet hits the ground, which is below zero too, it spreads out a little on landing, and then instantly freezes, encasing the surface in a layer of clear ice. The formation of freezing was shown in Figure 9.4.

Figure 9.3 Freezing rain. *Source:* Nicolas M. Perrault / Wikimedia Commons / Public Domain.

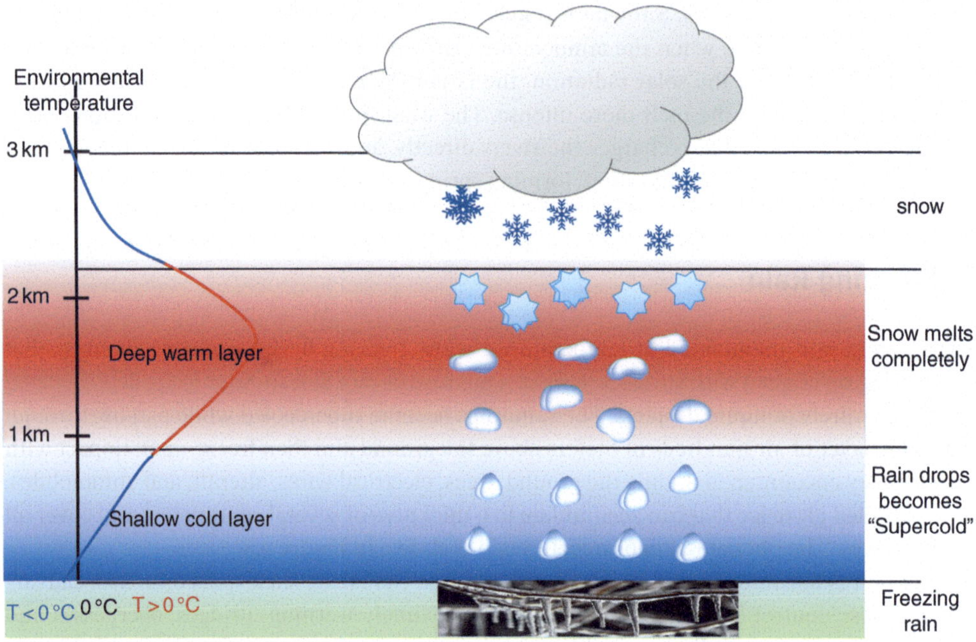

Figure 9.4 Freezing rain falls when snowflakes melt completely before reaching the surface, and refreeze upon contact with anything that is at or below 0°C. *Source:* United States Department of Commerce / Public Domain.

Water freezes at 0°C but in the same way that raindrops need a nucleus (microscopic particles of dust or dirt) to form around, water droplets can only freeze around an ice nucleus. However, it is possible that water droplets can exist below zero and remain in liquid form without a nucleus, which is called supercooled water droplets. Supercooled water droplets commonly occur in clouds, especially in winter, and they play a very important role in the way that rain forms. Freezing rain tends to start its life as snow, ice, sleet or hail, but passes through a layer of air that is above 0°C on the way down to the ground, melting into a liquid water droplet. If these droplets then fall through a zone of sub-zero air just above the ground, they become supercooled. When these supercooled droplets strike surfaces that are close to or below freezing, they freeze on impact forming a glaze of ice.

Freezing rain events cause major damage in the United States, which impact on various weather-sensitive user groups. Ice storms can cause significant damage to trees, telephone and power lines, and agriculture; even small amounts of ice can cause slippery conditions for humans and transportation (Changnon 2004). Associated weather conditions during an ice storm event also contribute to the amount of damage received. Precipitation amount can affect how much ice accumulates, while wind speed and direction can affect the load placed upon an object. During 1949–2000, freezing-rain events caused more than 16.3 billion dollars in total losses in the United States (Changnon 2004).

In the Great Lakes Region of North America, there is a general increase in the freezing-rain frequency from west to east; however, the distribution shows large spatial and temporal variability. Most freezing-rain events are short lived and occur near sunrise between the months of December and March. Continuous freezing rain typically lasts less than one hour, with 7% of events lasting longer than five hours. Freezing rain occurs primarily between December and March. It is usually short-lived, and is not mixed with other types of precipitation, and occurs most frequently near sunrise. There is large spatial and temporal variability in the annual frequency of freezing rain, with a general increase in frequency from west to east that appears related to the climatological track of surface cyclones, the proximity of the Atlantic Ocean to these parts of the lakes, and the presence of numerous valleys. This track favors freezing rain across the central and eastern regions of the Great Lakes. A relative minimum in frequency is observed west of Lakes Huron, Michigan, Ontario, and Erie, suggesting that the lakes may have a moderate influence on the frequency of freezing rain near the shoreline during some freezing-rain events (Cortinas 2000).

In North America, spatial variability in the annual frequency of freezing precipitation and ice pellets is considerable across the United States and Canada and occurs most frequently across the central and eastern portions of the United States and Canada, much of Alaska, and the northern shores of Canada. Freezing precipitation and ice pellets occur most often from December to March, except in northern Canada and Alaska where it occurs in the warm season as well. Freezing rain and freezing drizzle appear to be influenced by the diurnal solar cycle. Freezing precipitation is often short lived. Most freezing rain and freezing drizzle are not mixed with other precipitation types, whereas most ice pellets included other types of precipitation. Freezing precipitation and ice pellets occur most frequently with an air temperature that is slightly less than 0°C. In general, surface temperature remains at or below freezing for up to 10 hours, and for up to 25 hours for freezing drizzle after freezing rain events (Cortinas et al. 2004).

Zhou (2017) reported that, in China, based on the observations at 650 *in-situ* stations from 2000 to 2015, the freezing rain is generally in stripe-shape spatial distribution. Stations with freezing rain over 100 station-hours are mainly concentrated in the southeast areas of southwest China and the high-elevation mountainous areas of eastern China. Here, seven regions are divided including northwest China, north China, northeast China, southwest China, central China, east China, and south China according to common geographical partition method. Figure 9.5 shows the spatial distribution of freezing rain in China from 2000 to 2015. It can be seen that all of seven regions have

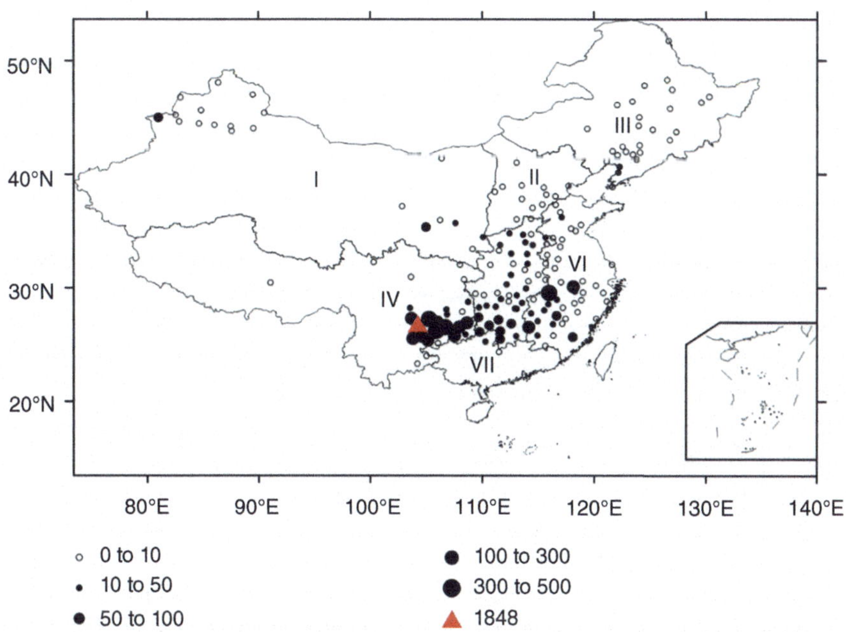

Figure 9.5 Distribution of freezing rain in China from 2000 to 2015 (unit: station-hour): (I) northwest China; (II) north China; (III) northeast China; (IV) southwest China; (V) central China; (VI) east China; (VII) south China. *Source:* Zhou et al. (2017) / Hindawi / CC BY 4.0.

the occurrence of freezing rain, wherein the freezing rain is more frequent in southwestern China, central China, and eastern China. Freezing rain occasionally happens in northwestern China, northern China, and northeastern China. The distribution of freezing rain shows a "strip-shaped" feature in China, which is consistent with the spatial distribution of temperature. The low temperature of three regions in the north of China makes freezing rain unlikely to occur. The excessive high temperature is also the reason why freezing rain would not occur in southern part of south China. In the three central regions, more freezing rain events are due to the exchange of warm and cold air masses, especially for the areas around 26°N. There are 175 stations of freezing rain throughout China, accounting for 26.9% of total observation stations. Among them, there are 113 stations with 0–10 station-hours, which mainly distributed in northwest areas of northwest China, north China, the southern areas of northeast China, central China, and central and northern areas of eastern China; there are 37 stations with 10–50 station-hours are distributed in the northern and southern areas of central China; there are ten stations with 50–100 station-hours, which are mainly distributed at southern areas of central China; there are ten stations with 100–300 station-hours, which are mainly distributed in the southeastern areas of southwest China and high-elevation mountainous areas of eastern China; there are four stations with 300–500 station-hours, which are mainly distributed in southeast areas of southwest China and high-elevation mountainous areas of eastern China. Located at altitude of 2236 m, Weining in Guizhou province observes the longest station-hours of freezing rain (1848 station-hours), which is significantly longer than those of the nearby stations. There are mainly two types of freezing-rain temperature profiles, including single-warm-layer profiles and double-warm-layer profiles. The freezing-rain profiles of mountainous areas and urban areas in southwest China are similar to each other, and the freezing rain in central and east China shares the similar single-warm-layer profile (Zhou et al. 2017).

Figure 9.6 shows the freezing-rain events in the United States reported by Houston and Changnon (2007), which is based on the maximum number of freezing-rain hours experienced in any single year during 1928–2001. Small parts of New England and a large section of Missouri and Illinois have experienced a year with 60–80 hours of freezing rain. Values in the 40–60 hour range appear across most of the Midwest, northeast, and along the foot of the mountains from North Carolina northwards. The fewest number of hours with freezing rain in any given year between 1928 and 2001 was also determined, revealing that most of the nation has gone for years without any freezing rain. However, locations in the Midwest and northeast have annual minimums of one hour and portions of New York State and Massachusetts have annual minimums of two hours (Houston and Changnon 2007).

The regional maxima of freezing precipitation exists across the contiguous United States (CONUS) and Canada. These maxima have been shown to be related to major storm tracks, topographic features, and moisture sources. The two mechanisms firmly established as the causes of freezing precipitation are the collision–coalescence (CCP, or "warm rain") and "classical" melting processes. The CCP typically occurs in situations where clouds are primarily composed of liquid water, at above and/or below freezing temperatures. For this to occur, cloud-top temperatures are typically greater than −10°C, where significant snow production is not expected (Politovich 1989; Rauber et al. 2000). Rasmussen et al. (1995) noted that low concentrations of cloud condensation nuclei combined with significant liquid water content ($>0.25\,g^{-3}$) may be important in the growth of cloud-sized drops to drizzle sizes by the CCP.

The classical melting process occurs when snow falls into a layer of above-freezing air, melts to form rain or drizzle, then falls into a layer of subfreezing air to become supercooled rain (FZRA) or freezing drizzle (FZDZ). It generally has been accepted that most FZRA forms via the melting process that about 62% of all FZRA was associated with warm layers above (Huffman and Norman 1988).

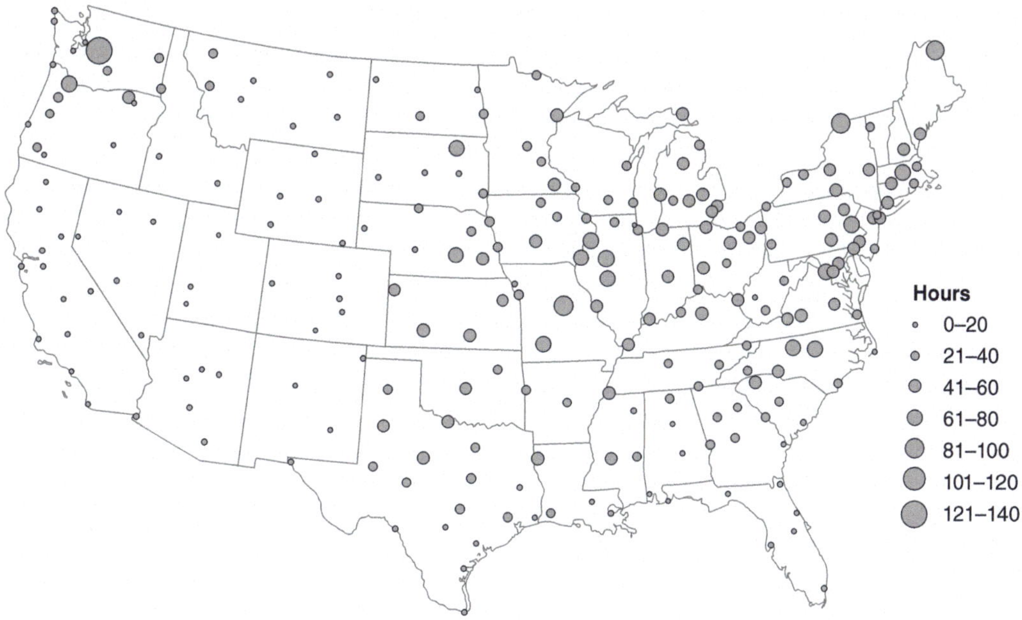

Figure 9.6 Maximum number of hours with freezing rain in a single year, 1928–2001. *Source:* Houston and Changnon (2007) / Springer Nature.

Ice pellets have also been shown to form by either process. The primary formation mechanism for ice pellets appears to be incomplete melting of snowflakes during the classical process (Bernstein 2000). Hanesiak and Stewart (1995) and Zerr (1997) concluded that melting zones were too cool and/or shallow to completely melt snowflakes, that ice particles remained suspended within the raindrops, and that subsequent supercooling caused the water drops to freeze and form ice pellets. Freezing of completely melted droplets in the subfreezing zone below was considered a secondary mechanism. Ice pellets can also form by the freezing of supercooled drizzle droplets that were formed via the CC process (Kajikawa et al. 1988).

The distribution of dry-bulb temperature in the United States shows that nearly 39% of all hours were with temperatures of −0.6 or 0°C, and 71% of all hours had values between −2.2 and 0°C. The southeast region had the greatest number of hours (72%) with dry-bulb temperatures between −1.7°C and 0°C. In the southwest, while the most frequent temperatures for freezing rain were −0.6°C and 0°C, the occurrence of freezing rain during temperatures below −1.7°C was more prevalent than the other regions. The lowest temperature reported during freezing rain was −19.4°C that occurred on January 17, 1997 at Mansfield, Ohio and the highest temperature reported during freezing rain was 5.6°C that occurred on 18 January 1959 at Sioux City, Iowa. Studies have shown that temperatures below freezing do not have an effect on the amount of ice accumulated during an event (i.e. the amount of ice accumulated at −4°C is the same as that accumulated at −1°C) (Houston and Changnon 2007).

Wind conditions with FZRA are very important since they affect the amount of ice formation as well as cause damage to structures that collect ice. Higher wind speeds enhance ice accretion while wind direction perpendicular to wires, for example, creates more damage than parallel winds. The most frequent hourly mean speeds were 4.0–5.9 $m s^{-1}$, representing 31.6% of all FZRA hours. Most hours with freezing rain (78%) had mean wind speeds between 2.0 and 7.9 $m s^{-1}$. Only 3.8% of all freezing-rain hours had mean wind speeds of $10.0\,m\,s^{-1}$ or higher. Regionally, winds with FZRA were higher in the eastern regions of United States than in the western regions. Freezing precipitation occurred most frequently with winds from between north and east, and least often with winds from between the south-southwest and west-northwest (Houston and Changnon 2007).

Relatively low topographic features and water bodies can play a significant role in the occurrence of freezing precipitation, especially FZDZ. Previous studies have shown that clean air is important for the growth of precipitation via the CCP, and there is evidence that enhanced FZDZ occurs with wind directions have a long fetch over the Great Lakes. Aome stations bordering the lakes do not show this trend, which indeed merits further investigation (Bernstein 2000).

An investigation of the conditions present during freezing-rain and ice-pellet events was undertaken by Penn (1957) and Stewart and King (1987), their results showed that during ice-pellet events, the ice particles falling into the melting layer would not completely melt before falling into the refreezing layer, whereas the ice particles falling during a freezing-rain event do completely melt in the melting layer. After studying the characteristics present during 34 freezing-rain, freezing-rain/ice pellet, and ice-pellet events, it was found that the precipitation type reported at the surface seemed to be more sensitive to the melting characteristics present in the temperature profile than to the refreezing characteristics.

9.4 Winter Storms

As shown in Figure 9.7, winter storm precipitation, both freezing rain and sleet, occur by the same general process: liquid raindrops in a layer of warm air well above the surface fall into a layer of freezing air hugging the ground. The difference between these two winter precipitation types depends on the thickness of the layer of freezing air. Freezing rain occurs when the layer of freezing air is so thin that the raindrops do not have enough time to freeze before reaching the ground. Instead, the water freezes on contact with the surface, creating a coating of ice on whatever the raindrops contact. A winter storm is a combination of heavy snow, blowing snow and/or dangerous wind chills. A winter storm is life-threatening. In the winter, most precipitation forms as snow within the clouds because temperatures at the top of the storm are cold enough to make snowflakes. Snowflakes are collections of frozen ice crystals that form as water vapor condenses into water droplets and freezes. These ice crystals stick together as they fall toward the ground, forming snowflakes. If the air temperature remains at or below 0°C between the cloud and the ground, the precipitation will fall as snow. If the air near the ground is above freezing, the precipitation will melt to form rain or freezing rain.

A storm where precipitation falls as snow is called a snowstorm, as shown in Figure 9.8. A snowstorm where there is no accumulation beyond a light dusting of snow is a snow flurry. A brief snowstorm with snow falling at varying intensity and some accumulation is a snow shower. If the snow showers are accompanied by strong gusts of wind and accumulation of lots of snow, they are called snow squalls. A blizzard is considered a severe kind of snowstorm and is described in more detail below.

A blizzard is a severe snowstorm defined by the strength of the winds rather than the amount of snow it brings. With wind speeds at or above 35 mph, blizzards create blowing snow conditions, where snow on the ground is picked up by the wind, causing reduced visibility and the accumulation of snowdrifts. A blizzard lasts for three or more hours and often leads to the accumulation of lots of snow, either as new snowfall or as redistribution of previously fallen snow as blowing snow.

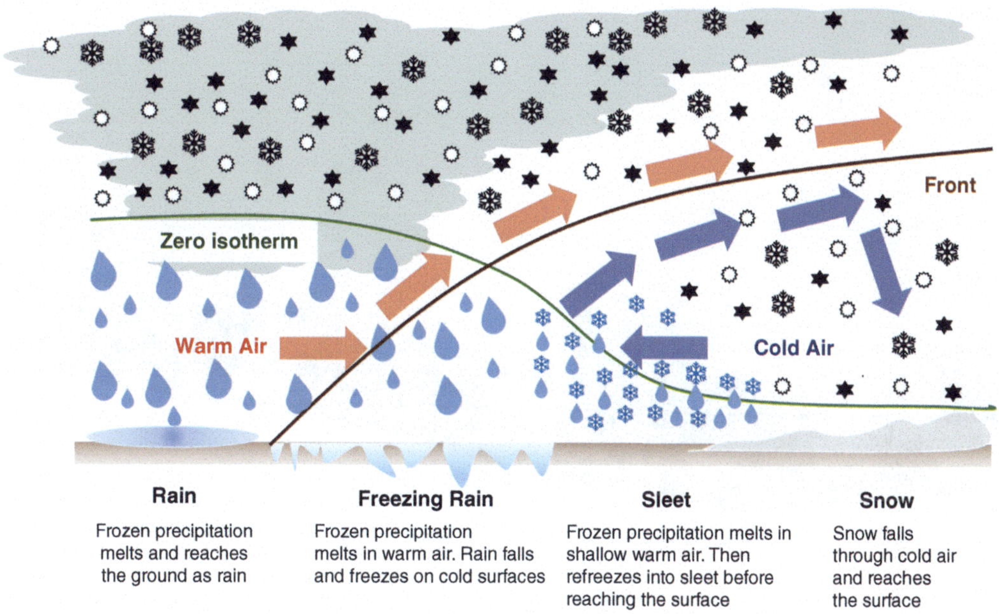

Figure 9.7 Winter storm precipitation from NOAA, both freezing rain and sleet occur by the same general process: liquid raindrops in a layer of warm air well above the surface fall into a layer of freezing air hugging the ground. The difference between these two wintry precipitation types depends on the thickness of the layer of freezing air. Freezing rain occurs when the layer of freezing air is so thin that the raindrops do not have enough time to freeze before reaching the ground. Instead, the water freezes on contact with the surface, creating a coating of ice on whatever the raindrops contact. *Source:* United States Department of Commerce / Public Domain.

An ice storm is a winter storm that has an accumulation of at least 0.25 in. (6.35 mm) of ice on all outdoor surfaces. The ice forms a slick layer on the ground that can make driving and walking conditions dangerous and can cause branches and powerlines to snap due to the weight of the ice. There are different types of icy winter weather, which are determined by the temperature of air masses in the storm. Sleet forms when the snowflakes falling toward the surface first pass through a layer of air that is above freezing, which causes the snowflakes to partially melt, and then pass through a layer of air that is below freezing. This can cause the snowflakes to refreeze into pellets of ice. Similar to sleet formation, freezing rain forms when precipitation (either rain or snow) passes through a layer of warmer air, allowing it to become rain, and then through a layer of much colder air. But this time, the rain is not able to refreeze as is falls through the shallow layer of freezing air near the surface. The rain is supercooled through this process and freezes instantly upon contact with cold surfaces (UCAR Center for Science Education 2022).

Severe snowstorms could be quite dangerous: a 6 in. (15 cm) snow depth will make some unplowed roads impassable, and it is possible for cars to get stuck in the snow. Snow depth exceeding 12 in. (30 cm) especially in southern or generally warm climates will make the roofs of some homes cave in and cause loss of electricity. Standing dead trees can also be brought down by the weight of the snow, especially if it is wet. Even a few inches of dry snow can form drifts many feet high under windy conditions. Heavy snow can immobilize a region and paralyze a city, stranding commuters, closing airports, stopping the flow of supplies, and disrupting emergency and medical services. The weight of snow can cause roofs to collapse and knock down trees and power lines.

Figure 9.8 Snow storm, heavy snow can immobilize a region and paralyze a city, stranding commuters, closing airports, stopping the flow of supplies, and disrupting emergency and medical services. The weight of snow can cause roofs to collapse and knock down trees and power lines. Homes and farms may be isolated for days and unprotected livestock may be lost. In the mountains, heavy snow can lead to avalanches. *Source:* Evgeni Dinev Photography / Moment / Getty Images.

Homes and farms may be isolated for days and unprotected livestock may be lost. In the mountains, heavy snow can lead to avalanches. The cost of snow removal, repairing damages, and the loss of business can have severe economic impacts on cities and towns.

In China, snow disasters occur nearly every year, especially in the four pastoral areas, namely, Inner Mongolia, Qinghai, the TP, and Xinjiang (Figure 9.9). Snow disasters vary in different areas: over the TP, even mild snowfall can cause detriment owing to the dwarf grass; in north part of Xinjiang, the grass is higher than that of the TP, so an excessively thick snow layer causes disasters; snow in Inner Mongolia is frequently accompanied by wind, and can thus cause snowstorms (Lv et al. 2010).

Shijin et al. (2019) suggested that, from 1961 to 2015, there have been 238 events recorded on the TP, in which it suffered serious snow disasters with a large-scale damage (more than 600 000 livestock deaths) in 1974, 1975, 1979, 1982, 1989, and 1995, respectively. In the past 20 years, the snow disasters with large-scale damage showed a clear decreasing trend, but these snow disasters caused serious devastation to animal husbandry. By contrast, the frequency of small-scale snow disasters with a death rate of less than 600 000 increased in recent years and there was a gradual increase in casualties and losses. In view of the frequency of small-scale disasters, the number of the snow disasters with livestock deaths and casualties increased from an average of two to three years in the 1960s and gradually increased to almost once a year in recent years. In view of the scale of livestock deaths, snow disasters have been mainly small-scale in the past 20 years, but the total cumulative number of deaths increased year by year. From October 1995 to the end of April 1996, there were five continuous major snowstorms in the Yushu Prefecture in the middle of the TP. The cumulative snow depth in the snow-covered area reached 60 cm and the mean temperature was between −15.8 and −26.5°C in most areas during this period. This snow disaster caused a large area of grassland

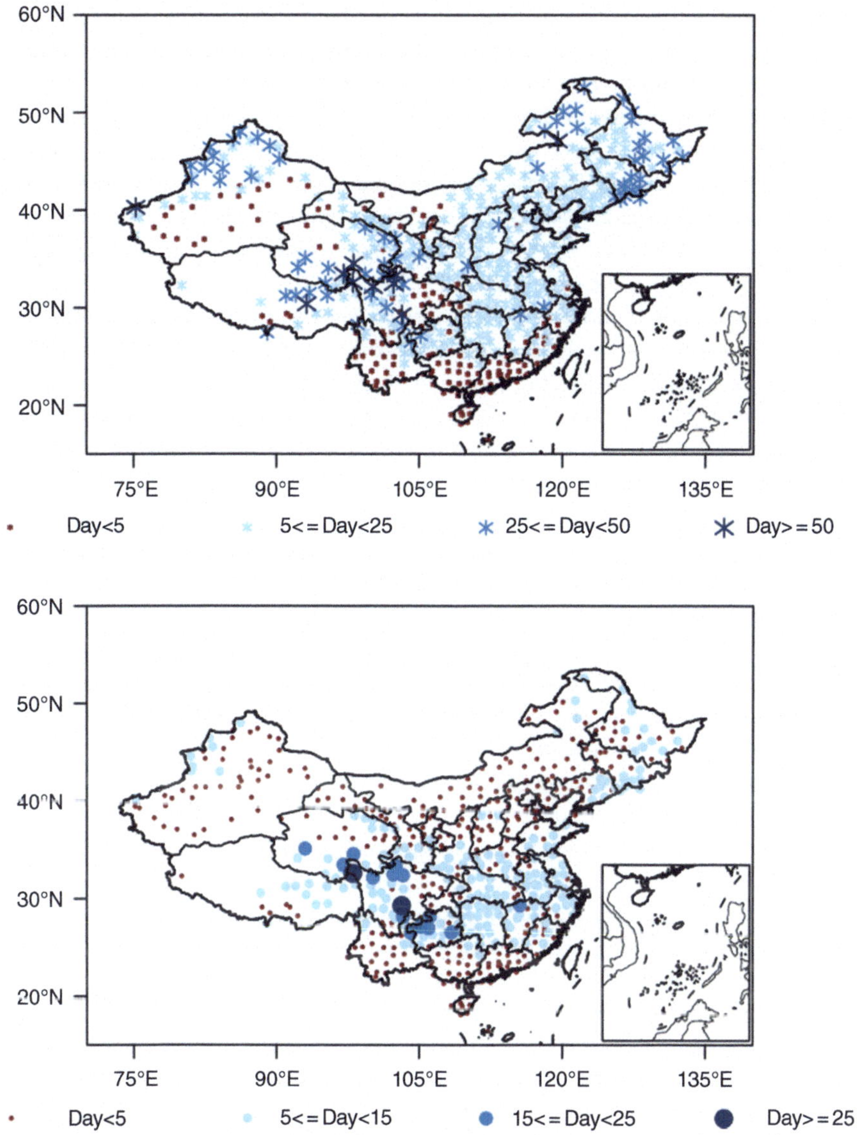

Figure 9.9 Annual mean snowfall days ($P > 0.1$ mm and $T < 0$°C) and freezing ($P > 0.1$ mm and $0 < T < 3$°C) days (period 1961–2018, no data for Taiwan).

covered by snow to be blocked to traffic and causing interruption in communication resulting in 630000 livestock death and a death rate of 23.19% and the loss of 287 million dollars in the past 40 years.

The annual mean snowfall distribution shows that the area with more than 50 days of snowfall was southern Qinghai, while the regions with 25–50 days were mainly high-latitude and high-altitude areas, such as Heilongjiang, Jilin province, the eastern part of Inner Mongolia, northern Xinjiang, Eastern Qinghai, and central Tibet. The regions with less than 5–25 days were in central China, and those with less than five days were in southern China. The annual mean freezing days show a similar pattern to snowfall days. The distribution of snowfall and freezing days in China is comparatively concentrated. There are three heavy snowfall areas in China: northeastern Inner Mongolia, Heilongjiang and Jilin, northern Xinjiang and the eastern TP (Gao 2016).

Over the TP, areas at high-risk of snow disaster are located in certain areas of the counties of Yecheng and Pishan in the Xinjiang region; areas at medium-risk of snow disaster are found between the Gangdise Mountains and the Himalayas in the central-western TP, and the southeastern TP; the risk of snow disaster is generally low throughout the large area to the south of 30°N and the region on the border of the eastern TP. Overall, the risk of snow disaster in high-altitude areas of the central TP is higher than that at the edge of the TP (Liu et al. 2014).

Winter snowstorms can basically be divided into two types: snowstorms formed by northerly winds and those formed by southeasterly winds (one to two days), weak intensity (winds below magnitude 6–7), and few occurrences. The wind direction of the last named is basically concentrated in the southeastern-easterly direction, and it lasts for a long time (usually three to four days) and has a strong intensity (usually winds of 6–8 magnitude), and there are many occurrences. Therefore, the southeastern-easterly snowstorm is the most prominent in the comparison of the two types of snowstorms.

According to Pei (2019), snowstorm weather in polar regions is a large-scale weather phenomenon, which is closely related to the surface pressure field. After statistical analysis, the two types of snowstorms have completely different characteristics of the pressure situation. The common feature of the northerly snowstorms is the pressure field of "high in the east and low in the west," that is, in the northeast of the Antarctic Peninsula, there is a high-pressure ridge extending south from the Atlantic subtropical region to the Weddell region, and in the west of the Antarctic Peninsula and the Weddell Sea is a low-pressure or cyclone area. The northerly snowstorm occurs in the northerly air flow with a large pressure gradient between high and low pressure. Xu et al. (2004) suggested that the common feature of the distribution of the pressure field of the southeastern snowstorms is that the pressure field is "high in the west and low in the east" or "high in the southwest and low in the northeast," that is, the western and southern parts of the Antarctic Peninsula are high-pressure control areas, and the eastern and northern parts of the Antarctic Peninsula are low-pressure areas. Snowstorm weather is generally matched with low-level temperature inversion. The appearance of this temperature inversion increases the stability of the atmosphere and blocks the vertical exchange of air. It may have the following two effects on the formation and maintenance of snowstorm weather: (i) suppressing the vertical dissipation of the aerodynamic energy in the lower atmosphere and confining the lower atmosphere to flow in the horizontal direction. This is like the tube wall effect of the rigid body, which is beneficial to the increase of the lower horizontal wind speed; (ii) the temperature inversion layer blocks the upward transport of the fine snowflake particles scraped by the strong wind, so that they concentrate in the bottom layer of the atmosphere and reduce the visibility of the atmosphere. Therefore, the temperature inversion layer is beneficial to the generation and maintenance of snowstorm weather. According to the statistics, the low-level temperature inversion is an important factor for the generation and maintenance of snowstorms.

Ice storms are significant cold-weather events that prove hazardous to multiple sectors. Ice storm occurred when ice accumulation on surfaces exceeds 0.25 in. (0.635 cm), creating major hazards. Ice storms usually occur as part of a wider outbreak of severe winter storms, which includes snow, sleet, and rain. The distinction between freezing rain and these other types of winter precipitation is usually difficult to predict as slight changes in the thermodynamic structure of the atmosphere can alter the precipitation type. Meanwhile, the ability to forecast ice storm events is very important due to the particular hazards they produce in comparison to other winter weather events.

The atmospheric conditions required for freezing rain can occur during the late fall, winter, and early spring and develop as a result of a variety of different synoptic weather patterns. A survey by Call (2008) showed that, ice storms were generally only third to flooding and tornadoes in the potential threat they cause to their respective regions. Impacts of ice storms are wide ranging and generally affect large portions of society. Damage from ice storms cost an annual average of 313 million dollars in the United States alone (Zarnani et al. 2012). The northeast region experiences the greatest frequency of "catastrophic" losses, as ice storms are also more frequent. Although numbers are low compared to some other severe weather events in the United States, ice storms can still lead to multiple injuries and fatalities.

Synoptic situation plays a key role in the development of ice storms. Analyses of synoptic situations discussed by Robbins and Cortinas (2002) showed large-scale variability between the regions with the most favorable conditions for ice storm events. The most significant similarity between these regions was the common occurrence of mid-level warm advection, seen as an elevated warm layer in freezing-rain atmospheric profiles.

9.5 Ice Avalanches

Ice-avalanche disasters are closely related to the movement and ablation of glaciers. The collapse can be divided into four types: ice avalanche, avalanche, rock avalanche or debris flow according to the composition of the debris (Alean 1985). The formation of ice avalanches is similar to a landslide, which refers to the process in which a part of the glacier falls off a steep ice bed and moves to a place without ice and snow. There is a clear distinction between ice avalanches and avalanche disasters. Ice avalanches are formed in two different ways: first the ice body on the steep glacier slope is continuously broken, forming ice avalanches falling on the surface of the glacier. The debris on the ice can reintegrate into the glacier or accumulate to form a new ice tongue, The second way is similar to a landslide. The ice avalanche is the process of the normal melting of the glaciers on high-altitudes slopes. After rolling off, the ice eventually breaks down into smaller pieces (Figure 9.10). The frequent glacier jump, volcanic activity, and abundant rainfall in summer can result in an ice-avalanche disaster. Ice avalanches are the rupture of large glacial blocks on steep glaciers. The ice tongues of the steep glaciers in alpine areas cannot reach the ablation zone for normal ablation. They then break off and cause ice avalanches. Sometimes ice avalanches are not caused by climate, rainfall and other factors, but caused by earthquakes (Hao et al. 2018).

According to Pudasaini and Hutter (2007), ice avalanches are different from avalanches, debris flows, landslide and other physical processes. They are composed of multiphase materials such as ice, snow, water, moraine debris, and air, and their internal structure is extremely complex and diverse. The solid phase of the avalanche body includes ice bodies and moraine debris, and the liquid phase is composed of ice and snowmelt water. However, during the movement of the avalanche body, due to the friction of the substrate and the internal liquefaction, the solid phase and the liquid phase are transformed. In addition, while moving, the mixture of ice and snowmelt water and loose moraine debris in the avalanche body will move up and down under the influence of the terrain and its own structure, taking in air, so that a small amount of gas is wrapped inside the avalanche body. In terms of physical properties, the avalanche body is a collection of fragmented materials, and there are countless material interfaces inside.

Figure 9.10 Three stages in the formation of an ice avalanche. First, it needs a low temperature that freezes water into ice. Second, there needs to be a lot of precipitation, so that the snowflakes turn into grain snow, and then from grain snow to glacial ice, which of course takes a long time. Third, there must be terrain suitable for snow and ice accumulation. Under such conditions, glacial ice accumulates to a certain thickness and then moves slowly along a certain slope of the ground under the action of gravity. The general movement rate is tens to hundreds of meters per year. *Source:* Bethan Davies.

From the collapse of the ice avalanche to the end of the accumulation, the material composition of the entire ice avalanche is not static. At the beginning, the entire ice avalanche is a solid ice body, and then the moraine in the ice bed and the ice and snow melt due to friction. Water and the substances in various phases are also transformed during movement, so that the material composition of the ice avalanche is constantly changing (Bennett and Glasser 2011).

Faillettaz et al. (2015) suggested that, the average speed of ice avalanches can reach more than 70 $\mathrm{km\,h^{-1}}$, which can directly destroy settlements and vegetation along the way, and cause great damage to the ecological environment in the surrounding areas. Such high-speed movement is mainly related to the potential energy of the ice avalanche, and may also be related to the excessive pore water pressure in the body of the ice avalanche; the speed of movement increases first and then decreases. The body of the ice avalanche is the collapse of the glacier or a fracture in the initial stage. The meltwater will enter the glacier through fissures formed by the collapse or fracture. After the friction between the avalanche body and the ice bed, a thin water film is formed between the ice-avalanche body and the ice bed, which increases the speed of movement. In addition, during part of an ice avalanche, a staged and stratified movement process can also occur. The upper layer of ice can break away from the ice bed and rush directly down while the lower layer of ice keeps stable. In the process of friction with the ice bed, the body slips downstream with the moraine debris in it.

The high-altitude and low-temperature characteristics of TP provide favorable geographical and climatic conditions for the development of glaciers (Yang et al. 2019). Mountain systems such as the Himalayas, the Hengduan Mountains, and the Nyainqentanglha Mountains located in the southern TP are affected by the Asian monsoon, and at the same time, they also obviously prevent these warm and humid monsoons from continuing to penetrate into the hinterland of the TP. Therefore, the rainfall in this area is the highest in Asia. Under the influence of the monsoon, the glaciers increased significantly, which provided the material basis for the development of glaciers, and made a large number of modern glaciers develop inside the TP. Therefore, the TP has become the largest distribution area of glaciers in China and the main storage area for solid freshwater resources (Liu et al. 2000).

Affected by global warming, the glaciers over the TP have experienced obvious ablation and retreat. From 1999 to 2015, the glaciers over the TP showed a decreasing trend. During this period, a total of 2333 glaciers changed, among which 150 glaciers disappeared and 2156 retreated significantly, with a total reduction area of 254.5 km^2. Only 27 glaciers expanded in area, with an enlarged area of 10.62 km^2 (Yang et al. 2008).

Sun et al. (2022)'s results indicated that, in terms of quantity changes, the Qilian Mountains have the most prominent changing glaciers, up to 605, accounting for 25.93% of the total number of changed glaciers; followed by the Karakoram Mountains, with a total of 409 glaciers changed, accounting for 17.53% of the total number of glaciers that changed; the changes of glaciers in the Kunlun Mountains ranked third in the number of changes, accounting for 12.56% of the total number of glaciers. The area of glaciers in the Nyainqentanglha Mountains, Gangdise Mountains, Qilian Mountains, and Himalayas have evidently retreated, and the sum of the retreated areas accounted for 60.48% of the total area of change, while the area of glaciers in other mountain systems retreated less. Among them, the glacier area in the Nyainqentanglha Mountains decreased the most, accounting for 56.31 km^2 for 21.24% of the total glacier change area; The Gangdise Mountains followed, with a total decrease of 43.58 km^2. The Qilian Mountains ranked second as far as change in the glacier area was concerned with a total reduction of 33.33 km^2. The reduction of glaciers in the Himalayas is 27.11 km^2. Glacier ablation will bring greater risk of ice avalanchse.

The frequency of ice avalanches in the TP is obviously increasing. The earliest recorded ice-avalanche disaster in the 20th century was in the Rum Lake Basin in July 1931 (Xia 2012), and three more ice-avalanche disasters occurred successively in the following 30 years. The 1960–1979 period is a slight increase in the overall number of ice avalanches compared to the previous 30-year time period. Between 1980 and 1999, the number of ice avalanches was roughly the same as in the previous 20-year time period. However, after entering the twenty-first century, the number of ice avalanches in the TP has increased significantly. As of December 2018, there have been 11 ice avalanches in less than 20 years (Barry and Gan, 2022).

According to Qin et al. (2021), from the geographical location of the distribution of disasters, the ice-avalanche disasters are mainly concentrated in the Himalayas and the Nyainqentanglha Mountains on the TP. A large number of glaciers develop on the TP, and their distribution is uneven. Most of the glaciers are concentrated in the Nyainqentanglha Mountains, the Himalayas, and the Gangdise Mountains. Although the global temperature increase will lead to the overall retreat of glaciers on the TP, the degree of ablation of glaciers in different regions of the TP is different. The average annual rainfall in the region is rich, so the frequency of ice avalanches in the Himalayas and Nyainqentanglha Mountains, which are located in the southern TP and have a large number of glaciers, will be relatively high. From the perspective of the administrative division of disaster distribution, ice avalanches are mainly distributed in the Tibet Autonomous Region with a large number of glaciers, and only two ice avalanches have been recorded in Qinghai Province. Judging from the specific time of the disaster, basically all ice avalanches occurred between June and September because it is the rainy season on the TP, with more rainfall and relatively high temperatures over the Tibet plateau.

According to Pei (2019), the characteristics of ice-avalanche disasters on the TP can be summarized as "two concentrations and one increase," that is, the occurrence time is concentrated, mainly between June and September each year; the occurrence location is concentrated, mainly in the southern TP near the Nyainqentanglha Mountains and the Himalayas, where glaciers are melting rapidly and in large numbers and the frequency of occurrence has increased. After entering the twenty-first century, the frequency of ice avalanches has increased significantly compared with the twentieth century.

Pei (2019) also suggested that, when considering the impact of ice-avalanche disasters from the two climatic factors of temperature and rainfall, it can be revealed that temperature is the long-term influence factor. When the overall temperature of the TP is higher, the glaciers developed inside it are gradually melting. Stability is also deteriorating, a change that can only be detected over long periods of time. Rainfall is an influencing factor in a short period of time. A large amount of rainfall in a short period of time will directly lead to the accelerated collapse of glaciers and ice-avalanche disasters.

The direct-ice avalanche disaster can be considered as ice-avalanche disaster in the traditional sense. Its process of formation and development, can generally be divided into three intervals: the formation area, the circulation area, and the accumulation area. The formation area of direct-ice avalanche disasters is generally in places with steep terrain and a large number of glaciers within the plateau; the circulation area is generally close to the formation area, and the formed debris basically moves along the valley, causing a direct-ice avalanche disaster. Before and after the occurrence, there will be a great impact on the landform in the valley, so the movement path of the avalanche debris can be clearly seen on the remote sensing images. The accumulation area is generally located below the circulation area, and the terrain in this area is relatively flat. The ice-avalanche debris accumulates together due to the loss of the driving force of sliding, generally forming a relatively obvious accumulation fan. The formation process of disasters can be mainly divided as follows: (i) under the background of increasing global temperatures, the internal consolidation of glaciers becomes worse, the stability decreases, and the sliding driving force becomes larger; (ii) under the influence of other factors such as rainfall or earthquakes, the internal friction of the glacier continues to decrease or the integrity of the glacier deteriorates, while the glacier activity continues to increase. When the driving force is finally greater than the static friction, the ice-avalanche disaster occurs. The scope of direct-ice avalanches is generally limited by terrain conditions. Due to the steep terrain on the TP, once an ice avalanche occurs, the fractured glaciers will generally obtain a great initial movement speed that is dangerous. The performance of the debris will also be greatly improved, but after the terrain conditions become smooth in the later stage of the movement, the movement speed of the debris will gradually decrease, and finally the movement will be completely stopped under the action of external forces such as friction, forming an accumulation fan.

Ice-lake outburst disasters in the traditional sense refer to sudden flood disasters caused by glaciers blocking lakes or moraines blocking the drainage of lakes, or dam collapses (Qin et al. 2014). In order to distinguish glacial-lake outburst disasters caused by other factors, a separate definition of glacial-lake outburst disasters caused by ice avalanches is made, that is, the ice-avalanche material formed during the ice avalanche is directly washed into the downstream glacial lake, causing flood surges. Eventually, a flood disaster results from the destruction of glacier-lake dams. Generally, this type of ice-avalanche disaster mainly causes damage to downstream villages or buildings such as roads and bridges through the burst flood, forming a disaster chain. The burst flood generally moves a long distance in the valley, and the vegetation, buildings, and villages where the flood passes are all washed away and submerged, which is extremely dangerous. Tong et al. (2018) proposed that the main process of burst-flood formation is "acceleration of glacier melting → increase in recharge of glacier meltwater → increase in area of ice lake → deterioration of stability of ice lake → occurrence of ice avalanche disaster, a large amount of debris pouring into the ice lake → destruction of the ice-lake dam.

The chain-blocking disaster is the most complicated type of disaster, and it can be regarded as a kind of debris flow and flood disaster in essence. According to the disaster-causing process, the chain-blocking disaster can take place during an ice avalanche. The broken glacier blocks carry scraped debris to block the downstream river channel to form a dammed lake, and after the collapse of the dam, a disaster occurs in the downstream area. Generally speaking, the the blockage chain disaster can be caused by avalanche debris and Quaternary loose deposits in the channel. Ice-avalanche debris mainly refers to the glacial block that breaks during the collapse of the glacier. This part of the debris will generate heat due to friction in the process of moving down the slope, and part of it will be converted into ice and snowmelt water. The loose deposits in the channel play the role of wetting, so that the deposits start to slide after creeping up. The unmelted part enters the river together with the debris, transforming the river into a lake. When the river water continuously erodes the dam, the remaining glaciers in the dam melt, and the remaining deposits are not enough to withstand the continuous impact of the river, and the dam lake bursts, resulting in corresponding flood or debris-flow disaster (Jia et al. 2019).

The chain-blocking disaster has certain similarities with the glacial-debris flow and the glacial-lake outburst disaster. Compared with glacial-debris flows, these two types of glacial disasters are ultimately caused by debris-flow flooding, inundation of surrounding or downstream areas, but there are certain differences in the causes and processes of these two types of disasters. Blocking-chain disasters are mainly caused by the collapse of glaciers. The main emphasis is on the ice avalanche and the two processes of "blocking" and "collapse." There are many triggering factors for glacial-debris flows, including heavy rainfall, ice and snow, and melt water, and so on. In addition, the two processes of "blocking" and "collapse" were not experienced during the development and formation of glacial-debris flows, so there are still some differences between the two types of disasters. Compared with glacial-lake outburst disasters, both types of ice-avalanche disasters share the process of dammed lake outburst. The collapsed dammed lake was formed during the disaster. Therefore, the ice-lake outburst disaster can be regarded as a part of the chain disaster (Ullah et al. 2020).

Due to the frequent occurrence of ice avalanches on the TP, there are certain differences in their respective causes. Even for the same ice avalanche, there may be some controversy about the cause of its formation.

9.6 Ice-Jam Floods

Ice-jam floods are caused by the blockage of ice in the river and the discharge of water storage in the river channel. They result from the comprehensive action of factors such as heat, power, and channel shape. As far as the causes of floods are concerned, they can be divided into ice-jam floods, ice-dam floods, and melting ice floods (Gao and Jin 2003).

The occurrence of floods is closely related to the geographical location of the river. The latitude of the lower section of the river is higher than that of the upper section. The lower section thaws later than the upper section. The upper section of the ice sheet is thin, and the lower section is thick. If the river channel is curved and narrow, ice dams are often formed when the river is thawing, causing harm (Leopold 1972). The lower reaches of the Yellow River and the upper reaches of the Ningmeng River. The flow direction is from low latitude to high latitude. The latitude difference between the upper and lower reaches is 3°00′ and 4°37′, respectively, and severe floods sometimes occur. The most serious situation during the flood season is the occurrence of ice blockages or ice dams in the river, causing the water level in the upper reaches to rise sharply. When the river thaws, the water storage capacity of the trough is released sharply, forming peaks and severe floods (Beltaos and Berrell 2021).

After the river was frozen, a large number of ice flowers and broken ice can accumulate under the ice cap, blocking part of the water-passing section, causing the water level in the upper reaches of the river to be significantly high. When the ice plug melts, the storage water discharges to form a flooding process (Björnsson 2004). Ice plugs often occur in river sections where the water-surface gradient changes from steep to slow, and a large number of ice splinters and crushed ice flow downstream. Ice flakes and crushed ice dive and accumulate under the ice sheet, forming ice jams. The size of the ice block varies. In large rivers, the ice block can be as long as several tens of kilometers, and the thickness of the ice block can reach more than 5 m. If the river channel is curved and narrow, and the upstream water has a large amount of ice, the downstream section of the water can be seriously blocked if the climate is cold, and the backwater level is high. The impact is far away and lasts a long time. When the ice plug gradually melts with the increase of temperature, the water storage capacity of the trough is released slowly, forming a gentle flood process (Beltaos 1990).

According to Niu (1997), there are situations when large amount of drift ice blocks the river channel. The ice blocks can accumulate to form a dam-like ice body across the cross-section, which severely blocks the water-passing cross-section and makes the upstream water level of the dam significantly higher. When the head is suddenly destroyed, the water storage capacity of the trough quickly drains, forming a peak that evolves downstream. Ice dams form in the lower bends and narrow reaches. According to the incomplete statistics, from 1875 to 1938 in the lower reaches of the Yellow River, a dike breach occupied 38%. In the Songhua River in 1981, there were 16 ice dams which were only in the 365 km section from Yilan to Fujin, and they were 6–13 m high. The main conditions for the formation of ice-dam floods are: (i) sufficient quantity and strength of ice blocks in the upstream section; (ii) water flow conditions for transporting a large number of ice blocks; (iii) narrow stretches such as at the end of the backwater of the reservoir, the area where the river flows into the lake or the sea, a sharp bend of the river, the narrow section of the river channel and the river section with ice blocks, and so on. Although ice-jam floods and ice-dam floods are both upstream flooding phenomena caused by ice blockage, there are obvious differences between the two: (i) ice plugs mostly occur in the early stage of freezing as well as ice dams; (ii) ice plugs are composed of ice flowers and broken ice, and ice dams are composed of large and firm ice blocks; (iii) after the ice dam is formed, the duration can last for a few hours or as long as ten days; (iv) when the ice block melts, no obvious peaks are generated, and when the ice dam bursts, peaks are often generated, and generally increase along the way; (v) the ice block disappears when the river bed is scoured over a long distance; the ice dam ruptures locally at the head of the river bed; (vi) the ice plug is gradually washed away by the water flow, and the remaining ice flakes can increase the thickness, resulting in a lot of ice remains, and melting slowly on the spot; (vii) the formation of ice plugs is slow, and the water rises very fast upstream of the ice dam and the resulting ice blocks often damage buildings and bridges, causing great losses.

When the temperature in the upper reaches rises, the ice sheet gradually melts due to thermal action, and the water storage in the river channel is discharged, leading to the continuous flood along the river reach. Before the opening of the river, due to the enhanced solar radiation, the duration of sunshine was prolonged and the temperature rose above 0°C. This caused the frozen ice caps to melt, and weirs appeared on the banks and in the center of the river one after another, flooded by slow release. The water potential of the melting ice flood is stable, and the peak flow is small.

Flood hazards refer to the disasters caused by the overrunning and flooding of river embankments caused by floods. The pressure and expansion force of the ice can lead to the damage of hydraulic engineering facilities and coastal buildings, and the ice can also force the interruption of shipping (Sayers et al. 2013).

Ice-jam flood can cause considerable damage because when the ice blocks the river, the water level rises rapidly, causing the dikes to burst and flooding cities, mines, villages, farmland, and so on. Disasters resulting from ice-jam floods can be categorized as follows: (i) ice block flood disasters usually occur in the initial sealing period, often in narrow curved sections of rivers with a large latitude difference from low latitude to high latitude, such as the transition section from steep slope to gentle slope, and the end of the backwater of a reservoir. For example, in January 1962, a huge ice block appeared between the Yanguoxia Reservoir and Liujiaxia Reservoir in the upper reaches of the Yellow River, and the water level rose sharply. (ii) Ice-dam flood disasters usually occur during the thawing period. When the temperature rises, the ice cap in the upper section disintegrates rapidly, and the flow along the way increases, causing the river water level to rise and swell the ice cap, and a large number of ice blocks flow rapidly downstream. When the ice dam develops to a certain stage, it cannot withstand the pressure of the upstream ice water and collapses to form a peak, causing downstream havoc. Rivers with large latitude differences, curved channels, large amounts of drift ice and gentle slopes flowing from low latitudes to high latitudes are prone to such disasters. Their characteristics are very high water levels. According to incomplete statistics, the Inner Mongolian reaches of the upper reaches of the Yellow River had flood disasters of varying degrees every year before 1952. In 1933, for example, where the river rose in Dengkou County, an ice dam was formed. Both sides of the river were inundated for 150 km, forming a wide flooded area.

Ice hazards are also serious threats to structures. In Xinjiang, hydroelectric stations are often shut down due to ice blockages. In 1965 and 1966, for example, the first-level power station of Aksu River Shengli Canal was closed down more than ten times due to ice damage, and the longest closure lasted for a month.

Ice pressure, which is the force of ice directly acting on the building, including the dynamic pressure caused by the impact of drift ice and the static pressure caused by the expansion of the entire ice sheet, can cause disasters as well. Dynamic pressure will affect the construction of sluices, bridge piers and ports. For example, when the Inner Mongolia section of the Yellow River thawed and opened in the spring of 1962, the pier of the Wuda Railway in Wuda City was hit by the floes. The body of the bridge shook violently, and people crossing it could hardly stand up. The width is $2.13\,\mathrm{mm\,s^{-1}}$, which is more than ten times larger than the allowable width of a train running over the bridge and stopping.

River ice is an integral component of the hydrological and hydraulic regimes of cold region rivers, and affects an extensive portion of the global hydrological system, particularly in the northern hemisphere (Yang et al. 2020). It regulates numerous hydrological, geomorphical, biological and chemical processes in rivers, and provides important socioeconomic and ecological benefits (Prowse 2001). However, river ice can also cause extensive damage to infrastructure, navigation, and hydroelectric production, and can result in detrimental effects to river ecology and water quality (Beltaos and Burrell 2015).

River ice impacts on local river hydraulics in many different ways. The presence of ice cover and ice jams obstruct river flows due to additional resistance and reduced cross-sectional area, which can occasionally lead to backwater flooding (Beltaos and Prowse 2001). Similarly, in some small river systems, freezing of riverbeds may block groundwater inflows, which may also generate flooding (Daly 2013). Ice jams can also change river channel properties and flow paths through ice scouring and erosion (Boucher et al. 2009).

A schematic diagram of a river-ice jam and ice-jam induced backwater level, compared with open-water or ice-cover conditions, shows that ice-jam induced backwater levels are substantially higher than backwater levels that arise from intact, stationary, floating ice cover or open-water conditions (Figure 9.11). Studies have also suggested that ice jams can result in two to three times higher water depths in river channels for the same or even lower discharges compared to open-water (ice free) conditions (e.g. Lindenschmidt and Rokaya 2019; Prowse and Beltaos 2002).

There is a decreasing trend in the overall duration of the ice-cover season across the northern hemisphere. In Europe, ice-cover duration was reported to decrease by 46 days per century in the Oder River, which forms the border between Germany and Poland, between 1956 and 2015. The lower part of the Danube River in Romania also experienced a reduction in ice-cover duration by 28 days per century in the 1937–2016 period, whereas the lower reaches of the Drava River in Hungary observed a decline of 14 days per century in the 1875–2014 period. In the Baltic region the major rivers (e.g. Narva, Parnu, Daugava, etc.) showed a declining trend in ice-cover duration from 28 to 63 days per century.

In the Yellow River in Asia, a reduction of 14.2 days per century in ice-cover duration between 1954 and 2015 was reported. In Canada, ice-cover duration decreased by 22 days per century in the Red River in Manitoba between 1798 and 1981 (Rannie 1983) and in Atlantic Canada, where warming trends are weak or even negative (cooling), mixed or later showing breakup trends. Similarly, in the United States, the ice-cover duration decreased from 15 to 37 days across four

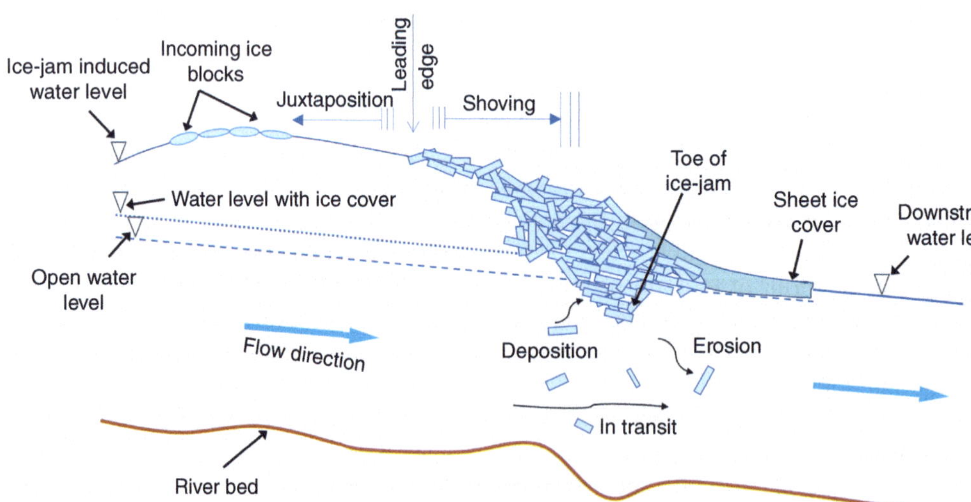

Figure 9.11 Schematic diagram of river-ice jam and associated backwater level. *Source:* Rokaya et al. (2022) / with permission of Elsevier.

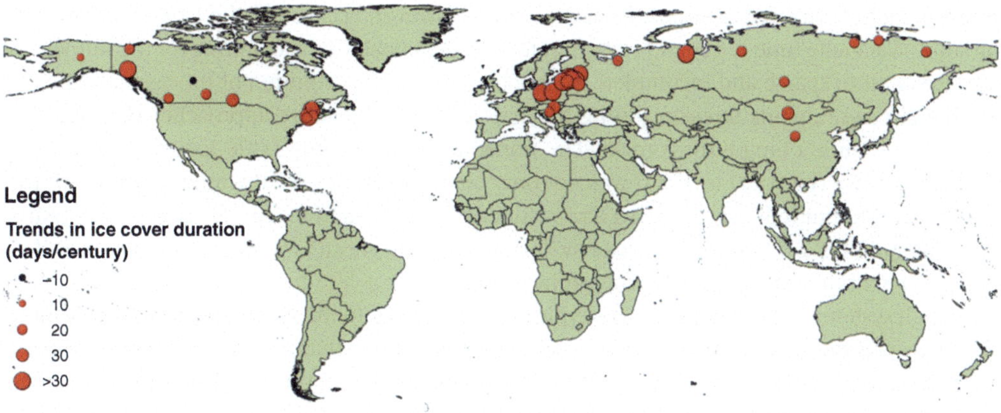

Figure 9.12 Reported historical decline in ice-cover duration in rivers in the NH. *Source:* Rokaya et al. (2022) / with permission of Elsevier.

rivers in Maine, whereas in the interior of Alaska, a decreasing trend in ice-cover duration by four days per century during the 1972–2016 period was discovered. These results suggest that ice-cover duration in northern rivers has decreased across the northern hemisphere in response to warming temperatures in the last few decades. Globally river-ice cover duration has decreased from 0.3 to 4.3 percentage points on a monthly scale between 2008 and 2018 and 1984 and 1994 (Beltaos and Prowse 2009).

The reported decline in ice-cover duration in rivers in the northern hemisphere from historical observations reveals that in most of the northern regions, as shown in Figure 9.12, a decreasing trend of a month or more per century in ice-cover duration is already recorded, but there are also indicators of regional disparities in trend direction and magnitude. These trends of later freeze-up and early breakups, resulting in shorter ice-cover durations over much of the northern hemisphere corresponds closely to the increasing air temperature trends observed over most of this region, which has been most pronounced at the end of the last century. There has been Aadecreasing trend of 26 cm per century between 1912 and 2001 (Huntington et al. 2003). Several large Russian rivers, had a decline in ice thickness by 2–14 cm in the 1980–2000 period compared to the 1950–1979 period.

9.7 Snow Avalanches

A snow avalanche is a rapid flow of snow down a mountain. Although snow avalanches can occur on any slope given the suitable conditions, certain times of the year and certain locations are naturally more dangerous than others. Wintertime, particularly from December to April, is when most snow avalanches tend to happen. However, avalanche fatalities have been recorded for every month of the year.

All that is necessary for a snow avalanche is a mass of snow and a slope for it to slide down. For example, while the temperature remains low, the snow sticks to the surface and does not slide off. After the temperature increases, however, the snow will slide down the windshield of a vehicle, often in small slabs. This is an avalanche on a miniature scale

Snow is highly sensitive to atmospheric warming. However, because of the lack of sufficiently long snow avalanche time series and statistical techniques capable of accounting for the numerous biases inherent to sparse and incomplete avalanche records, the evolution of process activity in a warming climate remains unclear. Filling this gap requires innovative approaches that put avalanche activity into a long-term context. The transition from the late Little Ice Age to the early twentieth century (i.e. 1850 to 1920) was not only characterized by local winter warming in the order of +1.35°C but that this warming also resulted in a more than sevenfold reduction in yearly avalanche numbers, a severe shrinkage of avalanche size, and shorter avalanche seasons as well as in a reduction of the extent of avalanche-prone terrain.

Although systematic avalanche records typically extend back to the 1950s, it has so far remained difficult to observe unequivocal trends on climatically driven changes in avalanche numbers, runout distances, or flow types (Eckert et al. 2010). Likewise, literature is also critically lacking regarding future avalanche activity in a warmer climate – the only study currently available (Castebrunet et al. 2014) points to a 20–30% reduction in avalanche numbers in the French Alps by the end of the twenty-first century as compared to the reference period 1960–1990. According to Ballesteros-Cánovas et al. (2018), several factors explain this lack of direct evidence: (i) homogeneous time series of snow avalanches have remained too short for trend analyses over timescales relevant for the detection of climate change impacts (Mock et al. 2017); (ii) interpretations may have suffered from confounding influences of climatic and socioeconomic drivers or biases in the proxy data (Laternser and Schneebeli 2002); (iii) with global warming, a temporary increase in extreme snowfall or wet-snow amounts could emerge at higher elevations (Le Roux et al. 2021), which in turn could enhance avalanche activity (Lavigne et al. 2015). For instance, during the exceptional avalanche episode of January 2018 in the European Alps, and despite extremely high air temperatures, numerous avalanches were triggered across the Alps (Ballesteros-Cánovas et al. 2018) (in a subregion of Switzerland alone, nearly 19 000 avalanches were observed on satellite images). Similarly, a drastic increase in avalanche activity has been evidenced lately on high-elevation slopes in the Himalayas and attributed to the increase in increasing air temperatures.

A snow avalanche has three main parts, in which the starting zone is the most volatile area of a slope, where unstable snow can fracture from the surrounding snow cover and begin to slide. Typical starting zones are higher up on slopes. However, given the right conditions, snow can fracture at any point on the slope. Several factors may affect the likelihood of an avalanche, including weather, temperature, slope steepness, slope orientation (whether the slope is facing north or south), wind direction, terrain, vegetation, and general snowpack conditions. Different combinations of these factors can create low, moderate, or extreme avalanche conditions. Some of these conditions, such as temperature and snowpack, can change on a daily or hourly basis.

9.8 Freezing and Thawing Damage

In cold regions, frost damage has a serious impact on the life of the building. In the process of cyclical freeze–thaw of subgrade due to periodic temperature changes, paved roads located in seasonally frozen areas are susceptible to damage, which poses a great threat to the safety of highway transportation (Zhang et al. 2019). Frozen soil is a kind of soil containing ice at temperatures below 0°C. It can be divided into permafrost and seasonally frozen soil if a seasonal freezing period exists.

Seasonally frozen areas are widely distributed, accounting for 53% of the world's land area, while in China, they account for 53.5% of the total land area (Zhan et al. 2018).

The mechanism of frost heave and thaw settlement of subgrade in seasonally frozen ground is complex. The heat transfer, moisture migration, and phase change all interact during the freeze–thaw process (Mo and Lou 2020).

During winter, the freeze–thaw cycle occurs that causes most of the damage to roads during winter. The first part of the cycle occurs when it snows or rains. The moisture falls through cracks in the road. While in summer, the water would drain away in the soil, in winter the moisture gets frozen in place. When water is frozen, it expands. When it expands it causes the cracks in the brittle road to grow larger. This cycle continues to repeat throughout winter, with the water being re-thawed sinking further under the road, then refreezing and causing the cracks to expand even further.

Another potentially more damaging effect of the freeze–thaw cycle is frost heave. Frost heave refers to upward movements of soil and rocks due to the freeze–thaw cycle. This process occurs because, as the ice freezes it attracts water to it through the soil. This sucks both water and dirt toward the surface. This process mostly affects fine soils in subfreezing temperatures. With the ability to destabilize buildings frost heave can create a huge amount of damage to roadways and asphalt surfaces.

The accumulated water in the capillary in the foundation expands after being frozen, which causes the foundation soil to generate a uniform expansion force, resulting in the destruction of the building. There are countless capillaries in the foundation soil. The groundwater mainly rises to the foundation ground through the capillaries in the foundation. When winter comes, when the negative atmospheric temperature is introduced into the ground, the free water in the surface soil first freezes into ice crystals, and as the temperature continues to drop, the outermost layer of bound water also begins to freeze, causing the ice crystals to gradually expand and form an ice interlayer in the soil layer. The more fine particles there are in the soil, the more capillaries are formed and have greater the impact on the foundation.

The freezing line in the cold area is about 2 m, and the shallow foundation located in the frost heave area. If the embedded depth is shallower than the freezing depth, it will be affected by the frost heave force on the foundation. If the frost heave force is greater than the load on the foundation, the foundation may be damaged. Generally, the ice layer of the foundation soil on the sunny side of a room is shallow, and the frozen layer of the foundation soil on the shady side is deeper. Therefore, the frozen layer on the shady side contains a lot of ice. The expansion of water changes greatly, and the arching upward is severe; the ice layer on the sunny side is shallow, the water content is small, the soil expansion change is small, and the building foundation is unevenly affected by the expansion force of the frozen soil. When the soil layer has thawed, the ice crystals accumulated in the foundation melt, which greatly increases the water content in the soil. In addition, the drainage capacity of the fine-grained soil is poor, or the soil layer under the foundation is not thawed, and the water in the thawed soil layer above cannot penetrate. The base soil layer is in a saturated and softened state, and its strength is greatly reduced, causing the building to sink, which is called subsidence.

Whether frost heaves or thaws, it is generally uneven, so that freezing and thawing alternate every year, resulting in cracks in buildings with shallow foundations. Frost damage is particularly serious for houses built in soils with poor bearing capacity which consist of many fine particles and high groundwater levels,.

9.9 What's Next

There are still deficiencies in disaster prevention in cold regions:

1) Prediction: observational data about mountain hazards are extremely scarce around the world. Avaielable data, such as the weather data obtained from the spots far away from the danger zones, however, cannot exactly reflect the real formation conditions of hazards at high altitudes. In addition, calculation from regional remote sensing data needs further validation. Data limitation has been a long stumbling block for the quantitative forecasting model.
2) Dynamics mechanisms: there is a lack of studies on the dynamics of hazard formation and mass movement in extremely cold areas. Existing analysis and simulations are still insufficient.
3) Risk analysis: attention should be paid to multi-process of hazard chain and multi-scale risk. Field experiments simulate the physical processes of structure destruction well but fail to satisfy the similarity rules; on the other hand, lab experiments apply to dynamical investigation but the results usually lack practical application. Therefore, the key to enhancing risk analysis is to determine the dynamic mechanism of interaction between different disasters and bearing bodies.

10

Climate Changes in Cold Regions

In the past few decades, global temperature have risen rapidly. In this context, climate changes in cold regions are more significant, which threaten the vulnerable human living and ecological environment in cold regions. Under global warming, changes of climate in cold regions occurred in both historical periods and future periods. This chapter will discuss the climate changes in cold regions, including historical changes in the past few decades and the projected changes for the next 50–100 years.

10.1 Climate Change in Cold Regions in the Past Decades

Global surface temperatures have generally been warming for the entire length of the instrumental record. The most significant and highest warming occurred in recent decades. The most up-to-date consensus from global climate models predicts warming during the boreal winter. The linear trends of January surface air temperature show increases in most of regions (Figure 10.1). Changes of surface air temperature will greatly affect the thermal state of the cryosphere components (such as permafrost, snow, etc.). Recent climatic changes are pronounced in permafrost regions. Since 1980, the Arctic has been warming at approximately twice the global rate. The strongest temperature changes (~1 °C decade^{-1}) occur in winter and spring, and the smallest changes occur in autumn (IPCC 2013).

The Tibetan Plateau (TP) has the largest extent of high-altitude frozen ground at middle and low latitudes in the northern hemisphere (NH) and is surrounded by dozens of seasonally frozen ground. Rising air temperatures have resulted in frozen ground degradation over the TP since the last century. Based on the daily frozen soil depth, annual mean daily minimum air temperatures and annual mean air temperatures obtained from 19 *in situ* observations in the TP for the period of 1960–2019, Wang et al. (2020) suggested that the maximum depth of frost penetration (MDFP), the beginning when the soil froze, and ending when the soil thawed have changed considerably. On average, the MDFP has reduced by 0.14–1.71 m at most stations. The annual frozen period has decreased by about 40 days. The changes in seasonal freezing of soils appear to be attributed to the rising of minimum air temperatures in winter, especially at higher elevations from 4500 to 5000 m. The MDFP presents a decreasing trend at most stations, with a much greater decline at higher elevations. MDFP declined much more rapidly at most stations especially after 1995, compared to only a slight decrease between 1960 and 1995. The smallest MDFP of all stations was about 0.17 m for the Chengdu station in 2007. The averaged MDFP is 1.05 m for stations below 3000 m, with a decrease of 0.21 m during 1960–2019. At the stations above 4500 m, the changing of the MDFP is distinctly higher than other stations. During 1960–2019, the average MDFP decreased by 1.37 m

Climatology in Cold Regions, First Edition. Chenghai Wang.

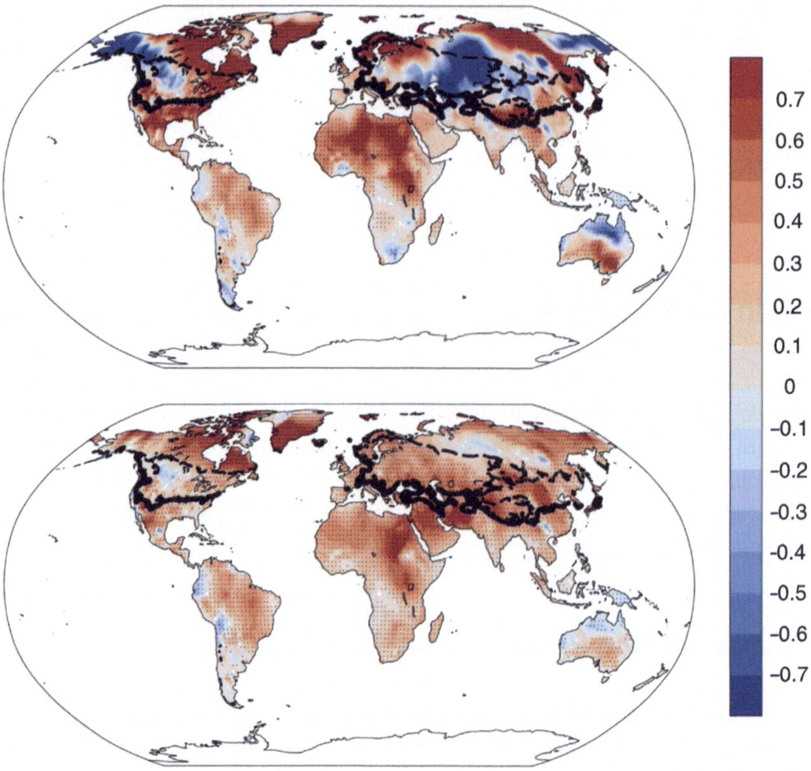

Figure 10.1 Linear trend of January (upper) and annual mean (lower) surface air temperature (units: °C decade^{-1}) during 1981–2010.

above 4500 m, while it decreased by 0.35 m below 4500 m. These results suggest that the variations of MDFP are correlated to the higher elevations (Figure 10.2).

10.2 Projection of Future Climate in Cold Regions

The cryosphere is a layer in which the Earth's surface water exists in solid form; it includes glaciers (e.g. mountain glaciers, ice caps, polar ice caps, and ice shelves), frozen ground (seasonal frozen ground and permafrost), snow cover, solid-state precipitation, sea ice, river ice, lake ice, etc. Various components of the cryosphere are mainly distributed in the high latitudes and high-elevation areas, which are quite sensitive to changing temperature, thus, the cryosphere is commonly known as the indicator of climate change. Historically, the cryosphere plays a key role in triggering changes in climate and gradually forces changes in the atmosphere. Global warming is the main characteristic of the Earth's climate over the past century. The results of the global mean linear trends in the land and oceanic surface temperatures indicate that the global average temperatures rose by 0.85°C between 1880 and 2012 (IPCC 2013). Over the past 30 years, temperature rise over every 10-years interval is higher than any time since 1850. In the northern hemisphere, 1983–2012 was likely the warmest period in the past 1400 years, and had an obvious warming trend compared with the previous decade or two (Qin et al. 2014). Under global warming, the cryosphere

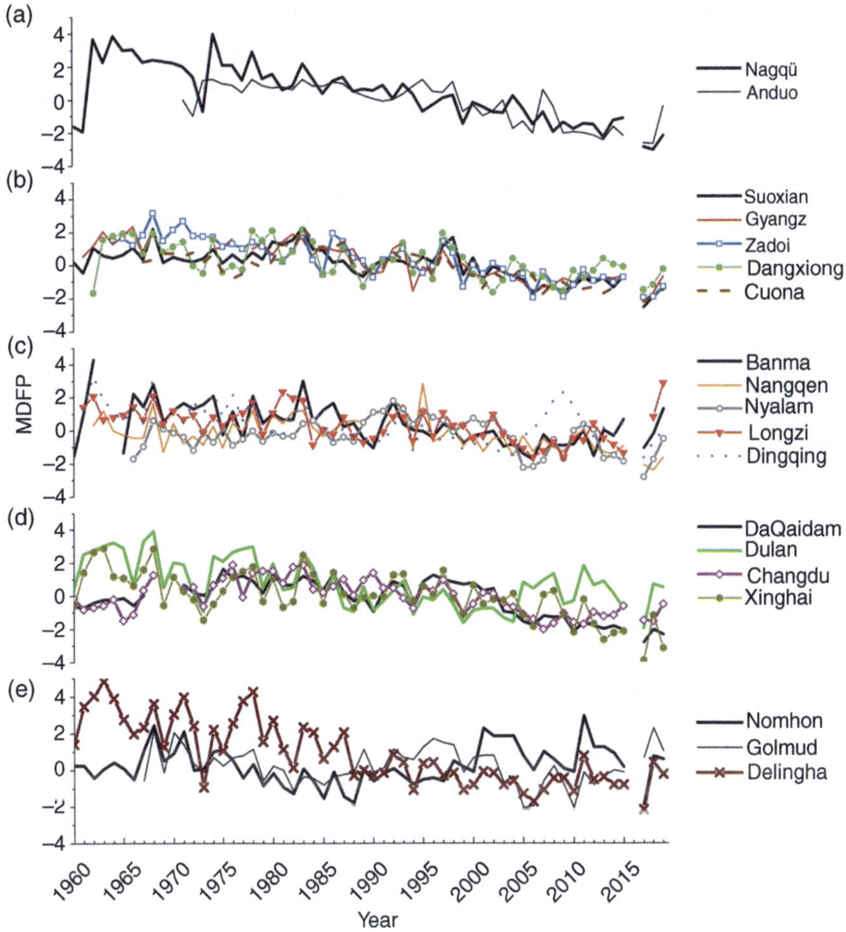

Figure 10.2 Time series of the maximum depth of frost penetration; normalized relative to 1981–2010 over the Tibetan Plateau during 1960–2019 (Wang et al. 2020). (a) Stations with altitudes of 4500–5000 m. (b) Stations with altitudes of 4000–4500 m. (c) Stations with altitudes of 3500–4000 m. (d) Stations with altitudes of 3000–3500 m. (e) Stations with altitudes of 2500–3000 m.

has responded very rapidly, and local glaciers, permafrost, and snow cover have changed markedly (Burke et al. 2013; Park et al. 2015; Rafiq and Mishra 2016).

Permafrost is highly sensitive to a warming climate. Its degradation is mainly based on the following three variables: the increase in permafrost temperature, the decrease in the extent of permafrost, and the deepening of the active layer thickness. Observations have shown that the permafrost in the TP, Canada, Alaska, Sweden, and other regions has degraded to various extents (Wang et al. 2002; Slater and Lawrence 2013). The temperature increases in permafrost have pronounced differences in different regions. The most significant increases in permafrost temperature were found in northern Alaska, northern European Russia, which increased by 2–3°C between 1971 and 2010 (Romanovsky et al. 2010; Oberman 2012). The permafrost temperature at a depth of 6 m increased at the rate of 0.02°C per year between 2006 and 2010 along the Qinghai-Tibet railway (Wu et al. 2012). The active layer thickness tended to deepen in the NH since the mid-1990s, with the most pronounced deepening found in northern European Russia, eastern Siberia, and

Chukotka. The active layer thickness declined slightly in Canada and western Siberia (IPCC 2013). The active layer thickness has increased at a rate of 0.71 cm per year in the eastern TP since 1980 (Zhang and Wu 2012). The permafrost area in the Arctic decreased at a rate of 0.55×10^6–0.81×10^6 km^2 per year between 1967 and 2000 (Burke et al. 2013).

Based on multi-model ensemble mean results from 17 global climate models in the Coupled Model Intercomparison Project Phase 5 (CMIP5), Wang et al. (2019) analyzed the changes in permafrost and snow cover in the northern hemisphere under a scenario in which the global surface temperature has risen by 1.5°C, relative to 1861–1880. Their results showed that the threshold of 1.5°C warming will be reached in 2027, 2026, and 2023 under representative concentration pathway (RCP) 2.6, RCP4.5, RCP8.5, respectively. When the global average surface temperature rises by 1.5 °C, the southern boundary of the permafrost will move 1–3.5° northward, particularly in the southern Central Siberian Plateau. The permafrost area will be reduced by 3.43×10^6 km^2 (21.12%), 3.91×10^6 km^2 (24.1%), and 4.15×10^6 km^2 (25.55%) relative to 1986–2005 in RCP2.6, RCP4.5 and RCP8.5, respectively (Figure 10.3). The snow-water equivalent (SWE) will decrease in over half of the regions in the northern hemisphere but increase only slightly in the Central Siberian Plateau. The SWE will decrease significantly (more than 40% relative to 1986–2005) in central North America, western Europe, and northwestern Russia. The permafrost area over the TP will decrease by 0.15×10^6 km^2 (7.28%), 0.18×10^6 km^2 (8.74%), and 0.17×10^6 km^2 (8.25%), respectively, in RCP2.6, RCP4.5, RCP8.5. The SWE in winter (December, January, February) and spring (March, April, May) over the TP will decrease by 14.9% and 13.8%, respectively.

Remarkable decreases in the area and/or depth of snow have been documented in many regions of the northern hemisphere including western North America (Groisman et al. 2004; Stewart et al. 2004), central Europe (Falarz 2002; Vojtek et al. 2003; Scherrer et al. 2004), and China (Ji and Kang 2012; Wang and Wang 2012). Satellite data show that terrestrial snow cover extent in the northern hemisphere has experienced significant decreases from earlier melt over the period since 1970 (Vaughan et al. 2013), and this trend is projected to continue in the future with an estimated projected decrease in northern hemisphere spring snow cover of 25% by the end of the twenty-first century for RCP8.5 (Collins et al. 2013).

SWE represents the water stored in the snowpack and is a key variable for climate and hydrological science. Local SWE responds to both precipitation and air temperature, which are both projected to increase in the twenty-first century (Collins et al. 2013) and the magnitude and seasonal character of the change represents a complex interplay between a shortened snow accumulation period, the fraction of precipitation that falls as snow (influenced by total precipitation and air temperature) and the frequency and intensity of winter thaw events (Räisänen 2008; Brown and Mote 2009).

Changes in SWE over the northern hemisphere landmasses have been investigated for the early (2016–2035), middle (2046–2065), and late (2080–2099) twenty-first century using a multi-model ensemble from 20 global climate models from the CMIP5. The multi-model ensemble was found to provide a realistic estimate of observed mean winter SWE in the northern hemisphere compared with the GlobSnow project. The multi-model ensemble project significant decreases in SWE over the twenty-first century for most regions of the northern hemisphere for RCP2.6, RCP4.5, and RCP8.5. This decrease is particularly evident in the TP and North America. The only region with projected increases is eastern Siberia. Projected reductions in mean annual SWE exhibit a latitudinal gradient with the largest relative changes over lower latitudes. SWE is projected to undergo the largest decreases in the spring where it is most strongly negatively correlated with air temperature. The reduction in snowfall amount from warming is shown to be the main contributor to projected changes in SWE during September to May in 2023 over the northern hemisphere (Figure 10.4).

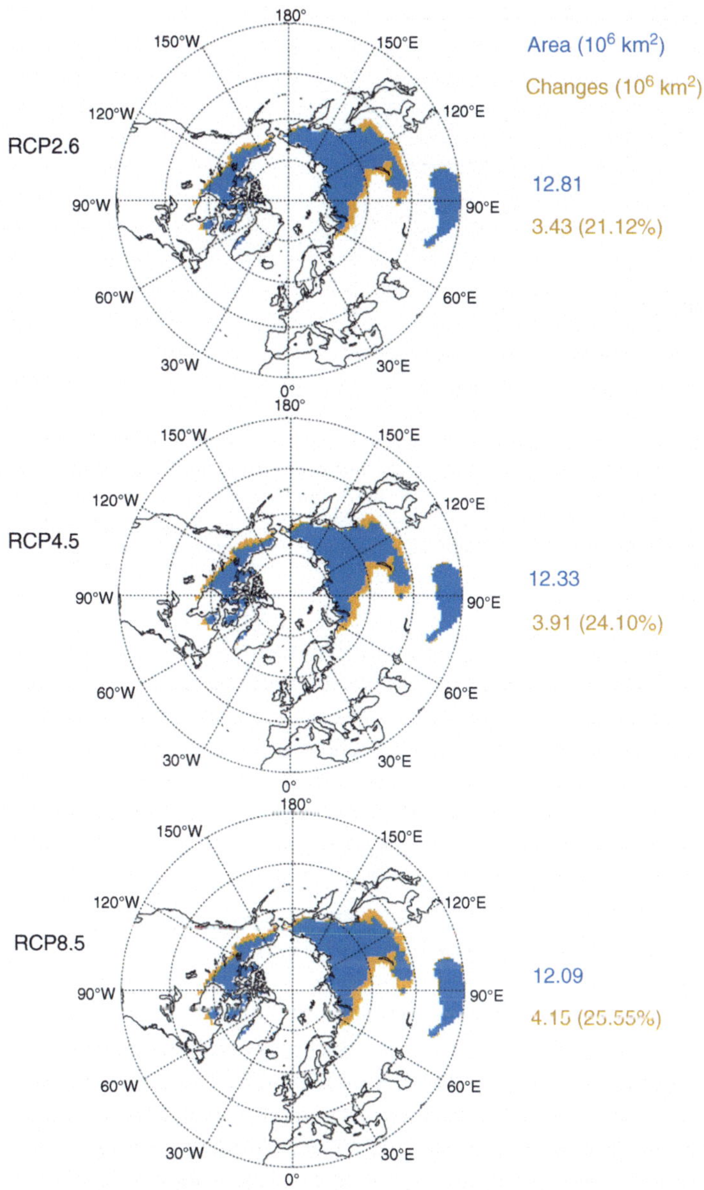

Figure 10.3 Absolute change in permafrost extent (relative to 1986–2005) in the northern hemisphere when the global average surface temperature rises by 1.5 °C (relative to 1861–1880) under the three representative concentration pathway (RCP) scenarios (Kong and Wang 2017) (orange: the permafrost extent in 1986–2005; blue: the permafrost extent when the global average surface temperature rises by 1.5 °C).

The spatial pattern of projected mean annual SWE change shows a clear north–south gradient with the largest relative reductions over more southern latitudes. The seasonal pattern of projected changes shows the largest changes occurring during the spring–summer period with the lowest changes for annual maximum SWE.

Permafrost occupies much of Earth's land area, that is, the area of permafrost underlying the exposed land surface is estimated to be $16\text{–}21 \times 10^6\,km^2$ (Gruber 2012), and it is sensitive to

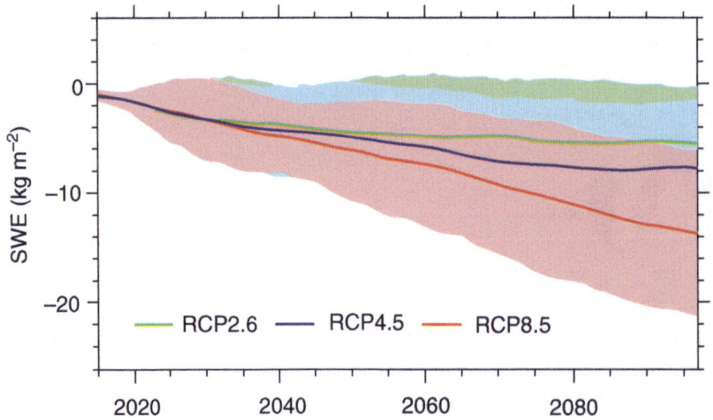

Figure 10.4 Projected changes in mean annual snow-water equivalent (SWE) over northern-hemisphere land during the twenty-first century for all three representative concentration pathways (RCP) (Shi and Wang 2015). The mean value for the 1986–2005 reference period is subtracted from all values. Also shown is the multi-model mean for all available models for each scenario. The ten-year running average was derived for each model before calculating the multi-model mean. Shaded areas denote the intermodal standard deviation for each ensemble mean.

climate change, especially in high-latitude and high-altitude regions (Walter et al. 2006; Li et al. 2008). A number of recent studies have suggested that permafrost is warming and degrading, and active layer thickness (ALT) is increasing (Jafarov et al. 2013; Park et al. 2013; Guglielmin et al. 2014; Sobota and Nowak 2014; Wang et al. 2014; Peng et al. 2015; Pogliotti et al. 2015; Yu et al. 2015). Widespread warming in permafrost regions will result in substantial changes in terrestrial hydrology, vegetation composition, and ecosystem function, as well as carbon dioxide and methane fluxes (Nelson and Anisimov 1993; McNamara et al. 1998; Jorgenson et al. 2001; Christensen et al. 2004; Hinzman et al. 2005; Smith et al. 2005). Most of these processes will be confined to the seasonally thawed active layer, across which exchanges of heat, moisture, and gases between the atmosphere and terrestrial systems will occur (Kane et al. 1991; Nelson and Anisimov 1993; Burn 1998).

The active layer is the part of ground that is subjected to annual thawing and freezing in areas underlain by permafrost, and it is an indicator of climate change in cold regions. Recent observations have detected a general increasing trend in the active layer thickness (ALT) in many regions, including Russia and northern Europe (Mazhitova 2008), eastern Siberia (Fyodorov-Davydov et al. 2008), the Canadian Arctic (Smith et al. 2010), northern Alaska (Hinkel and Nelson 2003), and the TP (Wu and Zhang 2010; Zhao et al. 2010; Li et al. 2012). The widespread increase in ALT may affect the local hydrology, ecology, and the infrastructure of northern communities (Kane et al.1991; Jin et al. 2000; Wu et al. 2003; Instanes et al. 2005). Although there are many uncertainties associated with estimates of the magnitude of these effects, there is an urgent need to resolve questions regarding the fate of permafrost and changes in the ALT (Walker 2007).

Soil moisture (SM) is a crucial variable that influences the mean annual ground temperature (MAGT) of permafrost and the thermal regime of the active layer (Goodrich 1982; Kane et al. 2001; Nelson et al. 1997a,1997b; Park et al. 2015). As a result of seasonal and interannual variations of SM associated with changes in climate and soil temperature (Nelson et al. 1997a,1997b; Park et al. 2013), it is important to study the impact of SM on permafrost. However, due to the lack of equivalent SM observations, studies generally assumed that SM is a constant factor in the spatial–temporal scale. Moreover, snow also plays an important role in the water and heat exchange processes between the ground and the atmosphere. In addition to these two influencing factors, ALT can also be affected by other factors such as vegetation, soil texture, and corresponding thermal properties, and topography. These variables interact in a complex way and result in large variations in the ALT (Hinzman et al. 1997; Zhang et al. 1997). Many numerical models have been developed to estimate ALT, such as analytic steady-state equations (Sazonova et al. 2004), heat conduction equations (Koichi 1986; Mackay 1995; Zhang et al. 2003). However, these models mainly depend on the temperature indicator. In view of the complex impact of the above-mentioned factors on ALT, Kudryavtsev's model (Kudryavtsev et al. 1974) is a more effective method for calculating ALT.

A more realistic SM could improve the estimation of ALT. In addition, one of the effective ways to improve the estimation of ALT in permafrost is to use reasonable datasets of SM. However, the spatial details in high mountainous areas are not sufficiently estimated, and the data used in mountainous areas or places with sparse weather stations must be used with caution (Ji et al. 2014). In recent years, land-surface data assimilation has become widely applied. The Global Land Data Assimilation System (GLDAS) generates optimal fields of land-surface states and provides improved SM products on a global scale, and it is available to climate research communities as observational data for the time series ranging from 1948 to 2012 (Rodell et al. 2004). Research shows that GLDAS can capture basic features of observed SM on TP (Bi and Ma 2015), and the root mean square error between GLDAS SM and the station-averaged SM in central TP is less than $0.06\,m^3\,m^{-3}$ (Chen et al. 2013).

The relationship between ALT and annual thawing index (ATI), snow depth (SND), and SM are different in both Eurasian and North American regions. Besides, ALT changes relative to the SND, soil thermal and moisture regimes formed in the pre-thaw season. The contribution of SND and SM to ALT with changing temperatures is different in different periods. Based on the simulation, Park et al. (2015) examined the insulation effects of SND on the underlying soil temperature and permafrost over the pan-Arctic area. Simulation results indicated that snow cover contributes to the warming of permafrost in northern regions and could play a prominent role under scenarios of future Arctic warming.

Snow density can be used as an input parameter in Kudryavtsev's model (Wang and Wang 2012), whereas SND is computed by temperature and precipitation using Eq. (10.1) (Anisimov and Nelson 1996).

$$\bar{Z}_s = \sin^2\phi \left\{ \sum_{i=1}^{k} P_i / \rho_r \left(k - i + 1\right) / k \right\} \tag{10.1}$$

$\bar{Z}_s$ is the SND, P_i is the water-equivalent precipitation for those months i ($i = 1, 2, \ldots, k$) in which the mean temperature is 0°C or below, ρ_r is the relative snow density, and φ is the latitude of the location.

Kudryavtsev's model is employed to estimate ALT in permafrost regions over the northern hemisphere landmass. This is a semi-empirical method that has been developed in Russia primarily to address the practical needs of cold-region engineering, and it is adjusted to operate on hemisphere and regional scales (Kudryavtsev et al. 1974). The required spatially explicit mineral soil textures come from the land-surface module of the Community Earth System Model (CESM), and the thermal properties follow Anisimov et al. (1997). Organic soil plays an important role in the responses of permafrost to warming (Yi et al. 2007, 2009). However, thickness of organic soil is controlled by several factors including vegetation growth, wildfire, drainage. The Kudryavtsev model has two major outputs: ALT and mean annual soil temperature at the depth of thawing, which can be used to determine whether permafrost exists or not (Figure 10.5).

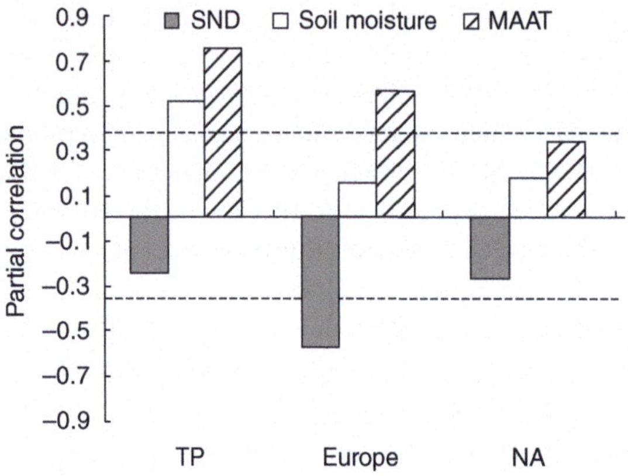

Figure 10.5 The partial correlations between active layer thickness in the persistent permafrost regions and snow depth (SND), soil moisture and mean annual air temperature (MAAT) in Europe, North America (NA), and the Tibetan Plateau (TP) during the period 1986–2005. The dashed lines indicate the 90% significant levels, the partial correlations between ALT and SND, SM, and MAAT are indicated by the black, white, and diagonally-striped bars, respectively.

The partial correlation coefficients (Eqs. (10.2) and (10.3)) can be used to quantify the relative impacts of SM, SND, and temperature on permafrost and ALT in the permafrost regions.

$$r_{12,34} = \frac{r_{12,3} - r_{14,3}r_{24,3}}{\sqrt{\left(1-r_{14,3}^2\right)\left(1-r_{24,3}^2\right)}} \tag{10.2}$$

$$r_{12,3} = \frac{r_{12} - r_{13,23}}{\sqrt{\left(1-r_{13}^2\right)\left(1-r_{23}^2\right)}} \tag{10.3}$$

where $r_{12,34}$ indicates the partial correlation between the first and second variable after eliminating the influence of the third and fourth variables on the first variable, and $r_{12,3}$ is the partial correlation between the first and second variable after eliminating the influence of the third variable on the first variable. The definitions for $r_{14,3}$ and $r_{24,3}$ follow that for $r_{12,3}$; and r_{12} is the correlation coefficient between the first and second variable. The definitions for r_{13} and r_{23} follow that for r_{12}.

10.2.1 Impact of Soil Moisture on Frozen Ground and its Indicators

Compared with SND and air temperature, SM is a relatively robust factor for frozen ground. When the SM remains constant, changes in heat conduction and heat capacity will have an evident effect on frozen ground (Xie et al. 2012; Yu and Wang 2012; Wang et al. 2014). To validate the link between SM and frozen ground, Figure 10.6a–c shows the relationship between GLDASs SM and MDFP observed at the Henan, Yushu, and Qamdo stations; and Figure 10.6d–f illustrates the correlation between GLDAS SM and ALT observed at FH1, TG1, and TM1 stations over the TP. There was a significant positive correlation between SM and ALT, and a significant negative correlation between SM and MDFP. This implies that SM contributes to the thickening of ALT and restrains the increase of MDFP.

An obvious difference is that the permafrost edge is more consistent with the 0°C isotherm than that with permafrost areas of $14.09 \times 10^6\,\text{km}^2$ and $16.71 \times 10^6\,\text{km}^2$, respectively. The distribution of permafrost extent is closer to the result from the International Permafrost Association (IPA) (Heginbottom et al. 1993; Brown et al. 1997). This implies that approximately realistic SM can improve the estimated performance of Kudryavtsev's model for permafrost extent, which also shows that SM plays an important role in changes of the permafrost (Figure 10.7).

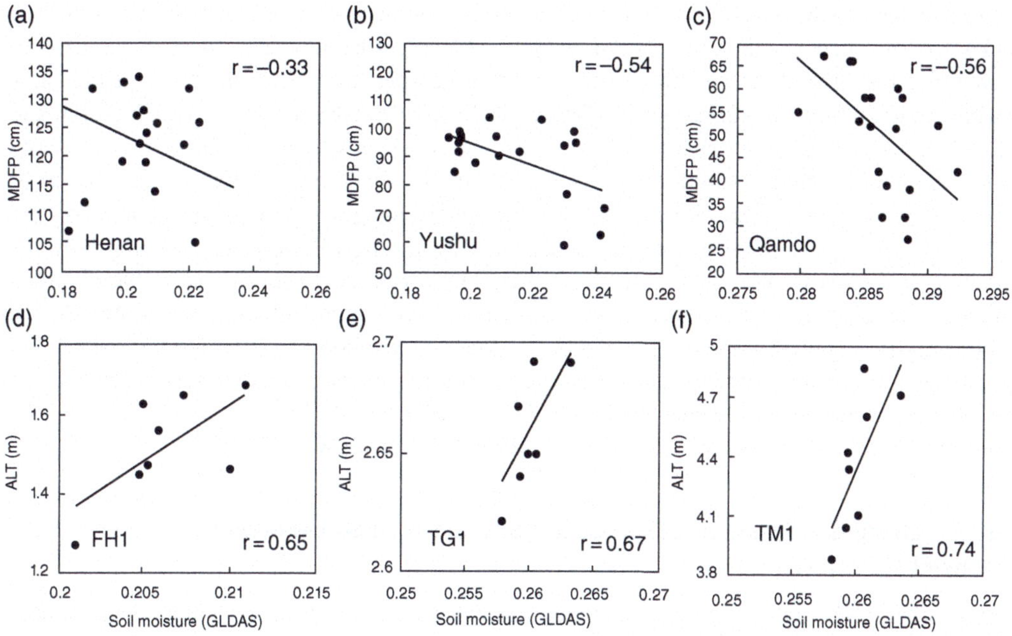

Figure 10.6 The correlation coefficients between soil moisture (SM) and observed maximum depth of frozen ground (MDFP) at (a) Henan, (b) Yushu, and (c) Qamdo observation sites during the period of 1986–2005; and active layer thickness (ALT) at (d) FH1, (e) TG1, and (f) TM1 during 1998–2005. These six observation stations are located over the Tibetan Plateau (TP). SM is from Global Land Data Assimilation System (GLDAS), over the Qinghai-Xizang (TP).

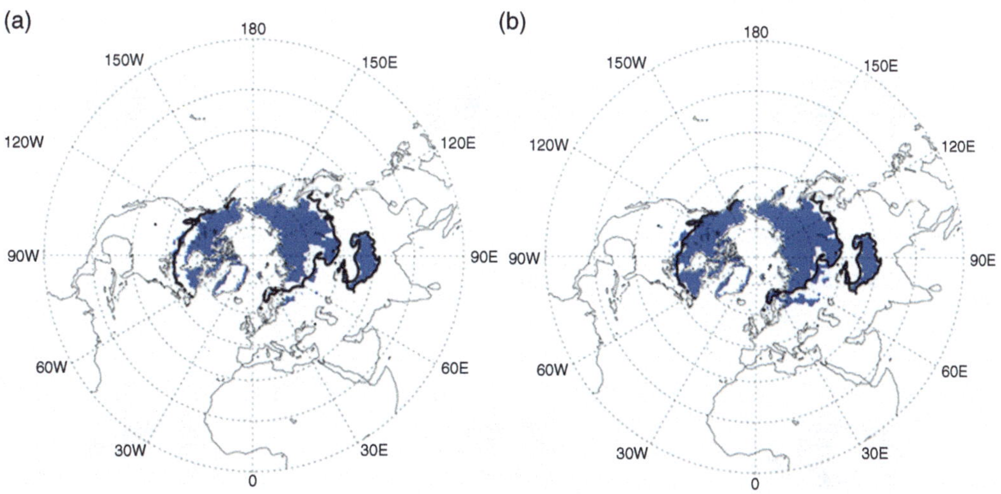

Figure 10.7 The estimated permafrost extent based on Climatic Research Unit (CRU) data and Kudryavtsev's method. The black lines indicate the 0°C isotherm in mean annual air temperature from the CRU data set during the period 1986–2005. (a) Soil moisture is a constant. (b) Soil moisture from the Global Land Data Assimilation System.

Except for SM, the mean annual air temperature (MAAT), and SND also play an important role in ALT variations. Both SM and MAAT have a significant impact on ALT over the TP, and with correlations exceeding the 90% significance level, this is likely because of the larger interannual variation of SM and the higher increase of MAAT over the TP, which is consistent with Intergovernmental

Panel on Climate Change (IPCC) (Stocker et al. 2013) (Figure 10.8). In Europe, the effect of SND and MAAT are stronger than that of SM on ALT. This might be attributed to heavy snowfall in winter and spring, and the resulting stronger insulating effect and high reflectivity of the snow having a more significant effect on ALT (Park et al. 2013). A notable feature is that SND, SM, and MAAT all have a weak relationship with ALT in North America, where permafrost characteristics differ from that in middle latitudes (such as the TP) and in Mediterranean Europe. This might be attributed to the Arctic alpine climate, and relatively small changes in ALT (Zhang et al. 2005).

The IPCC fifth assessment report suggested that temperature will increase continuously in the twenty-first century (Stocker et al. 2013), and warming will be more pronounced in high-latitude and high-altitude regions; the multi-model ensemble projects significant decreases in SWE during the twenty-first century for most regions of the northern hemisphere for RCPs 2.6, 4.5, and 8.5 (Shi and Wang 2015). Permafrost will degrade as well in a future climate-change scenario. An issue that deserves concern is what changes in the relative contribution of SND, SM, and MAAT to ALT would occur, which also involve the regime of thermal hydrology over permafrost.

10.2.2 Changes in Permafrost Extent and Active Layer Thickness in the Twenty-First Century

Temperature, SND, and SM are crucial climate variables for permafrost. To validate the reliability of SM, MAAT, and SND simulated by the CMIP5 models, Figure 10.8 shows that the correlations between the observation of SND (calculated by the Climatic Research Unit [CRU] temperature)

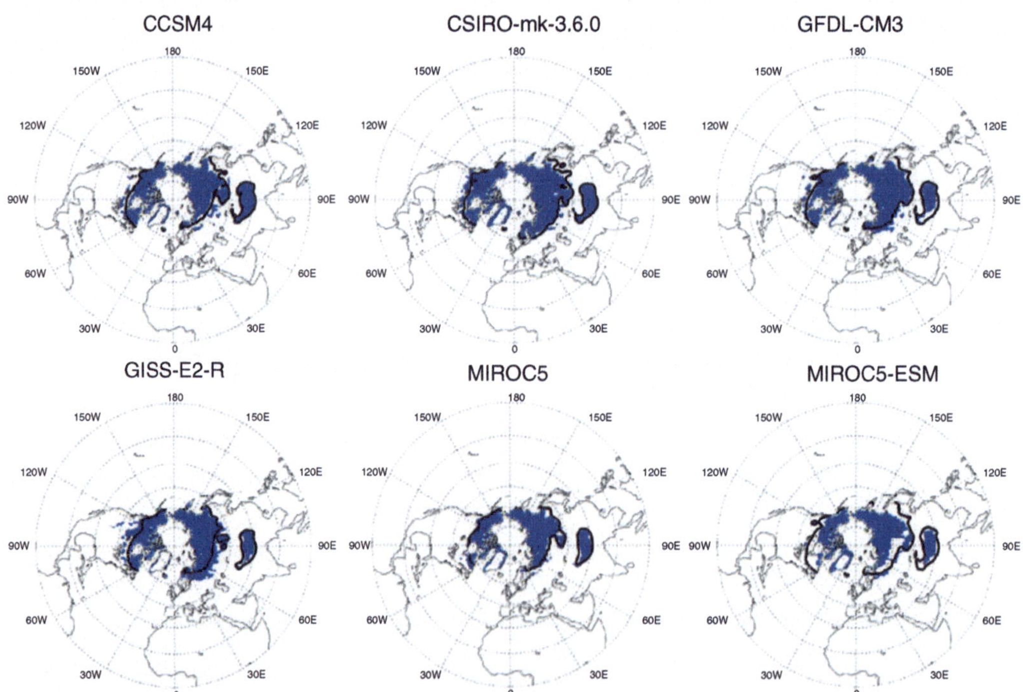

Figure 10.8 The partial correlations between active layer thickness (ALT) in the persistent permafrost regions and snow depth (SND), soil moisture (SM), and mean annual air temperature (MAAT) in Europe, North America, and the Tibetan Plateau during the period 1986–2005. The dashed lines indicate the 90% significant levels, the partial correlations between ALT and SND, SM, and MAAT are indicated by the black, white, and diagonally striped bars, respectively.

and precipitation by Eq. (10.1), MAAT (from CRU), SM (from GLDAS), agree with those from the CMIP5 multi-model ensemble simulation in Europe, North America, and TP, respectively. Here, the reanalysis data are regarded as the observations. Although there are some biases between the simulation and observation, however, almost all of the simulations generally captured the characteristics of the variables, and most of the correlations between them exceeded the 90% significance level, namely the CMIP5 models can well reproduce temperature, SND, and SM.

In general, the multi-model ensemble mean has better performance than each individual model, similar to the approach of Shi and Wang (2015). Figure 10.9 shows the estimated mean permafrost extent from six individual and ensemble means in CMIP5 by the Kudryavtsev method during the period 1986–2005; the six model simulations have good correlation with the GlobSnow and the National Snow and Ice Data Center SWE datasets. Compared with the permafrost extent estimated using the CRU data sets, the six models capture the distribution pattern of permafrost, and the range of permafrost is consistent with the 0°C isotherm of MAAT, which also agrees with the result from Brown et al. (1997). The estimated permafrost area ranges from 13.8–21.0 × 10^6 km^2 in 1986 to

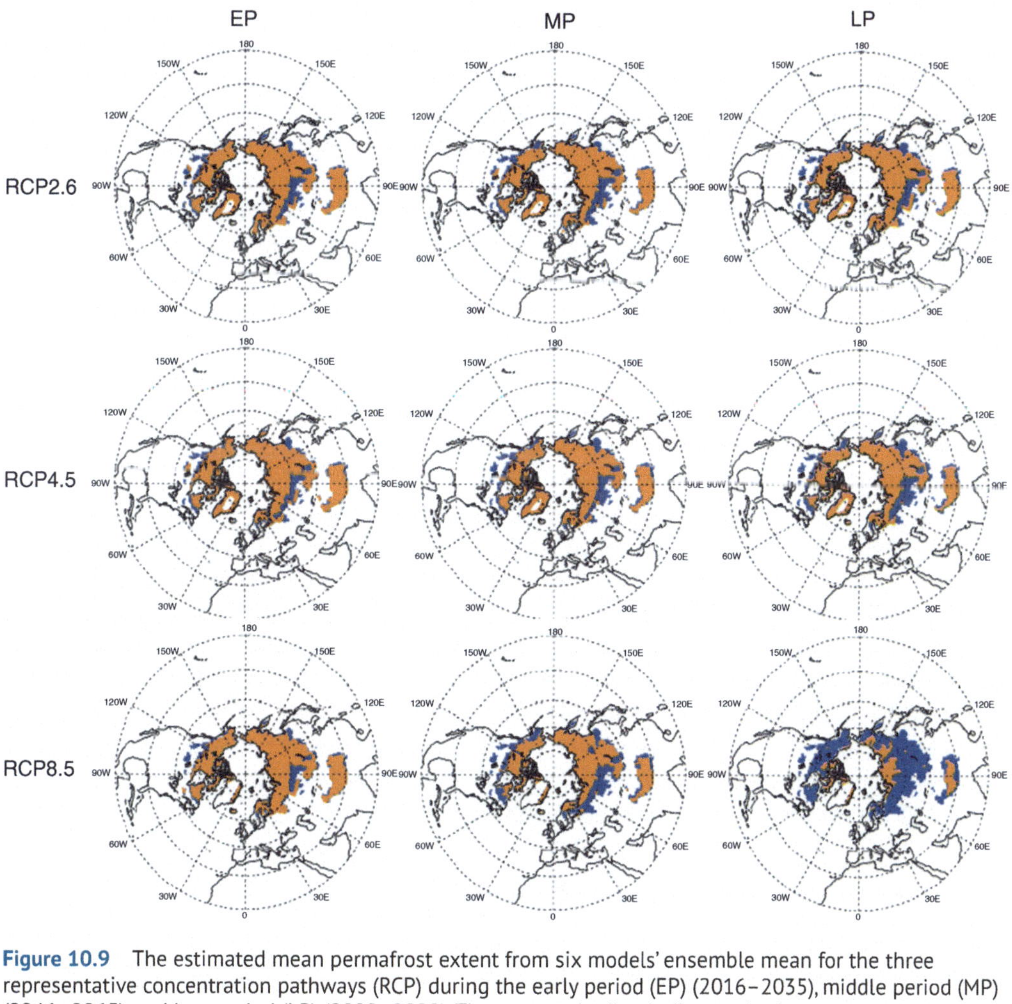

Figure 10.9 The estimated mean permafrost extent from six models' ensemble mean for the three representative concentration pathways (RCP) during the early period (EP) (2016–2035), middle period (MP) (2046–2065), and late period (LP) (2080–2099). The orange shading indicates the future permafrost extent, and the blue shading is the mean permafrost extent for 1986–2005 (Kong and Wang 2017).

$11.1–16.8 \times 10^6\,km^2$ in 2005. The decreasing trend of multi-model mean permafrost extent during the period 1986–2005 is about $-8 \times 10^5\,km^2\,10yr^{-1}$ ($P > 95\%$). Moreover, the spatial correlations between the permafrost extent calculated by the CMIP5 simulation and CRU data ranges from 0.75 to 0.80 ($P > 99.9\%$), whereas the multi-model ensemble results show that the spatial correlation coefficient reaches 0.82 ($P > 99.9\%$). This shows that the model ensemble results can well reproduce the permafrost distribution, and the multi-model ensemble is more effective for reproducing the permafrost distribution than the individual model.

Permafrost is a product of cold climate, and it has significantly responded to global warming. The multi-model ensemble shows the MAGT at the depth of seasonal thawing gradually increased during the three periods of the twenty-first century for the three RCPs. In most permafrost regions, MAGT is closer to −2°C, which indicates that most permafrost gradually changes into warm permafrost. Especially during the late period for RCP8.5, only in Arctic regions (north of 60°N) and the TP, MAGT is below 0°C in other regions, MAGT basically increases to above 0°C, which implies that most permafrost might change into warm permafrost or even disappear; the changes of permafrost in the twenty-first century mainly depend on warm permafrost (Peng et al. 2015).

Furthermore, the averaged permafrost distribution in the period of 1986–2005, and the early, middle, and late periods of the twenty-first century for the three RCPs indicated that, for RCP2.6 and RCP4.5, permafrost gradually degrades from the southern margins of permafrost, however, the decreased magnitude of permafrost extent was weak, relative to the reference period 1986–2005. For RCP8.5, the permafrost degradation gradually becomes significant. Especially during the late period, most permafrost completely disappears, and only exists in the Canadian Arctic, north European Russia, and the center of the TP. Relative to 1986–2005, mean decreased permafrost extent in 2099 is $1.38 \times 10^6\,km^2$ for RCP2.6, $4.34 \times 10^6\,km^2$ for RCP4.5, and $10.32 \times 10^6\,km^2$ for RCP8.5. In addition, the projected future evolution of permafrost extent mainly depends on the future periods and scenarios prescribed. The ensemble mean trend of the permafrost area during the period 1986–2005 is $(-0.9 \pm 1.09) \times 10^6\,km^2\,10yr^{-1}$. The decreasing trend of permafrost area during the early, middle, and late periods for RCP2.6 and RCP4.5 is less than that during 1986–2005, and the changing trend appears to stabilize during the late period, whereas for RCP8.5, the trend ranges from (-0.9 ± 0.83) to $(-1.5 \pm 1.07) \times 10^6\,km^2\,10yr^{-1}$ during the three periods in the twenty-first century. This implies that the permafrost area under RCP8.5 is on a declining channels during the entire twenty-first century.

The changes in permafrost are also reflected in the spatial distribution of ALT in persistent permafrost regions. Relative to the reference period of 1986–2005, ALT in persistent permafrost regions consistently shows an increasing trend for the three RCPs during the early, middle, and late periods. However, for RCP2.6 and RCP4.5, ALT in the persistent permafrost regions gradually becomes stable in 2099, while RCP8.5 is still showing an increasing trend. The most significant increase in ALT occurs in the TP, especially for RCP8.5 during the late period; the largest increment of ALT is greater than 0.8 m. Permafrost in the center of the TP belongs to continuous permafrost, and is regarded as cold permafrost with a MAGT of below −2 °C. As permafrost temperatures in cold permafrost regions increase faster than those in warm permafrost regions, the increasing rate of ALT at the center of the TP is larger than that at its edge.

10.2.3 The Soil Thermal and Hydraulic Regimes in Permafrost Regions

The IPCC fifth assessment report (Stocker et al. 2013) suggested that temperature warming will be amplified at high latitudes and high altitudes, which may further lead to permafrost rapidly degrading (Zhang et al. 2005). The increase in ALT in the TP is larger than that in the Arctic and sub-Arctic regions. A similar feature of MAAT and ALT implies that MAAT over the TP is likely to be an important contributor to ALT. However, the changes in ALT indicate that the response of permafrost to climate change varies regionally, just as soil thermal and hydraulic regimes vary in different permafrost regions.

Changes in soil thermal and hydraulic regimes, the partial correlations between ALT and MAAT, SND, and SM for the period 1986–2005 and the early, middle, and late periods for the three RCPs are close. During 1986–2005, ALT is significantly related to MAAT in the three main permafrost regions, and all of the partial correlations exceed the 90% significant level. However, the significant correlation between SM and ALT only occurs over the TP, and SND significantly correlates with ALT only in Europe. This implies that MAAT is the dominant factor for ALT in the persistent permafrost regions, while the impact of SND and SM on ALT varies regionally, which is similar to the results of Park et al. (2013).

As the temperature increases in the twenty-first century, ALT also shows an increasing trend in most regions, but the thermal and hydraulic regimes significantly vary with the regions and the periods. Relative to the reference period of 1986–2005, MAAT has an increasing, stable, and decreasing trend over the TP during the early, middle, and late periods for RCP2.6, respectively. This implies that the impact of MAAT on ALT gradually decreases from the early to the late period under RCP2.6. Under RCP4.5 and RCP8.5, the increasing temperature significantly contributes to ALT. Because of the effect of temperature, the relative significant impact of SM and SND on ALT alternates under RCP2.6 and RCP4.5, and gradually decreases under RCP8.5. Meanwhile, the soil thermal and hydraulic regime varies in middle and high latitudes. MAAT is significantly correlated with ALT under RCP2.6 and RCP8.5 in North America, and SND significantly contributes to ALT under RCP2.6 and RCP4.5. As snow shows a more significant decreasing trend in North America (north of 60°N) than that in other regions, this might mean that a remarkable temperature rise could result in a rapid snowmelt and later snow accumulation (Dyer and Mote 2006; Park et al. 2013), especially for RCP8.5. The interannual variation of snow will become small and the snow would even disappear completely in some regions, which results in the gradually decreasing contribution of SND to ALT from low- to high-emission scenarios (RCPs). However, the thermal and hydraulic regimes in Europe completely differ from those in North America and the TP. SND had strong effects on the permafrost in Europe, although this significant effect only occurs under RCP2.6 and 4.5. With the temperature increasing, the snow cover extent and SND quickly decrease; this affects SM in the shallow soil layer and results in a significant impact of SM and SND on ALT and alternately, occurs under RCP2.6 and RCP4.5, and gradually declines under RCP8.5. Under the high emission (RCP8.5), especially during the late period, most permafrost completely degrades, and ultimately the feature of soil thermal and hydraulic regimes under the lower RCPs (RCP2.6, RCP4.5) will disappear under RCP8.5.

Changes in ALT are mainly related to the increasing MAAT in persistent permafrost regions in the twenty-first century, and with the climate warming, the impact of SM and SND on ALT gradually decreases. In addition, both SND and SM will show a decreasing trend in the twenty-first century (Hua and Chen 2011; Shi and Wang 2015), which will have a significant impact on permafrost status.

The strong relationship between SM and the observed MDFP and ALT over the TP illustrates that SM has a significant effect on MDFP and ALT over the TP. Furthermore, employing Kudryavtsev's model, the estimated permafrost extent based on the GLDAS SM is closer to the result from the IPA than that with a constant SM (Anisimov et al. 1997), which implies that SM is a key factor for estimating permafrost extent. In addition, the relationship between ALT and SM, MAAT, and SND shows that soil thermal and hydraulic regimes varied with the increasing temperature rate and regions. Observations over the period 1986–2005 show that both SM and MAAT significantly contribute to ALT over the TP, and the correlations between them exceed the 90% significance level. In Europe, ALT is mainly related to SND and MAAT, which implies that the soil thermal and hydraulic regimes are different in middle and high latitudes. Over the TP, warming occurs earlier, and the amplitude of warming is larger than that in other regions (Liu and Chen 2000; Zhong et al. 2011); the significant correlation between ALT and MAAT illustrates that MAAT has an important influence on ALT over the TP. Furthermore, SM over the TP has a larger interannual variation than that in North America and Europe (Stocker et al. 2013), which suggests that SM has significant effect on ALT over the TP. However, in Europe, which is rich in snow in winter and spring, snow will gradually melt with the temperature increasing, which results in changes in the insulation effects of snow. Hence, snow and temperature both play an important role in ALT in Europe. In North America, especially in northern areas, ALT has large interannual variations, but few significant trends, which is consistent with results from the IPCC (Stocker et al. 2013). The relationship between ALT and MAAT, SND, and SM indicates that no single variable significantly contributed to ALT in North America. Therefore, the soil thermal and hydraulic regimes vary regionally.

As the global temperature rises, the near-surface permafrost extent will decrease by 81% for RCP8.5 by the end of the twenty-first century (Stocker et al. 2013). The permafrost extent estimated by Kudryavtsev models will disappear along its southern edge. By the end of 2099, there are only 14.7, 12.0, and $5.7 \times 10^6\,km^2$ of permafrost existing for RCP2.6, RCP4.5, and RCP8.5, respectively. Most permafrost completely disappears in the late period for RCP8.5. Relative to the permafrost extent investigated by the IPA, the permafrost area will decrease by 64% by the end of twenty-first century for RCP8.5; this is less than the projected results from IPCC (Stocker et al. 2013), for which the results are based on the surface frost index without considering the impact of SM on permafrost (Slater and Lawrence 2013). This implies that SM is indeed a key factor for permafrost. Although the results have some uncertainty, it seems certain that permafrost will degrade, and that the soil thermal and hydraulic regimes in permafrost regions will undergo significant changes in the twenty-first century.

Relative to the reference period of 1986–2005, ALT consistently shows an increasing trend in persistent permafrost regions under the three RCPs during the twenty-first century, whereas the relationships between ALT and MAAT, SND, and SM illustrate that soil thermal and hydraulic regimes vary with the regions and periods (Figure 10.8). With increasing temperature, ALT is significantly related to MAAT over the TP under RCP4.5 and RCP8.5, whereas the relative significant contribution of SM and SND to ALT only occurs before the middle period under RCP2.6 and RCP4.5, which implies that MAAT is the main contributor to changes in ALT over the TP in the twenty-first century. However, in NA, MAAT only has a significant impact on ALT under the RCP2.6 and RCP8.5. With temperature increasing, snow generally begins to melt, and this leads to a significant contribution of snow to ALT for RCP2.6 and RCP4.5. The most significant changes in snow occurred in northern NA (north of 60°N), with the exception of the TP (Shi and Wang 2015), hence, snow may completely disappear under high emission, hence the impact of SND on ALT gradually decreases under RCP8.5. These imply that the thermal and hydraulic regimes in Europe differ from those in NA and the TP. Although MAAT and snow contributes significantly to ALT over the period 1986–2005, all of the contributions of MAAT, SND, and SM gradually decline in the three periods of the twenty-first century. In Europe, winter and spring are rich in snow; with the climate warming, the significant impact of SND on ALT gradually declines and disappears. In general, with increasing temperature, MAAT will still contribute to changes in ALT, whereas the impact of SM and SND will gradually decline. These changes of relationship among ALT, SM, SND, and MAAT reflect the differences of the thermal and hydraulic regimes over permafrost regions (Figure 10.10).

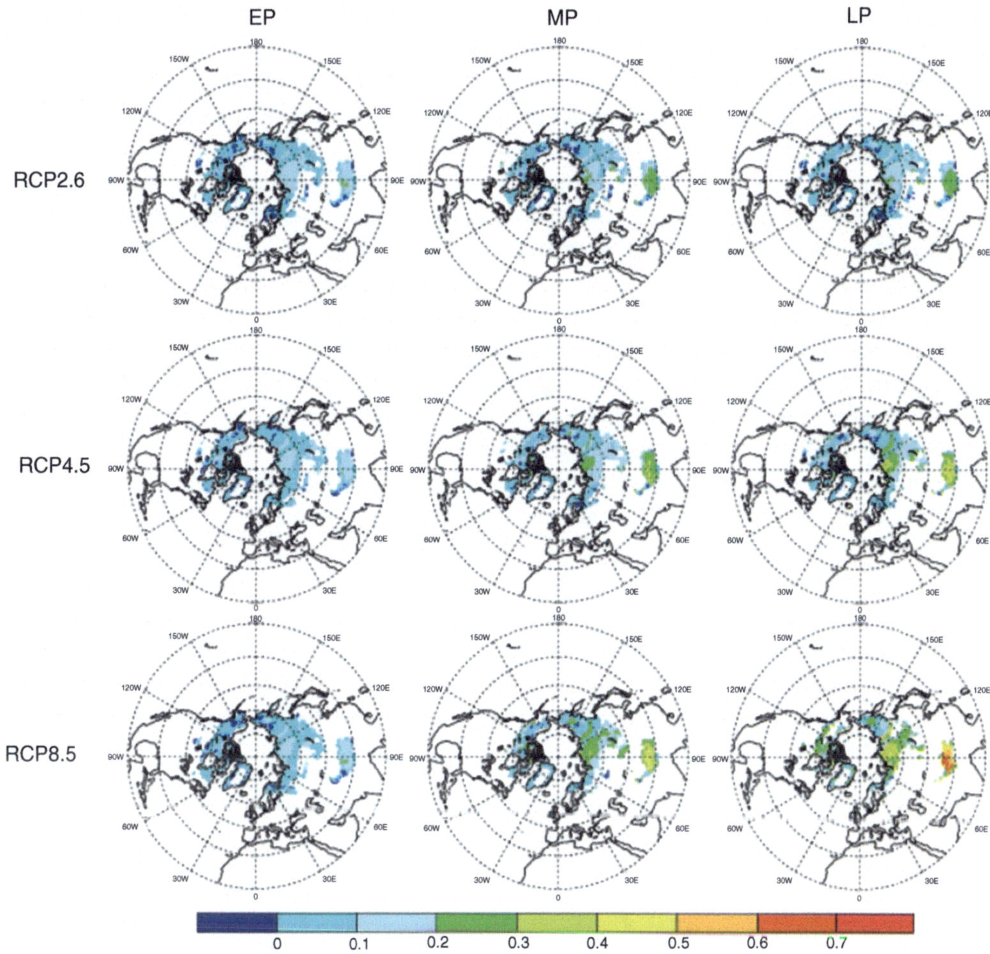

Figure 10.10 The changes in active layer thickness for the three representative concentration pathways (RCP) during the early (EP) (2016–2035), middle (MP) (2046–2065), and late (LP) (2080–2099) periods, relative to the reference period of 1986–2005.

10.3 Changes of Related Environment Response to Climate Change in Cold Regions

10.3.1 Changes of Hydrological Processes

In future, snow and ice cover worldwide is projected to shrink down, and water resources will likely be modified (Meybeck et al. 2001; Barnett et al. 2005; Hagg and Braun 2005; Bocchiola et al. 2010, 2011; Immerzeel et al. 2010; Soncini et al. 2015; Migliavacca et al. 2015). Countries that depend upon water supplied by mountains may face an extinction of permanently cryo-spheric areas, and decrease of the available water because of climate change (Barnett et al. 2005; Viviroli et al. 2007; Kääb et al. 2012; Diolaiuti et al. 2012a, 2012b). Water and food security are therefore at stake (e.g. Aase et al. 2009; Groppelli et al. 2011; Viganò et al. 2015). Climate change in the mountainous regions changes the spatial and temporal distribution of water (e.g. Rohrer et al. 1994;; Laternser and Schneebeli 2002), including the frequency of extreme floods and droughts (e.g. Liu et al. 2003; Bocchiola et al. 2011; Confortola et al. 2013; Bocchiola 2014), further increasing natural hazards.

Snow melt contributes for 20% of the total flow yearly, with monthly variation up to 100%. With an ice cover of 41% of the area, ice melt contributes for 55% of the total flow, and during monsoon ice melt contribution reaches 60% or so, with snow melt reaching 12%. Ragettli et al. (2015) estimated for upper Langtang Basin (350 km^2, 27% ice cover), a 40% share of snow melt (18% during monsoon), and a 26% of ice melt (and 38% during monsoon), maybe due to smaller ice-cover share than here. Palazzoli et al. (2015) studied projected (until 2100) stream flows within the Indrawati Basin in Nepal, closed at Dolalghat, displaying no ice cover. At 2050 they found an average increase of stream flows of +12% during the monsoon season, and +11% at 2090. Excluding the Community Climate System Model version 4, depicting very large increases of precipitation, one has −3%, and −5% at 2050, and 2090, respectively, as due to joint effects of decreasing precipitation, and increasing evapotranspiration. In this sense, in the Dudh Koshi Basin ice-cover presence may delay in time-flow decrease by buffering water resources for 50 years. Immerzeel et al. (2013) investigated future ice volume, and runoff from the Langtang Basin, Nepal, using outputs of all general circulation models from CMIP5 (RCP4.5, RCP8.5). At 2100, in front of a projected decrease of ice volume of −37%, and −53% under RCP4.5 and RCP8.5, they found consistently increasing yearly runoff (vs. 1961–1990) until 2100, from +31% in 2021–2050 for RCP4.5, to +88% in 2071–2100 for RCP8.5 (Figure 10.11).

The ongoing and expected future acceleration of ice melt under global warming will initially increase available water in summer. However, in the wake of large ice mass loss, potentially increased precipitation would not be able to offset the negative water balance. The evolution of such situation toward a reduction of available water is visible elsewhere worldwide. For

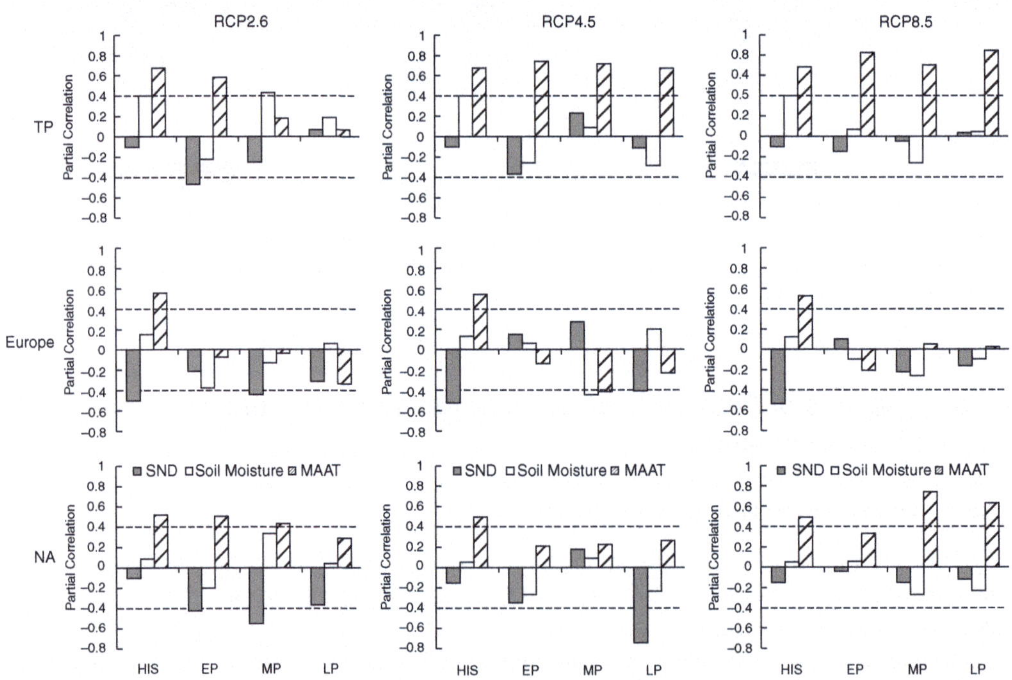

Figure 10.11 The partial correlations between active layer thickness in persistent permafrost regions and mean annual air temperature (MAAT), snow depth (SND), and soil moisture for the three representative concentration pathway (RCP) scenarios (RCP2.6, RCP4.5, and RCP8.5) over the Tibetan Peninsula, Europe, and North America. HIS, EP, MP, and LP refer to the periods 1986–2005, 2016–2035, 2046–2065, and 2080–2099, respectively.

instance, in the central Andes largely shrunken ice cover has already led to decreasing stream flows, and more decrease is expected for the future (e.g. Migliavacca et al. 2015). In the Italian Alps, glaciers shrinking in the last few decades (Diolaiuti et al. 2012a,b), together with the visible decrease of seasonal snow cover (Bocchiola and Diolaiuti 2010) is likely to have decreased summer flows at the highest altitudes (Bocchiola 2014). Projections for Alpine catchments indicated a reduced summer precipitation, and a decreased snowfall during winter, which would be the main reason for large prospective reduction of summer stream flows until 2100 (Confortola et al. 2013; Viganò et al. 2015).

10.3.2 Changes of Ecology

Arctic vegetation distribution is largely controlled by climate, especially by summer temperatures. Summer temperatures have been increasing in the Arctic and this increasing trend is expected to continue. Arctic vegetation has been shown to change in response to the increase of summer temperatures, which in turn affects Arctic fauna, human communities, and industries. In addition, the latitudinal distribution of vegetation is largely determined by the north-south temperature gradient. Climate warming will change the North-South climate gradient and, as a result, the distribution of vegetation will change accordingly. For the whole Arctic, a 5°C increase in summer warmth index (SWI) along the climate gradient corresponded to an increase in normalized differences vegetation index (NDVI) of approximately 0.07. The strongest positive relationship between NDVI and SWI occurred in partially vegetated and graminoid vegetation types. Recently deglaciated areas, areas with many water bodies, carbonate soil areas, and high mountains had lower NDVI values than those predicted by SWI. Plant growth in these areas was limited by substrate factors as well as temperature, and thus is likely to respond less to climate warming than other areas.

Climate change is one of the main factors that will affect biodiversity in the future and may even cause species extinctions. If global temperature increases, then both species turnover will increase, and mean stable area of species will decrease in all biomes (Alkemade et al. 2011). Some studies have suggested that, the most dramatic changes will occur in northern Europe, where more than 35% of the species composition in 2100 will be new for that region, and in southern Europe, where up to 25% of the species now present will have disappeared under the climatic circumstances forecasted for 2100. In Mediterranean scrubland and natural grassland/steppe systems, Arctic and tundra systems species turnover is high, indicating major changes in species composition in these ecosystems. The mean stable area of species decreases mostly in Mediterranean scrubland, grassland/steppe systems and warm mixed forests.

10.3.3 Changes in Disasters

Over the last decades, global warming has led to widespread shrinking of the cold regions. The reduction of snowfall and amounts of snow on the ground are increasingly observed, notably in Europe. Likewise, climate models project further drastic reductions in snow cover extent and duration, especially in low-elevation mountain environments. Changes in snowfall and snow cover characteristics already have direct consequences on the occurrence and magnitude of natural hazards. Among these, snow avalanches are not only a major threat to mountain populations and their assets, but also the climate change is considered as a major role through changes in rain–snow partitioning of precipitation and snow cover characteristics.

In order to have a deeper understanding of the formation process of ice avalanches and provide a corresponding basis for the prediction of ice avalanches in the future, You et al. (2020)

summarized the causes of ice avalanches in the TP as follows: (i) global warming and increasing rainfall are a major factor causing ice avalanches; (ii) most of the ice avalanches in the TP are caused by climate change, so climate change is the main reason for the frequent occurrence of ice avalanches in the plateau; (iii) the increase in the annual mean temperature over the TP causes the glaciers to begin to melt at an accelerated rate; (iv) the glacier ablation increases the meltwater between the contact surfaces of the glaciers, which further increases the fissures of the glaciers and destroys the integrity of the glaciers; (v) the increase in rainfall is conducive to the accumulation of substances inside the glacier, which increases the mass of the glacier and increases the driving force of the glacier.

In recent years, global warming caused by human activities and related environmental changes has led to the melting of glaciers in high mountain areas, and the probability of ice avalanches has increased. Glacier jumping is a phenomenon of rapid movement of glaciers, often accompanied by the occurrence of related disasters, so it is also called "catastrophic glacier advance." With the warming of the global climate, the frequency of glacier surges over the TP has been increasing. After the glacier jumps, there are generally two results. One is that the glacier still maintains a relatively complete structure after a large advance, and there is no phenomenon of avalanche, but the advancing glacier may cause damage to downstream villages, towns, transportation facilities, etc. The second result is that in the process of glacier jumping, its internal structure changes, and the ice body is scattered, forming a series of glacial disasters, including ice-avalanche disasters, debris-flow disasters. In addition, when the disaster occurred in summer, cracks appeared in the glacier due to ablation, and finally it gradually became unstable under the action of gravity. The throbbing glaciers carry a large amount of ice water and a mixture of rocks and move in the valley at extremely high speeds, eventually causing heavy-casualty ice avalanches.

11

Challenges and Outlook

This chapter will discuss some challenging issues that are involved but have not been entirely solved in models and modeling. These include determinant signals in cold regions for seasonal climate prediction, land-surface assimilation in a climate model, and the possible impacts of climate change on economic and social development, which are worth investigating in the future.

11.1 Challenges for Future Study

It is difficult to understand the climate in cold regions, since it is difficult to describe the climate as a whole. Cold regions are discontinuous in space, and the warm–cold duration is also different. Defining the extent of cold regions has to consider many factors together.

Interaction between land and atmosphere in cold regions is certainly a stationary external force for the global atmosphere. For instance, most of the middle-latitude zone in the northern hemisphere can be classified as a cold region, while an air parcel leaves the ocean (e.g. Atlantic Ocean), which must pass across the Eurasian continent, and will be influenced by the interaction between land and atmosphere in cold regions. Because the cold region has distinct annual variations, and the annual variations of snow cover, frozen ground, and soil moisture are different, thus, intensity and onset time of land–atmosphere interaction in cold regions are also different. The land–atmosphere interaction in the cold region also depends on the transition of atmospheric patterns from winter to summer. However, how to quantitatively describe the physical processes regarding the interaction between land and atmosphere is an open field.

It is undoubtable that the external forcing from cold region is essential for general circulation. Although separating the internal variability of the atmosphere and external forcing for general circulation is complex, land–atmosphere interaction in the cold region is undoubtedly crucial for the climate system around the globe.

Due to cold regions being located in the transition zone between cropping areas and nomadic areas, the changes in cropping areas or nomadic areas must influence the interaction between land and atmosphere. As a result, meteorological elements in cold regions must be changed accordingly, which will change the annual and seasonal mean atmospheric status in the transition zone of cropping and nomadic areas. Therefore, vegetation changes in cold regions should be noted.

As the temperature is generally low in cold regions, the evaporation is relatively weak in them; thus, the hydrological process in cold regions is unique, and estimation of hydrological elements in hydrology balance is generally underestimated, such as the surface runoff and soil moisture.

Climatology in Cold Regions, First Edition. Chenghai Wang.

Some cold regions are separated in geospatially. Where is the "hot pot" of the interaction between land and atmosphere which still lacks enough proof, especially for different spatial scales, time sequences, and coupling of land and atmosphere? So far, mechanisms of the coupling of land, as well as the parameterization in some atmospheric physical process have not been resolved; how the land-surface processes interact with the atmosphere through the boundary layer is unclear because of the insufficient observations and unresolved simulations uncertainty.

It is worth noting that many results come from numerical experiments. Due to a general lack of observations in cold regions, it is difficult to carry out meticulous research; developing an optimal physical parameterization scheme is also tricky. To some extent, the optimal solution of the interaction between land and atmosphere in cold regions can remarkably improve the performances of numerical models. Cold regions are mainly distributed in high latitude and altitudes areas over the globe, where *in situ* observations are relatively sparse; obtaining reliable and abundant observations is a challenging issue. In the current stage, satellite data assimilation in cold regions might be a better choice.

11.2 Conclusion

This volume has introduced the essential characteristics of climate, land-surface processes, land-air interactions, model and modeling, hydrological and ecological changes, and disasters in cold regions. The main conclusions are as follows:

1) The climate in cold regions should be classified separately from the traditional climate category. In the current climate-change scenario, climate change in cold regions has attracted much attention, which is also the last refuge from climate warming disasters for humans.
2) Interactions between land and atmosphere in the cold regions have their particularity. Snow cover and frozen ground dominate the land-surface process in cold regions; these processes affect the general circulation and climate at either regional or global scales by inducing soil moisture anomalies. The high albedo of abundant snow cover is generally the source of cooling, which has an effect on the atmosphere that cannot be ignored. Surface runoff and infiltration drive the variations of soil moisture, which would promote soil saturation, and then lead to a slower impact on the atmosphere through changing surface evaporation. We proposed and demonstrated that soil moisture anomalies in cold regions could persist for two to three seasons by the freezing–thawing process in the winter half-year. These imply that the land surface in cold regions can provide a "signal" ahead for two to three seasons for climate prediction.
3) A large extent of the cold region is the distinct external force for the climate system. The diabatic heating of the land surface will affect the air that passes across the cold region. Land–atmosphere interaction in cold regions might produce a stationary wave effect on atmospheric movement, and the combination of impacts will change the general circulation patterns at 200 and 500 hPa. In other words, land–atmosphere interaction in cold regions will influence the duration and the propagation of stream patterns in the transitional period from winter to summer.
4) The performance of the land-surface model in cold regions evidently influences the general circulation model's performance. In the current stage, the isothermal land-surface model fails to describe the coupling process of soil water and heat transport. It also ignores the movement of water vapor in soil; a fully coupled water–heat transport land-surface model is essential to understand soil's hydrothermal regime in cold regions.

5) There is a general lack of modeling for most physical processes and geographic components. The parameterizations of snow in the cold region are still defective; further observation and theoretical studies are expected. Retrievals of satellite/radar observation are the preferred approach. Combining multi-source observations such as satellite/radar retrievals and *in situ* observations can deepen our understanding of the snow process.
6) Parameterizations of the frozen ground process have been significantly improved, especially portraying the beginning of soil freezing and thawing. The simulation of soil temperature and soil moisture is still worse in land-surface models; the quick changes of soil moisture depend on the daily changes of soil temperature, which might be resolved through a technical method, not only by theoretical solutions.

Appendix

A. Soil Hydraulic and Thermal Properties

The isothermal hydraulic conductivity, $K_{L\psi}$ ($\mathrm{m\,s^{-1}}$), for liquid-phase flux due to a hydraulic potential gradient was calculated in the same manner as in Community Land Model version 4.5 (CLM 4.5) (Swenson et al. 2012; Oleson et al. 2013):

$$K_{L\psi} = 10^{-' F_{ice}} K_{sat}\left(\frac{\theta_l}{\theta_{sat}}\right)^{2B+3} \tag{A.1}$$

where $F_{ice} = \frac{\theta_i}{\theta_{sat}}$, $\Omega = 6$, K_{sat} is the saturated hydraulic conductivity, θ_{sat} is the saturated volumetric water content, and B is a parameter defined by Clapp and Hornberger (1978) and Cosby et al. (1984).

The thermal hydraulic conductivity K_{LT} ($\mathrm{m^2\,K^{-1}\,s^{-1}}$) for liquid-phase flux due to a temperature gradient that was calculated as

$$K_{LT} = K_{L\psi}\left(\psi G_{wT}\frac{1}{\gamma_0}\frac{d\gamma}{dT}\right) \tag{A.2}$$

where G_{wT} is the gain factor, γ ($\mathrm{g\,s^{-2}}$) is the surface tension of soil water, $\gamma = 75.6 - 0.1425 \cdot T - 2.38 \times 10^{-4} \cdot T^2$, and $\gamma_0 = 71.89\,\mathrm{g\,s^{-2}}$.

The isothermal water vapor hydraulic conductivity, $K_{v\psi}$ ($\mathrm{m\,s^{-1}}$), was calculated according to Fayer (2000) as:

$$K_{v\psi} = \frac{D\rho_{vs}Mg}{\rho_w RT}H_R = \alpha\left(\theta_s - \theta\right)\frac{D_a\rho_{vs}Mg}{\rho_w RT}H_R \tag{A.3}$$

where α is the tortuosity factor, the quantity $(\theta_s - \theta)$ represents the air-filled porosity, M is the molecular weight of water ($M = 0.018\,\mathrm{kg\,mol^{-1}}$), g is the gravitational constant ($g = 9.8\,\mathrm{m\,s^{-2}}$), and R is the gas constant ($R = 8.3145\,\mathrm{J\,mol^{-1}\,K^{-1}}$).

D_a is the diffusivity of water vapor in air calculated as

$$D_a = 2.3 \times 10^{-5}\left(\frac{T}{273.16}\right)^{1.75} \tag{A.4}$$

Climatology in Cold Regions, First Edition. Chenghai Wang.

ρ_{vs} (kg m^{-3}) is the saturated vapor density calculated by the modified empirical equation

$$\rho_{vs} = \exp\left[46.440973 - \frac{6790.4985}{T} - 6.02808 \ln T\right] \times 1000 \quad (A.5)$$

the density of water $\rho_w = 1000$ kg m^{-3}.

Relative humidity, H_R (%), can be determined (Campbell 1985) using

$$H_R = \exp\left[\frac{\psi M g}{RT}\right] \quad (A.6)$$

The thermal water vapor hydraulic conductivities, K_{vT} (m^2 K^{-1} s^{-1}), were calculated according to Fayer (2000) as:

$$K_{vT} = \frac{D\eta H_R}{\rho_w}\frac{d\rho_{vs}}{dT} = \alpha\left(\theta_s - \theta\right)\frac{D_a \eta H_R}{\rho_w}\frac{d\rho_{vs}}{dT} \quad (A.7)$$

where η is an enhancement factor and the derivative of ρ_{vs} with respect to temperature is:

$$\frac{d\rho_{vs}}{dT} = \rho_{vs}\left[\frac{6790.4985}{T} - 6.02808\right]\frac{1}{T} \quad (A.8)$$

The volumetric latent heating (LH) of vaporization of water L_0 (J m^{-3}) was calculated as

$$L_0 = \left[2.501 \times 10^6 - 2369.2\left(T - T_f\right)\right]\rho_w \quad (A.9)$$

where T_f is the freezing temperature.

B. Numerical Implementation

The structure of soil layer in the Full Couple Scheme (FCS) is consistent with the CLM4.5 (Oleson et al. 2012), which is divided into 15 layers, the node depths of the upper ten layers are 0.0071, 0.0279, 0.0623, 0.1189, 0.2122, 0.3661, 0.6198, 1.0380, 1.7276, and 2.8646 m.

In FCS, the equations are solved as one by one as shown in CLM4.5. the soil temperature calculation including the phase change process is performed first, and then soil moisture calculation is executed.

The numerical solution to the modified Richards equation was the same as the current scheme in CLM4.5 with implicit expression for time discretization and backward difference for space discretization (details can be found in Section 7.4 of the technical description of CLM4.5) (Oleson et al. 2012). The finite-difference forms of the water fluxes and partial derivatives were calculated as:

$$\begin{aligned} q_{i-1} = &-\left(K_{L\psi,i-1}\left[\frac{\left(\psi_{i-1} - \psi_i\right) - \left(\psi_{E,i-1} - \psi_{E,i}\right)}{z_i - z_{i-1}}\right] + K_{LT,i-1}\frac{T_{i-1} - T_i}{z_i - z_{i-1}}\right) \\ &-\left(K_{v\psi,i-1}\frac{\psi_{i-1} - \psi_i}{z_i - z_{i-1}} + K_{vT,i-1}\frac{T_{i-1} - T_i}{z_i - z_{i-1}}\right) \end{aligned} \quad (A.10)$$

$$q_i = -\left(K_{L\psi,i}\left[\frac{(\psi_i - \psi_{i+1}) - (\psi_{E,i} - \psi_{E,i+1})}{z_{i+1} - z_i}\right] + K_{LT,i}\frac{T_i - T_{i+1}}{z_{i+1} - z_i}\right) - \left(K_{v\psi,i}\frac{\psi_i - \psi_{i+1}}{z_{i+1} - z_i} + K_{vT,i}\frac{T_i - T_{i+1}}{z_{i+1} - z_i}\right) \tag{A.11}$$

$$\frac{\partial q_{i-1}}{\partial \theta_{i-1}} = -\left(\frac{K_{L\psi,i-1}}{z_i - z_{i-1}}\frac{\partial \psi_{i-1}}{\partial \theta_{i-1}} + \frac{\partial K_{L\psi,i-1}}{\partial \theta_{i-1}}\left[\frac{(\psi_{i-1} - \psi_i) - (\psi_{E,i-1} - \psi_{E,i})}{z_i - z_{i-1}}\right] + \frac{\partial K_{LT,i-1}}{\partial \theta_{i-1}}\frac{T_{i-1} - T_i}{z_i - z_{i-1}}\right) - \left(\frac{K_{v\psi,i-1}}{z_i - z_{i-1}}\frac{\partial \psi_{i-1}}{\partial \theta_{i-1}} + \frac{\partial K_{v\psi,i-1}}{\partial \theta_{i-1}}\frac{\psi_{i-1} - \psi_i}{z_i - z_{i-1}} + \frac{\partial K_{vT,i-1}}{\partial \theta_{i-1}}\frac{T_{i-1} - T_i}{z_i - z_{i-1}}\right) \tag{A.12}$$

$$\frac{\partial q_{i-1}}{\partial \theta_i} = -\left(-\frac{K_{L\psi,i-1}}{z_i - z_{i-1}}\frac{\partial \psi_i}{\partial \theta_i} + \frac{\partial K_{L\psi,i-1}}{\partial \theta_i}\left[\frac{(\psi_{i-1} - \psi_i) - (\psi_{E,i-1} - \psi_{E,i})}{z_i - z_{i-1}}\right] + \frac{\partial K_{LT,i-1}}{\partial \theta_i}\frac{T_{i-1} - T_i}{z_i - z_{i-1}}\right) - \left(-\frac{K_{v\psi,i-1}}{z_i - z_{i-1}}\frac{\partial \psi_i}{\partial \theta_i} + \frac{\partial K_{v\psi,i-1}}{\partial \theta_i}\frac{\psi_{i-1} - \psi_i}{z_i - z_{i-1}} + \frac{\partial K_{vT,i-1}}{\partial \theta_i}\frac{T_{i-1} - T_i}{z_i - z_{i-1}}\right) \tag{A.13}$$

$$\frac{\partial q_i}{\partial \theta_i} = -\left(\frac{K_{L\psi,i}}{z_{i+1} - z_i}\frac{\partial \psi_i}{\partial \theta_i} + \frac{\partial K_{L\psi,i}}{\partial \theta_i}\left[\frac{(\psi_i - \psi_{i+1}) - (\psi_{E,i} - \psi_{E,i+1})}{z_{i+1} - z_i}\right] + \frac{\partial K_{LT,i}}{\partial \theta_i}\frac{T_i - T_{i+1}}{z_{i+1} - z_i}\right) - \left(\frac{K_{v\psi,i}}{z_{i+1} - z_i}\frac{\partial \psi_i}{\partial \theta_i} + \frac{\partial K_{v\psi,i}}{\partial \theta_i}\frac{\psi_i - \psi_{i+1}}{z_{i+1} - z_i} + \frac{\partial K_{vT,i}}{\partial \theta_i}\frac{T_i - T_{i+1}}{z_{i+1} - z_i}\right) \tag{A.14}$$

$$\frac{\partial q_i}{\partial \theta_{i+1}} = -\left(-\frac{K_{L\psi,i}}{z_{i+1} - z_i}\frac{\partial \psi_{i+1}}{\partial \theta_{i+1}} + \frac{\partial K_{L\psi,i}}{\partial \theta_{i+1}}\left[\frac{(\psi_i - \psi_{i+1}) - (\psi_{E,i} - \psi_{E,i+1})}{z_{i+1} - z_i}\right] + \frac{\partial K_{LT,i}}{\partial \theta_{i+1}}\frac{T_i - T_{i+1}}{z_{i+1} - z_i}\right) - \left(-\frac{K_{v\psi,i}}{z_{i+1} - z_i}\frac{\partial \psi_{i+1}}{\partial \theta_{i+1}} + \frac{\partial K_{v\psi,i}}{\partial \theta_{i+1}}\frac{\psi_i - \psi_{i+1}}{z_{i+1} - z_i} + \frac{\partial K_{vT,i}}{\partial \theta_{i+1}}\frac{T_i - T_{i+1}}{z_{i+1} - z_i}\right) \tag{A.15}$$

The heat conduction equation was solved using the Crank–Nicolson method, which is the same procedure used in the current scheme in CLM4.5 (details can be found in Section 6.1 of the technical description of CLM4.5) (Oleson et al. 2012). The finite-difference forms of heat fluxes between two layers were calculated as:

$$F_{i-1} = -\lambda_{i-1}\left(\frac{T_{i-1} - T_i}{z_i - z_{i-1}}\right) + C_w q_{L,i-1}\left(\frac{T_{i-1} + T_i}{2}\right) + C_v q_{v,i-1}\left(\frac{T_{i-1} + T_i}{2}\right) + \left(\frac{L_{0,i-1} + L_{0,i}}{2}\right)q_{v,i-1} \tag{A.16}$$

$$F_i = -\lambda_i\left(\frac{T_i - T_{i+1}}{z_{i+1} - z_i}\right) + C_w q_{L,i}\left(\frac{T_{i+1} + T_i}{2}\right) + C_v q_{v,i}\left(\frac{T_{i+1} + T_i}{2}\right) + \left(\frac{L_{0,i+1} + L_{0,i}}{2}\right)q_{v,i} \tag{A.17}$$

Bibliography

Aase, T.H., Chaudhary, R.P., and Vetaas, O.R. (2009). Farming flexibility and food security under climatic uncertainty: Manang, Nepal Himalaya. *Area* 42 (2): 228–238. https://doi.org/10.1111/j.1475-4762.2009.00911.x.

Abdalati, W. and Steffen, K. (1998). Accumulation and hoar effects on microwave emission in the Greenland ice-sheet dry-snow ezones. *J. Glaciol.* 44 (148): 523–531. https://doi.org/10.3189/S0022143000002045.

Ablimitjan-Ablikim, C.C., Yusup, A., and Barigul, W. (2015). The temporal and spatial distribution features of snowmelt flood events in Xinjiang from 2001 to 2012. *J. Glaciol. Geocryol.* 37 (1): 226–232. (in Chinese).

ACIA (2004). *Impacts of a Warming Arctic: Arctic Climate Impact Assessment*. New York: Cambridge University Press http://www.acia.uaf.edu.

Adam, J.C. and Lettenmaier, D.P. (2008). Application of new precipitation and reconstructed streamflow products to streamflow trend attribution in Northern Eurasia. *J. Clim.* 21 (8): 1807–1828. https://doi.org/10.1175/2007JCLI1535.1.

Adam, J.C., Haddeland, I., Su, F., and Lettenmaier, D.P. (2007). Simulation of reservoir influences on annual and seasonal streamflow changes for the Lena, Yenisei and Ob' Rivers. *J. Geophys. Res.* 112: D24: 114. https://doi.org/10.1029/2007JD008525.

Adam, J.C., Hamlet, A.F., and Lettenmaier, D.P. (2009). Implications of global climate change for snowmelt hydrology in the twenty-first century. *Hydrol. Process.* 23: 962–972. https://doi.org/10.1002/hyp.7201.

Aggarwal, P.K., Joshi, P.K., Ingram, J.S.I., and Gupta, R.K. (2004). Adapting food systems of the Indo-Gangetic plains to global environmental change: key information needs to improve policy formulation. *Environ. Sci. Policy* 7: 487–498. https://doi.org/10.1016/j.envsci.2004.07.006.

Albergel, C., P. de Rosnay, C. Gruhier, J.Munoz-Sabater, S. Hasenauer, L. Isaksen, Y. Kerr, and W. WagnAlbergel, C., de Rosnay, P., Gruhier, C., et al. (2012). Evaluation of remotely sensed and modelled soil moisture products using global ground-based in situ observations. *Remote Sens. Environ.* 118: 215–226.

Albrecht, B.A. (1989). Aerosols, cloud microphysics, and fractional cloudiness. *Science* 245: 1227–1230. https://doi.org/10.1126/science.245.4923.1227.

Alean, J. (1985). Ice avalanches: some empirical information about their formation and reach. *J. Glaciol.* 31 (109): 324–333.

Algesten, G., Sobek, S., Bergström, A.-K. et al. (2003). Role of lakes for organic carbon cycling in the boreal zone. *Glob. Chang. Biol.* 10: 141–147. https://doi.org/10.1111/j.1365-2486.2003.00721.x.

Climatology in Cold Regions, First Edition. Chenghai Wang.

Alkemade, R., Bakkenes, M., and Eickhout, B. (2011). Towards a general relationship between climate change and biodiversity: an example for plant species in Europe. *Reg. Environ. Change* 11 (1): 143–150.

Ambaum, M.H.P., Hoskins, B.J., and Stephenson, D.B. (2002). Arctic oscillation or North Atlantic oscillation? *J. Clim.* 14: 3495–3507. https://doi.org/10.1175/1520-0442(2001)014<3495:AOONA O>2.0.CO;2.

Andrews, D.G., Holton, J.R., and Leovy, C.B. (1987). *Middle Atmosphere Dynamics*. Orlando, FL: Academic Press.

Anisimov, O.A. and Nelson, F.E. (1996). Permafrost distribution in the northern hemisphere under scenarios of climatic change. *Glob. Planet. Chang.* 14: 59–72. https://doi.org/10.1016/0921-8181(96)00002-1.

Anisimov, O.A., Shiklomanov, N.I., and Nelson, F.E. (1997). Global warming and active-layer thickness: results from transient general circulation models. *Glob. Planet. Chang.* 15: 61–77. https://doi.org/10.1016/S0921-8181(97)00009-X.

Arndal, M.F., Illeris, L., Michelsen, A. et al. (2009). Seasonal variation in gross ecosystem production, plant biomass, and carbon and nitrogen pools in five high arctic vegetation types. *Arct. Antarct. Alp. Res.* 41 (2): 164–173. https://doi.org/10.1657/1938-4246-41.2.164.

Aurela, M., Tuovinen, J.P., and Laurila, T. (2001). Net CO_2 exchange of a subarctic mountain birch ecosystem. *Theor. Appl. Climatol.* 70: 135–148. https://doi.org/10.1007/s007040170011.

Bai, J., Ouyang, H., Xiao, R. et al. (2010). Spatial variability of soil carbon, nitrogen, and phosphorus content and storage in an alpine wetland in the Qinghai-Tibet plateau, China. *Soil Res.* 48: 730–736. https://doi.org/10.1071/SR09171.

Bajtes, N.H. (1996). Total carbon and nitrogen in the soils of the world. *Eur. J. Soil Sci.* 47: 151–163. https://doi.org/10.1111/j.1365-2389.1996.tb1386.x.

Baker, D.F., Law, R.M., Gurney, K.R. et al. (2006). TransCom 3 inversion intercomparison: impact of transport model errors on the interannual variability of regional CO2 fluxes, 1988–2003. In: *Global Biogeochemical Cycles*, vol. 20, GB 1002. https://doi.org/10.1029/2004GB002439.

Baldwin, M.P. and Dunkerton, T.J. (1999). Propagation of the Arctic Oscillation from the stratosphere to the troposphere. *J. Geophys. Res.* 104: 30937–30946. https://doi.org/10.1029/1999JD900445.

Baldwin, M.P. and Dunkerton, T.J. (2001). Stratospheric harbingers of anomalous weather regimes. *Science* 294: 581–584. https://doi.org/10.1126/science.1063315.

Baldwin, M.P. and Holton, J.R. (1988). Climatology of the stratosphere polar vortex and planetary wave breaking. *J. Amos. Sci.* 45: 1123–1142. https://doi.org/10.1175/1520-0469(1988)045<1123:COTSP V>2.0.CO;2.

Ballesteros-Cánovas, J.A., Trappmann, D., Madrigal-González, J. et al. (2018). Climate warming enhances snow avalanche risk in the Western Himalayas. *Proc. Natl. Acad. Sci.* 115 (13): 3410–3415. https://doi.org/10.1073/pnas.1716913115.

Balshi, M.S., McGuire, A.D., Zhuang, Q. et al. (2007). The role of historical fire disturbance in the carbon dynamics of the pan-boreal region: a process-based analysis. *J. Geophys. Res.* 112: G02029. https://doi.org/10.1029/2006JG000380.

Bamzai, A.S. and Shukla, J. (1999). Relation between Eurasian snow cover, snow depth, and the Indian summer monsoon: an observational study. *J. Clim.* 12: 3117–3132. https://doi.org/10.1175/1520044 2(1999)012<3117:RBESCS>2.0.CO;2.

Ban-Weiss, G.A., Bala, G., Cao, L. et al. (2011). Climate forcing and response to idealized changes in surface latent and sensible heat. *Environ. Res. Lett.* 6 (3): 034032. https://doi.org/10.1088/174 8-9326/6/3/034032.

Bao, H., Koike, T., Yang, K. et al. (2016). Development of an enthalpy-based frozen soil model and its validation in a cold region in China. *J. Geophys. Res. Atmos.* 121: 5259–5280. https://doi.org/ 10.1002/2015JD024451.

Bao, H., Yang, K., and Wang, C. (2017). Characteristics of GLDAS soil-moisture data on the Tibet Plateau. *Sci. Cold Arid Reg.* 9 (2): 127–141. https://doi.org/10.3724/SP.J.1226.2017.00127.

Barnett, T.P., Dümenil, L., Schlese, U. et al. (1989). The effect of Eurasian snow cover on regional and global climate variations. *J. Atmos. Sci.* 46: 661–686. https://doi.org/10.1175/1520-0469(1989)046<0661:TEOESC>2.0.CO;2.

Barnett, T.P., Adam, J.C., and Lettenmaier, D.P. (2005). Potential impacts of a warming climate on water availability in snow-dominated regions. *Nature* 438: 303–309. https://doi.org/10.1038/nature04141.

Barnston, A.G., Tippett, M.K., L'Heureux, M.L. et al. (2012). Skill of real-time seasonal ENSO model predictions during 2002–11: is our capability increasing? *Bull. Am. Meteorol. Soc.* 93 (5): 631–651.

Barry, R.G. and Gan, T.Y. (2022). *The Global Cryosphere: Past, Present, and Future*. Cambridge University Press.

Bastviken, D., Cole, J., Pace, M., and Tranvik, L. (2004). Methane emissions from lakes: dependence of lake characteristics, two regional assessments, and a global estimate. *Glob. Biogeochem. Cycles* 18: GB4009. https://doi.org/10.1029/2004GB002238.

Bates, B.C., Kundzewicz, Z.W., Wu, S., and Palutik, J.P. (2008). Climate change and water. IPCC Secretariat, Geneva, Technical Paper of the Intergovernmental Panel on Climate Change, 210 pp.

Baumann, S. and Stefan Winkler, S. (2010). Parameterization of glacier inventory data from Jo Tunheimen/Norway in comparison to the European Alps and the Southern Alps of New Zealand. *Erdkunde* 64: 2155–2177.

Beltaos, S. (1990). Fracture and breakup of river ice cover. *Can. J. Civ. Eng.* 17 (2): 173–183.

Beltaos, S. and Burrell, B.C. (2015). Hydrotechnical advances in Canadian river ice science and engineering during the past 35 years. *Can. J. Civil Eng.* 42 (9): 583–591. https://doi.org/10.1139/cjce-2014-0540.

Beltaos, S. and Burrell, B.C. (2021). Effects of river-ice breakup on sediment transport and implications to stream environments: a review. *Water*. 13 (18): 2541.

Beltaos, S. and Prowse, T.D. (2001). Climate impacts on extreme ice-jam events in Canadian rivers. *Hydrol. Sci. J.* 46 (1): 157–181. https://doi.org/10.1080/02626660109492807.

Beltaos, S. and Prowse, T. (2009). River-ice hydrology in a shrinking cryosphere. *Hydrol. Processes* 23 (1): 122–144. https://doi.org/10.1002/hyp.7165.

Benn, D.I. and Owen, L.A. (1998). The role of the Indian summer monsoon and the mid-latitude westerlies in Himalayan glaciation: review and speculative discussion. *J. Geol. Soc. Lond.* 155: 353–363. https://doi.org/10.1144/gsjgs.155.2.0353.

Bennett, M. and Glasser, N.F. (2011). *Glacial Geology: Ice Sheets and Landforms*. John Wiley & Sons.

Bense, V.F., Kooi, H., Ferguson, G., and Read, T. (2012). Permafrost degradation as a control on hydrogeological regime shifts in a warming climate. *J. Geophys. Res.* 117: F03036. https://doi.org/10.1029/2011JF002143.

Bergamaschi, P., Frankenberg, C., Meirink, J.F. et al. (2007). Satellite chartography of atmospheric methane from SCIAMACHY on board ENVISAT: 2. Evaluation based on inverse model simulations. *J. Geophys. Res.* 112: D02304. https://doi.org/10.1029/2006JD007268.

Bernstein, B.C. (2000). Regional and local influences on freezing drizzle, freezing rain, and ice pellet events. *Weather Forecasting* 15 (5): 485–508. https://doi.org/10.1175/1520-0434(2000)015<0485:RALIOF>2.0.CO;2.

Bhatt, U.S., Walker, D.A., Raynolds, M.K. et al. (2010). Circumpolar Arctic tundra vegetation change is linked to sea ice decline. *Earth Interact.* 14 (8): 1–20. https://doi.org/10.1175/2010EI315.1.

Bi, H.Y. and Ma, J.W. (2015). Evaluation of simulated soil moisture in GLDAS using in-situ measurements over the Tibetan Plateau. In: *Geosciences and Remote Sensing Symposium*, 4825–4828. https://doi.org/10.1109/IGARSS.2015.7326910.

Bian, L.G. and Lin, X.C. (2005). Changes of Antarctic sea ice in recent 30 years. *Polar research* (in Chinese).

Bittelli, M., Ventura, F., Campbell, G.S. et al. (2008). Coupling of heat, water vapor, and liquid water fluxes to compute evaporation in bare soils. *J. Hydrol.* 362 (3): 191–205. https://doi.org/10.1016/j.jhydrol.2008.08.014.

Björnsson, H. (2004). Glacial lake outburst floods in mountain environments. *Mountain Geomorphology*, 155–184.

Black, M.L. and Willoughby, H.E. (1992). The concentric eyewall cycle of hurricane Gilbert. *Mon. Weather Rev.* 120: 947–957. https://doi.org/10.1175/1520-0493(1992)120<0947:TCECOH>2.0.CO;2.

Black, P.B. and Tice, A.R. (1989). Comparison of soil freezing and soil water curve data for Windsor sandy loam. *Water Resour. Res.* 25: 2205–2210. https://doi.org/10.1029/WR025i010p02205.

Blanford, H.F. (1884). On the connection of the Himalaya snowfall with dry winds and seasons of drought in India. *Proc. Roy. Soc.* 37: 3–22. https://doi.org/10.1098/rspl.1884.0003.

Bo, Y., Li, X., and Wang, C. (2014). Seasonal characteristics of the interannual variations centre of the Tibetan Plateau snow cover. *Journal of Glaciol. Geocryol.* 36: 1353–1362. https://doi.org/10.7522/j.issn.1000-0240.2014.0162.(in Chinese).

Bocchiola, D. (2007). Use of scale recursive estimation for multisensor rainfall assimilation: a case study using data from TRMM (PR and TMI) and NEXRAD. *Adv. Water Resour.* 30: 2354–2372. https://doi.org/10.1016/j.advwatres.2007.05.012.

Bocchiola, D. (2014). Long term (1921–2011) changes of Alpine catchments regime in Northern Italy. *Adv. Water Resour.* 70: 51–64. https://doi.org/10.1016/j.advwatres.2014.04.017.

Bocchiola, D. and Diolaiuti, G. (2010). Evidence of climate change within the Adamello Glacier of Italy. *Theor. Appl. Climatol.* 100 (3–4): 351–369. https://doi.org/10.1007/s00704-009-0186-x.

Bocchiola, D. and Diolaiuti, G. (2013). Recent (1980–2009) evidence of climate change in the upper Karakoram, Pakistan. *Theor. Appl. Climatol.* 113 (3–4): 611–641. https://doi.org/10.1007/s00704-012-0803-y.

Bocchiola, D. and Rosso, R. (2007). The distribution of daily snow water equivalent in the Central Italian Alps. *Adv. Water Resour.* 30: 135–147. https://doi.org/10.1016/j.advwatres.2006.03.002.

Bocchiola, D., Mihalcea, C., Diolaiuti, G. et al. (2010). Flow prediction in high altitude ungauged basins: a case study in the Italian Alps (Pantano Basin, Adamello Group). *Adv. Water Resour.* 33: 1224–1234. https://doi.org/10.1016/j.advwatres.2010.06.009.

Bocchiola, D., Diolaiuti, G., Soncini, A. et al. (2011). Prediction of future hydrological regimes in poorly gauged high altitude basins: the case study of the upper Indus, Pakistan. *Hydrol. Earth Syst. Sci.* 15: 2059–2075. https://doi.org/10.5194/hess-15-2059-2011.

Bocchiola, D., Senese, A., Mihalcea, C. et al. (2015). An ablation model for debris covered ice: the case study of Venerocolo Glacier (Italian Alps). *Geogr. Fis. Din. Quat., GFDQ* 38 (2): 113–128.

Boccolari, M. and Parmiggiani, F. (2018). Trends and variability of cloud fraction cover in the Arctic, 1982–2009. *Theor. Appl. Climatol.* 132: 739–749. https://doi.org/10.1007/s00704-017-2125-6.

Boer, D.G., Morrison, H., Shupe, M.D., and Hildner, R. (2011). Evidence of liquid-dependent ice nucleation in high-latitude stratiform clouds from surface remote sensors. *Geophys. Res. Lett.* 38: L01803.

Bohn, T.J., Livneh, B., Nijssen, B. et al. (2013). Global validation of MTCLIM and related algorithms for uncoupled ecological and hydrological model meteorological forcings. *Agric. For. Meteorol.* 176: 38–49. https://doi.org/10.1016/j.agrformet.2013.03.003.

Bonan, G. (2008). *Ecological Climatology*. Cambridge: Cambridge University Press.

Bookhagen, B. (2012). Hydrology: Himalayan groundwater. *Nat. Geosci.* 5 (2): 97–98. https://doi.org/10.1038/ngeo1366.

Boos, W.R. and Kuang, Z. (2010). Dominant control of the South Asian monsoon by orographic insulation versus plateau heating. *Nature* 463: 218–222. https://doi.org/10.1038/nature08707.

Boos, W.R. and Kuang, Z. (2013). Sensitivity of the South Asian monsoon to elevated and non-elevated heating. *Sci. Rep.* 3: 590–595. https://doi.org/10.1038/srep01192.

Boriah, P.C., Steinbach, M., Kumar, V., and Klooster, S. (2008). Terrestrial vegetation dynamics and global climate controls in North America: 2001–05. *Earth Interact.* 12 (8): 1. https://doi.org/10.1175/2008EI249.1.

Bosilovich, M.G., Schubert, S.D., and Walker, G.K. (2004). Global changes of the water cycle intensity. *J. Clim.* 18: 1591–1608. https://doi.org/10.1175/JCLI3357.1.

Boucher, É., Bégin, Y., and Arseneault, D. (2009). Impacts of recurring ice jams on channel geometry and geomorphology in a small high-boreal watershed. *Geomorphology* 108 (3–4): 273–281. https://doi.org/10.1016/j.geomorph.2009.02.014.

Boucher, O., Randall, D., Artaxo, P. et al. (2013). Clouds and aerosols. In: *Climate Change (2013). The Physical Science Basis. Contribution of Working Group I to the Fifth Assessment Report of the Intergovernmental Panel on Climate Change* (ed. T.F. Stocker, D. Qin, G.-K. Plattner, et al.), 1–121. Cambridge and New York: Cambridge University Press.

Bounoua, L., Collatz, G.J., Los, S.O. et al. (2000). Sensitivity of climate to changes in NDVI. *J. Clim.* 13 (13): 2277–2292. https://doi.org/10.1175/1520-0442(2000)013<2277:soctci>2.0.co;2.

Bousquet, P., Peylin, P., Ciais, P. et al. (2000). Regional changes in carbon dioxide fluxes of land and oceans since 1980. *Science* 290: 1342–1346. https://doi.org/10.1126/science.290.5495.1342.

Bousquet, P., Ciais, P., Miller, J.B. et al. (2006). Contribution of anthropogenic and natural sources methane emissions variability. *Nature* 443: 439–443. https://doi.org/10.1038/nature05132.

Bridgham, S.D., Cadillo-Quiroz, H., Keller, J.K., and Zhuang, Q. (2013). Methane emissions from wetlands: biogeochemical, microbial, and modeling perspectives from local to global scales. *Glob. Chang. Biol.* 19: 1325–1346. https://doi.org/10.1111/gcb.12131.

Bristow, K.L. and Campbell, G.S. (1984). On the relationship between incoming solar radiation and daily maximum and minimum temperature. *Agric. For. Meteorol.* 31 (2): 159–166. https://doi.org/10.1016/0168-1923(84)90017-0.

Brown, J. (1997). Disturbance and recovery of permafrost terrain. In: *Disturbance and Recovery in Arctic Lands: An Ecological Perspective* (ed. R.M.M. Crawford), 167–178. Dordrecht: Kluwer.

Brown, R.D. (2000). Northern hemisphere snow cover variability and change, 1915–97. *J. Clim.* 13: 2339–2355. https://doi.org/10.1175/1520-0442(2000)013<2339:NHSCVA>2.0.CO;2.

Brown, R.D. and Mote, P.W. (2009). The response of northern hemisphere snow cover to a changing climate. *J. Clim.* 22: 2124–2145.

Brown, R.D. and Robinson, D.A. (2011). Northern Hemisphere spring snow cover variability and change over 1922–2010 including an assessment of uncertainty. *Cryosphere* 5: 219–229. https://doi.org/10.5194/tc-5-219-2011.

Brown, J., Ferrians, O.J. Jr., Heginbottom, J.A., and Melnikov, E.S. (1997). Circum-Arctic map of permafrost and ground-ice conditions. U.S. Geological Survey in Cooperation with the Circum-Pacific Council for Energy and Mineral Resources, Circum-Pacific Map Series CP-45, scale 1:10,000,000, 1 sheet.

Brubaker, K.L., Entekhabi, D., and Eagleson, P.S. (1993). Estimation of continental precipitation recycling. *J. Clim.* 6 (6): 1077–1089. https://doi.org/10.1175/1520-0442(1993)006<1077:EOCPR>2.0.CO;2.

Bruhwiler, L., Parmentier, F., Crill, P. et al. (2021). The Arctic carbon cycle and its response to changing climate. *Curr. Clim. Change Rep.* 7: 14–34. https://doi.org/10.1007/s40641-020-00169-5.

Brutel-Vuilmet, C., Ménégoz, M., and Krinner, G. (2013). An analysis of present and future seasonal Northern Hemisphere land snow cover simulated by CMIP5 coupled climate models. *Cryosphere* 7: 67. https://doi.org/10.5194/tc-7-67-2013.

Budyko, M.I. (1974). *Climate and Lige [M]*, vol. 508. New York: Academic Press.

Burke, E.J., Dankers, R., Jones, C.D. et al. (2013). A retrospective analysis of pan Arctic permafrost using the JULES land surface model. *Clim. Dyn.* 41 (1025): https://doi.org/10.1007/s00382-012-1648-x.

Burn, C.R. (1998). The active layer: two contrasting definitions. *Permafr. Periglac. Process.* 9: 411–416. https://doi.org/10.1002/(SICI)1099-1530(199810/12)9:4<411::AID-PPP292>3.0.CO;2-6.

Businger, S. and Reed, R.J. (1989). Cyclogenesis in cold air masses. *Weather Forecasting* 4: 133–156. https://doi.org/10.1175/1520-0434(1989)004<0133:CICAM>2.0.CO;2.

Butler, J.L., Bottomley, P.J., Griffith, S.M., and Myrold, D.D. (2004). Distribution and turnover of recently fixed photosynthate in ryegrass rhizospheres. *Soil Biol. Biochem.* 36: 371–382. https://doi.org/10.1016/j.soilbio.2003.10.011.

Butler, A.H., Sjoberg, J.P., Seidel, D.J., and Rosenlof, K.H. (2017). A sudden stratospheric warming compendium. *Earth Syst. Sci. Data* 9: 63–76. https://doi.org/10.5194/essd-9-63-2017.

Cai, C.X. (1978). Characteristics of polar high and its relationship with cold wave in Xinjiang. *Meteorol. Mon.* 4 (12): 13–16. (in Chinese).

Cai, Y., Qian, Z.A., Wu, T.W. et al. (2004). Distributionchanges of atmospheric precipitable water over Qinghai-Xizang Plateau and its surroundings and their changeable precipitation climate. *Plateau Meteorol.* 23 (1): 1–10. (in Chinese).

Call, D.A. (2009). An assessment of national weather service warning procedures for ice storms. *Weather Forecast.* 24 (1): 104–120. https://doi.org/10.1175/2008WAF2007111.1.

Callaghan, T.V., Tweedie, C.E., and Webber, P.J. (2011). Multi-decadal changes in tundra environments and ecosystems: The International Polar Year-Back to the Future Project (IPYBTF). *Ambio* 40: 555–557. https://doi.org/10.1007/s13280-011-0179-8.

Campbell, G.S. (1985). *Soil Physics with BASIC*. New York: Elsevier.

Cao, G., Xingliang, X., Long, R. et al. (2008). Methane emissions by alpine plant communities in the Qinghai-Tibet Plateau. *Biol. Lett.* 4: 681–684. https://doi.org/10.1098/rsbl.2008.0373.

Carleton, A.M. (1996). Satellite climatological aspects of cold air mesocyclones in the Arctic and Antarctic. *Global. Atmos. Ocean Syst.* 5: 1–42.

Castebrunet, H., Eckert, N., Giraud, G. et al. (2014). Projected changes of snow conditions and avalanche activity in a warming climate: the French Alps over the 2020–2050 and 2070–2100 periods. *Cryosphere* 8 (5): 1673–1697. https://doi.org/10.5194/tc-8-1673-2014.

Changnon, S.A. (2004). *Climate Atlas: Freezing Rain and Ice Storms*. Mont Fleuri: Printec Press.

Charlson, R.J., Schwartz, S.E., Hales, J.M. et al. (1992). Climate forcing by anthropogenic aerosols. *Science* 255: 423–430. https://doi.org/10.1126/science.255.5043.423.

Charlton, A.J., O'Neill, A., Lahoz, W.A., and Berrisford, P. (2005). The splitting of the stratospheric polar vortex in the Southern Hemisphere, September 2002: dynamical evolution. *J. Atmos. Sci.* 62: 590–602. https://doi.org/10.1175/JAS-3318.1.

Charney, J.G. (1975). Dynamics of deserts and drought in the Sahel. *Quart. J. Roy. Meteor. Soc.* 101: 193–202. https://doi.org/10.1002/qj.49710142802.

Che, H.Z., Shi, G.Y., Zhang, X.Y. et al. (2005). Analysis of 40 years of solar radiation data from China, 1961–2000. *Geophys. Res. Lett.* 32: L06803. https://doi.org/10.1029/2004GL022322.

Che, T., Xin, L., Jin, R. et al. (2008). Snow depth derived from passive microwave remote-sensing data in China. *Ann. Glaciol.* 49: 145–154. https://doi.org/10.3189/172756408787814690.

Chen, G. and Li, G.P. (2014). Dynamic and numerical study of waves in the Tibetan Plateau vortex. *Adv. Atmos. Sci.* 31 (1): 131–138. https://doi.org/10.1007/s00376-013-1035-5.

Chen, L. (1997). The influence of winter–spring snow cover over Tibetan Plateau on atmospheric circulation and flood season precipitation in south China. In: *Middle and Long-Term Water Temperature Forecast Corpus*, vol. 1, 194–195. Beijing: China Waterpower Press.

Chen, L. (1998). Test and application of the relationship between anomalous snow cover in winter-spring over Qinghai-Xizang Plateau and the first summer precipitation in South China. *Q. J. Appl. Meteorol.* 9: 1–32. (in Chinese).

Chen, L. (2001). The role of the anomalous snow cover over the Qinghai-Xizang Plateau and ENSO in the great floods of 1998 in the Changjiang River valley. *Chin. J. Atmos. Sci.* 25: 184–192. https://doi.org/10.3878/j.issn.1006-9895.2001.02.04. (in Chinese).

Chen, L. and Wu, R. (2000). Interannual and decadal variations of snow cover over Qinghai-Xizang Plateau and their relationships to summer monsoon rainfall in China. *Adv. Atmos. Sci.* 17: 18–30. https://doi.org/10.1007/s00376-000-0040-7.

Chen, L.T. and Yan, Z.X. (1979). *Effects of Winter and Spring Snow Cover on the General Circulation of the Atmosphere and the Precipitation in Southern China during the Flood Season [A]. Collection of Long-Term Hydrometeorological Forecast in China* (Part 1) [M], 185–194. Beijing: Water Conservancy and Electric Power Press (in Chinese).

Chen, Q., Gao, B., Li, W.J., and Liu, Y. (2000). Studies on relationships among snow cover winter over the Tibetan Plateau and droughts/floods during Meiyu season in the middle and lower reaches of the Yangtze River as well as atmosphere/ocean. *Acta. Meteor. Sin.* 58: 582–595. (in Chinese).

Chen, R., Kang, E., Wu, L. et al. (2005). Cold regions in China. *J. Glaciol. Geocryol.* 27 (4): 470–475. https://doi.org/10.1016/j.coldregions.2006.03.001.

Chen, S.B., Liu, Y.F., and Thomas, A. (2006). Climatic change on the Tibetan Plateau: potential evapotranspiration trends from 1961–2000. *Clim. Change* 76: 291–319. https://doi.org/10.1007/s10584-006-9080-z.

Chen, S.T., Guo, B., Yang, F. et al. (2020). Spatial and temporal patterns of NPP and its response to climate change in the Qinghai-Tibet Plateau from 2000 to 2015. *J. Nat. Resour.* 35 (10): 2511–2527. https://doi.org/10.31497/zrzyxb.20201016.

Chen, X. and Kaung, W.H. (2019). Analysis of the characteristics of ecosystem changes on the Qinghai-Tibet plateau from 1990 to 2015. *J. Southwest Minzu Univ. (Nat. Sci. Edn.)* 45 (3): 233–242.

Chen, Y.-H. and Prinn, R.G. (2006). Estimation of atmospheric methane emissions between 1996 and 2001 using a three-dimensional global chemical transport model. *J. Geophys. Res.* 111: D10307. https://doi.org/10.1029/2005JD006058.

Chen, J., Wu, X., Yin, Y., Huang, Q., and Xiao, H. (2017). Characteristics of Cloud Systems over the Tibetan Plateau and East China during Boreal Summer. *J. Climate.* 30: 3117–3137.

Chen, Y.Y., Yang, K., Qin, J. et al. (2013). Evaluation of AMSR-E retrievals and GLDAS simulations against observations of a soil moisture network on the central Tibetan Plateau. *J. Geophys. Res.* 118 (10): 4466–4475. https://doi.org/10.1002/jgrd.50301.

Cheng, G. and Jin, H. (2013). Permafrost and groundwater on the Qinghai-Tibet Plateau and in northeast China. *Hydrgeol. J.* 21: 5–23. https://doi.org/10.1007/s10040-012-0927-2.

Cherkauer, K.A. and Lettenmaier, D.P. (1999). Hydrological effects of frozen soils in the upper Mississippi River basin. *J. Geophys. Res.* 104 (D16): 19599–19610. https://doi.org/10.1029/1999JD900337.

Cherkauer, K.A. and Lettenmaier, D.P. (2003). Simulation of spatial variability in snow and frozen soil. *J. Geophys. Res.* 108 (D22): 8858. https://doi.org/10.1029/2003/JD003575.

Christensen, T.R., Johansson, T., Akerman, H.J. et al. (2004). Thawing sub-arctic permafrost: effects on vegetation and methane emissions. *Geophys. Res. Lett.* 31: L04501. https://doi.org/10.1029/2003GL018680.

Christensen, T.R., Johansson, T., Olsrud, M. et al. (2007). A catchment-scale carbon and greenhouse gas budget of a subarctic landscape. *Phil. Trans. R. Soc. A* 365: 1643–1656. https://doi.org/10.1098/rsta.2007.2035.

Clapp, R.B. and Hornberger, G.M. (1978). Empirical equations for some soil hydraulic properties. *Water Resour. Res.* 14: 601–604.

Clein, J.S. and Schimel, J.P. (1995). Microbial activity of tundra and taiga soils at sub-zero temperatures. *Soil Biol. Biochem.* 27: 1231–1234. https://doi.org/10.1016/0038-0717(95)00044-F.

Clow, D.W. (2010). Changes in the timing of snowmelt and streamflow in Colorado: A response to recent warming. *J. Clim.* 23 (9): 2293–2306. https://doi.org/10.1175/2009JCLI2951.1.

Cohen, J., Barlow, M., Kushner, P.J. et al. (2007). Stratosphere-troposphere coupling and links with Eurasian land surface variability. *J. Clim.* 20 (21). 5335–5343. https://doi.org/10.1175/2007JCLI1725.1.

Cohen, J., Pfeiffer, K., and Francis, J.A. (2018). Warm Arctic episodes linked with increased frequency of extreme winterweather in the United States. *Nat. Commun.* 9: 869. https://doi.org/10.1038/s41467-018-02992-9.

Cole, J.J., Caraco, N.F., Kling, G.W., and Kratz, T.K. (1994). Carbon dioxide supersaturation in the surface waters of lakes. *Science* 265: 1568–1570. https://doi.org/10.1126/science.265.5178.1568.

Cole, J.J., Prairie, Y.T., Caraco, N.F. et al. (2007). Plumbing the global carbon cycle: integrating inland waters into the terrestrial carbon budget. *Ecosystems* 10: 171–184. https://doi.org/10.1007/s10021-006-9013-8.

Colliander, A., and Coauthors (2017). Validation of SMAP surface soil moisture products with core validation sites. *Remote Sens. Environ.* 191: 215–231.

Collins, M., Knutti, R., Arblaster, J. et al. (2013). Long-term climate change: projections, commitments and irreversibility. In: *Climate Change 2013: The Physical Science Basis. Contribution of Working Group I to the Fifth Assessment Report of the Intergovernmental Panel on Climate Change* (ed. T.F. Stocker, D. Qin, G.-K. Plattner, et al.), 1031–1136. Cambridge: Cambridge University Press.

Colony, R., Radionov, V., and Tanis, F.L. (1998). Measurements of precipitation and snow pack at the Russian North Pole drifting station. *Polar Rec.* 34: 3–14. https://doi.org/10.1017/S0032247400014923.

Confortola, G., Soncini, A., and Bocchiola, D. (2013). Climate change will affect hydrological re gimes in the Alps: a case study in Italy. *J. Alp. Res.* 101 (3): https://doi.org/10.4000/rga.2176.

Conil, S., Douville, H., and Tyteca, S. (2009). Contribution of realistic soil moisture initial conditions to boreal summer climate predictability. *Clim. Dyn.* 32 (1): 75–93. https://doi.org/10.1007/s00382-008-0375-9.

Conway, J.P., Cullen, N.J., Spronken-Smith, R.A., and Fitzsimons, S.J. (2015). All-sky radiation over a glacier surface in the Southern Alps of New Zealand: characterizing cloud effects on incoming shortwave, longwave and net radiation. *Int. J. Climatol.* 35: 699–713.

Cook, B.I., Bonan, G.B., and Levis, S. (2006). Soil moisture feedback to precipitation in South Africa. *J. Clim.* 19: 4198–4206.

Cortinas, J. (2000). A climatology of freezing rain in the Great Lakes region of North America. *Mon. Weather Rev.* 128 (10): 3574–3588. https://doi.org/10.1175/1520-0493(2001)129<3574:ACOFRI>2.0.CO;2.

Cortinas, J.V. Jr., Bernstein, B.C., Robbins, C.C. et al. (2004). An analysis of freezing rain, freezing drizzle, and ice pellets across the United States and Canada: 1976–90. *Weather Forecasting* 19 (2): 377–390. https://doi.org/10.1175/1520-0434(2004)019<0377:AAOFRF>2.0.CO;2.

Cosby, B.J., Hornberger, G.M., Clapp, R.B., and Ginn, T.R. (1984). A statistical exploration of the relationships of soil moisture characteristics to the physical properties of soils. *Water Resour. Res.* 20: 682–690.

Craig, R.A. and Hering, W.S. (1959). The stratospheric warming of January-February 1957. *J. Atmos. Sci.* 16 (2): 91–107. https://doi.org/10.1175/1520-0469(1959)016<0091:TSWOJF>2.0.CO;2.

Cuffey, K.M. and Paterson, W.S.B. (2010). *The Physics of Glaciers*, 4e. San Diego: Academic Press.

Cuntz, M. and Haverd, V. (2018). Physically accurate soil freeze-thaw processes in a global land surface scheme. *J. Adv. Model. Earth Syst.* 10: https://doi.org/10.1002/2017MS001100.

Cuo, L., Zhang, Y., Gao, Y. et al. (2013a). The impacts of climate change and land cover/use transition on the hydrology in the upper Yellow River basin, China. *J. Hydrol.* 502: 37–52. https://doi.org/10.1016/j.jhydrol.2013.08.003.

Cuo, L., Zhang, Y., Wang, Q. et al. (2013b). Climate change on the Northern Tibetan Plateau during 1957–2009: spatial patterns and possible mechanisms. *J. Clim.* 26: 85–109. https://doi.org/10.1175/JCLI-D-11-00738.1.

Cuo, L., Zhang, Y., Zhu, F., and Liang, L. (2014). Characteristics and changes of streamflow on the Tibetan Plateau: a review. *J. Hydrol.: Reg. Stud.* 2: 49–68. https://doi.org/10.1016/j.ejrh.2014.08.004.

Cuo, L., Zhang, Y., Bohn, T.J. et al. (2015). Frozen soil degradation and its effects on surface hydrology in the northern tibetan plateau. *J. Geophys. Res.-Atmos.* 120 (16): 8276–8298. https://doi.org/10.1002/2015JD023193.

Curio, J., Schiemann, R., Hodges, K., and Turner, A. (2018). Climatology of Tibetan Plateau vortices in reanalysis data and a high-resolution global climate model. *J. Clim.* 32 (6): 1933–1950. https://doi.org/10.1175/JCLI-D-18-0021.1.

Dai, A.G. and Wigley, T.M.L. (2000). Global patterns of ENSO-induced precipitation. *Geophys. Res. Lett.* 27: 1283–1286. https://doi.org/10.1029/1999GL011140.

Dai, J.X. (1990). *Plateau Climate*. Beijing: Meteorological Press (in Chinese).

Daly, S.F. (2013). Aufeis. In: *River Ice Formation* (ed. S. Beltaos), 1–320. Edmonton, AB: Committee on River Ice Processes and the Environment, Canadian Geophysical Union Hydrology Section.

Das, A. and Lindenschmidt, K.E. (2020). Current status and advancement suggestions of ice-jam flood hazard and risk assessment. *Environ. Rev.* 28 (6): 373–379. https://doi.org/10.1139/er-2019-0063.

Davidson, E.A. and Janssens, I.A. (2006). Temperature sensitivity of soil carbon decomposition and feedbacks to climate change. *Nature* 440: 165–173. https://doi.org/10.1038/nature04514.

Day, C.A. (2009). Modelling impacts of climate change on snowmelt runoff generation and stream flow across western US mountain basins: a review of techniques and applications for water resource management. *Prog. Phys. Geogr.* 33: 614–633. https://doi.org/10.1177/0309133309343131.

Decker, M. and Zeng, X. (2006). An empirical formulation of soil ice fraction based on in situ observations. *Geophys. Res. Lett.* 33: L05402. https://doi.org/10.1029/2005GL024914.

Degelia, S.K., Christian, J.I., Basara, J. et al. (2016). An overview of ice storms and their impact in the United States. *Int. J. Climatol.* 36 (8): 2811–2822. https://doi.org/10.1002/joc.4525.

Dell'Osso, L. and Chen, S.J. (1986). Numerical experiments on the genesis of vortices over the Qinghai-Tibet plateau. *Tellus A* 38 (3): 236–250. https://doi.org/10.1111/j.1600-0870.1986.tb00468.x.

Dewalle, D.R. and Rango, A. (2009). *Principles of Snow Hydrology*. Cambridge: Cambridge University Press https://doi.org/10.1017/CBO9780511535673.

Ding, F., Sun, Z.B., and Liu, J.L. (2010). Correlations of snow depth over the Tibetan Plateau in winter and spring with precipitation in summer over China. *Trans. Atmos. Sci.* 32: 783–791.

Ding, Y., Zhang, S., Chen, R. et al. (2020). *Hydrological Sciences of Cryosphere*. Beijing: Science Press.

Ding, Y.B., Xu, J.T., Wang, X.W. et al. (2020). Spatial and temporal effects of drought on Chinese vegetation under different coverage levels. *Sci. Total Environ.* 716: 137166. https://doi.org/10.1016/j.scitotenv.2020.137166.

Ding, Y.B., Xu, J.T., Wang, X.W. et al. (2021). Propagation of meteorological to hydrological drought for different climate regions in China. *J. Environ. Manag.* 283: 111980. https://doi.org/10.1016/j.jenvman.2021.111980.

Ding, Y.J. (2017). *Introduction to Cold Region Hydrology [M]*. Beijing: Science Press (in Chinese).

Diolaiuti, G., Bocchiola, D., D'agata, C., and Smiraglia, C. (2012a). Evidence of climate change im pact upon glaciers' recession within the Italian Alps: the case of Lombardy glaciers. *Theor. Appl. Climatol.* 109 (3–4): 429–445. https://doi.org/10.1007/s00704-012-0589-y.

Diolaiuti, G., Bocchiola, D., Vagliasindi, M. et al. (2012b). The 1975–2005 glacier changes in Aosta Valley (Italy) and the relations with climate evolution. *Prog. Phys. Geogr.* 36 (6): 764–785. https://doi.org/10.1177/0309133312456413.

Dirmeyer, P.A., Gao, X., Zhao, M. et al. (2006). GSWP-2: Multimodel analysis and implications for our perception of the land surface. *Bull. Amer. Meteor. Soc.* 87: 1381–1397.

Dirmeyer, P.A., Schlosser, C.A., and Brubaker, K.L. (2008). Precipitation, recycling, and land memory: an integrated analysis. *J. Hydrometeorol.* 10 (1): 278–288. https://doi.org/10.1175/2008JHM1016.1.

Dobinski, W. (2011). Permafrost. *Earth Sci. Rev.* 108 (3–4): 158–169. https://doi.org/10.1016/j.earscirev.2011.06.007.

Dominguez, F., Kumar, P., Liang, X.Z. et al. (2006). Impact of atmospheric moisture storage on precipitation recycling. *J. Clim.* 19 (8): 1513–1530. https://doi.org/10.1175/JCLI3691.1.

Dong, X., Xi, B., and Minnis, P. (2006). A climatology of midlatitude continental clouds from the ARM SGP Central Facility: part II: cloud fraction and surface radiative forcing. *J. Clim.* 19: 1765–1783. https://doi.org/10.1175/JCLI3710.1.

Douville, H. (2002). Influence of soil moisture on the Asian and African monsoons. Part II: interannual variability. *J. Clim.* 15 (7): 701–720. https://doi.org/10.1175/1520-0442(2002)015<0701:IOSMOT>2.0.CO;2.

Douville, H. (2004). Relevance of soil moisture for seasonal atmospheric predictions: is it an initial value problem? *Clim. Dyn.* 22 (4): 429–446. https://doi.org/10.1007/s00382-003-0386-5.

Douville, H. and Royer, J.-F. (1996). Sensitivity of the Asian summer monsoon to an anomalous Eurasian snow cover within the Meteo-France GCM. *Clim. Dyn.* 12: 449–466. https://doi.org/10.1007/BF02346818.

Douville, H., Chauvin, F., and Broqua, H. (2001). Influence of soil moisture on the Asian and African monsoons. Part I: mean monsoon and daily precipitation. *J. Clim.* 14 (11): 2381–2403. https://doi.org/10.1175/1520-0442(2001)014<2381:IOSMOT>2.0.CO;2.

Du, M., Li, G.P., and Ding, C.C. (2018). The waves of the plateau transverse shear line and its possible connection with the low vortex. *Plateau Meteorol.* 6: 16505–11615. https://doi.org/10.7522/j.

Du, Y.G., Cao, G.M., Wang, Q.L., and Wang, C.T. (2007). Effect of grazing on surface character and soil physical property in alpine meadow. *J. Mt. Sci.* 25: 338–343. https://doi.org/10.4067/S0718-95162011000400003.

Duan, A. and Wu, G. (2008). Weakening trend in the atmospheric heat source over the Tibetan Plateau during recent decades. Part I: observations. *J. Clim.* 21: 3149–3164. https://doi.org/10.1175/2007JCLI1912.1.

Duan, A., Wang, M., Lei, Y., and Cui, Y. (2013). Trends in summer rainfall over China associated with the Tibetan Plateau sensible heat source during 1980–2008. *J. Clim.* 26: 261–275. https://doi.org/10.1175/JCLI-D-11-00669.1.

Duan, W. and Wei, C. (2013). The "spring predictability barrier" for ENSO predictions and its possible mechanism: results from a fully coupled model. *Int. J. Climatol.* 33: 1280–1292. https://doi.org/10.1002/joc.3513.

Dyer, J.L. and Mote, T.L. (2006). Spatial variability and trends in observed snow depth over North America. *Geophys. Res. Lett.* 33: L16503. https://doi.org/10.1029/2006GL027258.

Eckert, N., Baya, H., and Deschâtres, M. (2010). Assessing the response of snow avalanche runout altitudes to climate fluctuations using hierarchical modeling: application to 61 winters of data in France. *J. Clim.* 23 (12): 3157–3180. https://doi.org/10.1175/2010JCLI3312.1.

Emanuel, K.A. and Rotunno, R. (1989). Polar lows as Arctic hurricanes. *Tellus* 41A: 1–17. https://doi.org/10.3402/tellusa.v41i1.11817.

Eugster, W., Kling, G., Jonas, T. et al. (2003). CO_2 exchange between air and water in an Arctic Alaskan and midlatitude Swiss lake: importance of convective mixing. *J. Geophys. Res.* 108: 4362. https://doi.org/10.1029/2002JD002653.

Euskirchen, E.S., Mcguire, A.D., Kicklighter, D.W. et al. (2006). Importance of recent shifts in soil thermal dynamics on growing season length, productivity, and carbon sequestration in terrestrial high-latitude ecosystems. *Glob. Chang. Biol.* 12 (4): https://doi.org/10.1111/j.1365-2486.2006.01113.x.

Faillettaz, J., Funk, M., and Vincent, C. (2015). Avalanching glacier instabilities: review on processes and early warning perspectives. *Rev. Geophys.* 53 (2): 203–224.

Fairlie, T.D.A. and O'Neill, A. (1988). The stratospheric major warming of winter 1984/85: observations and dynamical inferences. *Quart. J. Roy. Meteorlog. Soc.* 114 (481): 557–577. https://doi.org/10.1002/qj.49711448102.

Falarz, M. (2002). Long-term variability in reconstructed and observed snow cover over the last 100 winter seasons in Cracow and Zakopane (South Poland). *Clim. Res.* 19: 247–256. https://doi.org/10.3354/cr019247.

Fan, G., Zhou, D., Li, H. et al. (2006). Numerical simulation of carbon cycle characteristics and climate impact on the Qinghai-Tibet Plateau. In: *The 6th National Symposium for Young Meteorological Scientists*, 804–816.

Fang, J.Y., Liu, G.H., and Xu, S.L. (1996). Soil carbon pool in China and its global significance. *J. Environ. Sci.* 8: 249–254.

FAO/IIASA/ISRIC/ISS-CAS/JRC (2008). *Harmonized World Soil Database (Version 1.0).* Rome and Laxenburg: FAO.

Farouki, O.T. (1981). The thermal properties of soils in cold regions. *Cold Reg. Sci. Technol.* 5 (1): 67–75. https://doi.org/10.1016/0165-232X(81)90041-0.

Fasullo, J. (2004). A stratified diagnosis of the Indian monsoon-Eurasian snow cover relationship. *J. Clim.* 17: 1110–1122. https://doi.org/10.1175/1520-0442(2004)017<1110:ASDOTI>2.0.CO;2.

Fayer, M.J. (2000). *UNSAT-H version 3.0: Unsaturated soil water and heat flow model. Theory, user manual, and examples.* Richland, WA: Pacific Northwest National Laboratory.

Fekete, B.M., Vörösmarty, C.J., Roads, J.O., and Willmott, C.J. (2004). Uncertainties in precipitation and their impacts on runoff estimates. *J. Clim.* 17: 294–304. https://doi.org/10.1175/1520-0442(2004)017<0294:UIPATI>2.0.CO;2.

Feng, D., Gao, X., Zhou, Y. et al. (2017). Characteristics of spatial and temporal distribution of atmospheric methane concentration on the Qinghai-Tibet Plateau. *Clim. Environ. Res.* 22 (3): 9.

Feng, X., Liu, C., Rasmussen, R., and Fan, G. (2014). A 10-yr climatology of Tibetan Plateau vortices with NCEP Climate Forecast System reanalysis. *J. Appl. Meteorol. Climatol.* 53 (1): 34–46. https://doi.org/10.1175/JAMC-D-13-014.1.

Field, W. O. (1975). *Mountain Glaciers of the Northern Hemisphere* (2 vols). Hanover, New Hampshire: U. S. Army CRREL.

Fisher, J.A., Jacob, D.J., Wang, Q.Q. et al. (2011). Sources, distribution, and acidity of sulfate–ammonium aerosol in the Arctic in winter–spring. *Atmos. Environ.* 45 (39): 7301–7318. https://doi.org/10.1016/j.atmosenv.2011.08.030.

Flanner, M.G. (2013). Arctic climate sensitivity to local black carbon. *J. Geophys. Res. Atmos.* 118: 1840–1851. https://doi.org/10.1002/jgrd.50176.

Flerchinger, G.N. and Saxton, K.E. (1989a). Simultaneous heat and water model of a freezing snow residue-soil system: I. Theory and development. *Trans. Am. Soc. Agric. Eng.* 32: 565–571.

Flerchinger, G.N. and Saxton, K.E. (1989b). Simultaneous heat and water model of a freezing snow-residues-soil system II, field verification. *Trans. ASAE* 32: 573–578.

Flohn, H. (1960). Recent investigation on the mechanism of the 'summer monsoon' of southern and eastern Asia. *Proc. Syrup* 75–88.

Forzieri, G., Castelli, F., and Vivoni, E.R. (2011). Vegetation dynamics within the North American monsoon region. *J. Clim.* 24 (6): 1763–1783. https://doi.org/10.1175/2010JCLI3847.1.

Fransson, A., Chierici, M., Yager, P. et al. (2011). Antarctic Sea ice carbon dioxide system and controls. *J. Geophys. Res. Oceans* 116: C12035, 1–18. https://doi.org/10.1029/2010JC006844.

Frauenfeld, O.W. and Zhang, T. (2011). An observational 71-year history of seasonally frozen ground changes in the Eurasian high latitudes. *Environ. Res. Lett.* 6: https://doi.org/10.1088/1748-9326/6/4/044024.

French, H.M. (2007). *The Periglacial Environment*, 3e. Chichester: Wiley.

Fu, G., Niino, H., Kimura, R., and Kato, T. (2004a). Multiple polar mesocyclones over the Japan Sea on 11 February 1997. *Mon. Weather Rev.* 132: 793–814. https://doi.org/10.1175/1520-0493(2004)132<0793:MPMOTJ>2.0.CO;2.

Fu, G., Niino, H., Kimura, R., and Kato, T. (2004b). A polar low over the Japan Sea on 21 January 1997. Part I: observational analysis. *Mon. Weather Rev.* 132: 1537–1551. https://doi.org/10.1175/1520-0493(2004)132<1537:APLOTJ>2.0.CO;2.

Fu, Y. and Liu, G. (2007). Possible misidentification of rain type by TRMM PR over Tibetan Plateau. *J. Appl. Meteorol. Climatol.* 46 (5): 667–672.

Fueglistaler, S., Dessler, A.E., Dunkerton, T.J. et al. (2009). Tropical tropopause layer. *Rev. Geophys.* 47: RG1004. https://doi.org/10.1029/2008RG000267.

Fyodorov-Davydov, D.G., Kholodov, A.L., Ostroumov, V.E. et al. (2008). Seasonal thaw of soils in the North Yakutian ecosystems. In: *Proceedings of the 9th International Conference on Permafrost* (29 June–3 July 2008) (ed. D.L. Kane and K.M. Hinkel), 481–486. Fairbanks: Institute of Northern Engineering, University of Alaska.

Gaigerov, S.S. (1967). *Aerology of the Polar Region (Moscow, 1964)*. Jerusalem: Israel Program of Scientific Translations.

Galvin, J., Kendon, M., and McCarthy, M. (2019). Snow cover and low temperatures in February and March 2018. *Weather* 74: 104–110. https://doi.org/10.1002/wea.3469.

Gao, D.Y., Zhou, H., and Wang, W. (1985). Influence of water vapor pass along the Yarlungzangbo River on precipitation. *Mount. Res.* 3 (4): 239–249. (in Chinese).

Gao, H., Wang, J., Yang, Y. et al. (2021). Permafrost hydrology of the Qinghai-Tibet plateau: a review of processes and modeling. *Front. Earth Sci.* 8: 2296–6463. https://doi.org/10.3389/feart.2020.576838.

Gao, J. (2016). Analysis and assessment of the risk of snow and freezing disaster in China. *Int. J. Disaster Risk Reduct.* 19: 334–340. https://doi.org/10.1016/j.ijdrr.2016.09.007.

Gao, J., Williams, M.W., Fu, X. et al. (2012). Spatiotemporal distribution of snow in eastern Tibet and the response to climate change. *Remote Sens. Environ.* 121: 1–9. https://doi.org/10.1016/j.rse.2012.01.006.

Gao, P.S. and Jin, G.H. (2003). Investigation and analysis of river ice disaster in cold regions. *Journal of China Institute of Water Resources and Hydropower Research of North China*. (In Chinese)

Gao, X., Ye, B.S., Zhang, S.Q. et al. (2010). Glacial meltwater variation and its impact on runoff in the Tarim River Basin during 1961–2006. *Sci. China Earth Sci.* 5 (12): (in Chinese).

Gao, Y., Xiao, L., Chen, D. et al. (2016). Quantification of the relative role of land-surface processes and large-scale forcing in dynamic downscaling over the Tibetan Plateau. *Clim. Dyn.* 48: 1–17. https://doi.org/10.1007/s00382-016-3168-6.

Gao, Y., Chen, F., Lettenmaier, D.P. et al. (2018). Does elevation-dependent warming hold true above 5000 m elevation? Lessons from the Tibetan Plateau. *Clim. Atmos. Sci.* 1: 19.

Gao, Y., Yao, X.P., Li, S.S., and Wang, X.F. (2022). Dynamic and thermodynamic effects on the evolution of the transverse shear line over the Tibetan Plateau in boreal summer (in Chinese). *Chin. J. Atmos. Sci.* 46 (2): 486–500. https://doi.org/10.3878/j.issn.1006-9895.2108.21053.

Gerdel, R.W. (1969). Cold regions. Science and Engineering Monograph 1-A: Characteristics of the Cold Regions.

Gerrard, A.J., Kane, T.L., Thayer, J.P. et al. (2002). Synoptic scale study of the Arctic polar vortex's influence on the middle atmosphere. I. Observations. *J. Geophys. Res.* 107 (D16): https://doi.org/10.1029/2001JD000681.

Giacona, F., Eckert, N., Corona, C. et al. (2021). Upslope migration of snow avalanches in a warming climate. *Proc. Natl. Acad. Sci.* 118 (44): e2107306118, 1–18. https://doi.org/10.1073/pnas.2107306118.

Goetz, S.J., Bunn, A.G., Fiske, G.J., and Houghton, R.A. (2005). Satellite-observed photosynthetic trends across boreal North America associated with climate and fire disturbance. In:, vol. *Proceedings of the National Academy of Sciences (USA)*, 102, 13521–13525. https://doi.org/10.1073/pnas.0506179102.

Goodrich, L.E. (1982). The influence of snow cover on the ground thermal regime. *Can. Geotech. J.* 19: 421–432. https://doi.org/10.1139/t82-047.

Gordon, N.D. and Klein, S.A. (2014). Low-cloud optical depth feedback in climate models. *J. Geophys. Res. Atmos.* 119: 6052–6065. https://doi.org/10.1002/2013JD021052.

Gorham, E. (1990). Biotic impoverishment in northern peat-lands. In: *The Earth in Transition: Patterns and Processes of Biotic Impoverishment* (ed. G.M. Woodwell), 65–98. Cambridge: Cambridge University Press.

Gorham, E. (1991). Northern peatlands: role in the carbon cycle and probable responses to climatic warming. *Ecol. Appl.* 1 (2): 182–195. https://doi.org/10.2307/1941811.

Goritz, V. and Fung, I. (1994). Potential distribution of methane hydrate in the world's oceans. *Global Biogeochem. Cycles* 8: 335–347. https://doi.org/10.1029/94GB00766.

Gouttevin, I., Krinner, G., Ciais, P. et al. (2012). Multi-scale validation of a new soil freezing scheme for a land-surface model with physically-based hydrology. *Cryosphere* 6: 407–430. https://doi.org/10.5194/tc-6-407-2012.

Govorukha, L. S. (1988). *Sovremennoe Nazemnoe Oledenenie Sovyetskoi Arktiki (Modern Terrestrial Glaciation of the Soviet Arctic).* Leningrad: Gidrometeoizdat.

Graybeal, D.Y. and Leathers, D.J. (2006). Snowmelt-related flood risk in Appalachia: first estimates from a historical snow climatology. *J. Appl. Meteorol. Climatol.* 45 (1): 178–193. https://doi.org/10.1175/JAM2330.1.

Greening, K. and Hodgson, A. (2019). Atmospheric analysis of the cold late February and early March 2018 over the UK. *Weather* 74: 79–85. https://doi.org/10.1002/wea.3467.

Grenfell, T.C., Warren, S.G., and Mullen, P.C. (1994). Reflection of solar radiation by the Antarctic snow surface at ultraviolet, visible, and near-infrared wavelengths. *J. Geophys. Res.* 99 (D9): 18669–18684. https://doi.org/10.1029/94JD01484.

Groendahl, L., Friborg, T., and Soegaard, H. (2007). Temperature and snow-melt controls on interannual variability in carbon exchange in high Arctic. *Theor. Appl. Climatol.* 88: 111–125. https://doi.org/10.1007/s00704-005-0228-y.

Groisman, P.Y., Knight, R.W., Karl, T.R. et al. (2004). Contemporary changes of the hydrological cycle over the contiguous United States: trends derived from in situ observations. *J. Hydrometeorol.* 5: 64–85. https://doi.org/10.1175/1525-7541(2004)005<0064:CCOTHC>2.0.CO;2.

Groppelli, B., Bocchiola, D., and Rosso, R. (2011a). Spatial downscaling of precipitation from GCMs for climate change projections using random cascades: a case study in Italy. *Water Resour. Res.* 47: W03519. https://doi.org/10.1029/2010WR009437.

Groppelli, B., Soncini, A., Bocchiola, D., and Rosso, R. (2011b). Evaluation of future hydrological cycle under climate change scenarios in a mesoscale Alpine watershed of Italy. *Nat. Hazards Earth Syst. Sci.* 11: 1769–1785. https://doi.org/10.5194/nhess-11-1769-2011.

Grosse, G., Harden, J., Turetsky, M. et al. (2011). Vulnerability of high-latitude soil organic carbon in North America to disturbance. *J. Geophys. Res.* 116: G00K06. https://doi.org/10.1029/2010JG001507.

Gruber, S. (2012). Derivation and analysis of a high-resolution estimate of global permafrost zonation. *Cryosphere* 6: 221–233. https://doi.org/10.5194/tc-6-221-2012.

Guéguen, C., Guo, L., and Tanaka, N. (2005). Distributions and characteristics of colored dissolved organic matter in the western Arctic Ocean. *Cont. Shelf Res.* 25: 1195–1207. https://doi.org/10.1016/j.csr.2005.01.005.

Guéguen, C., Guo, L., Yamamoto-Kawai, M., and Tanaka, N. (2007). Colored dissolved organic matter dynamics across the shelf/basin interfaces in the western Arctic Ocean. *J. Geophys. Res.* 112: C05038. https://doi.org/10.1029/2006JC003584.

Guglielmin, M., Dalle Fratte, M., and Cannone, N. (2014). Permafrost warming and vegetation changes in continental Antarctica. *Environ. Res. Lett.* 9: 045001. https://doi.org/10.1088/1748-9326/9/4/045001.

Gui, S., Liang, S., and Li, L. (2010). Evaluation of satellite-estimated surface longwave radiation using ground-based observations. *J. Geophys. Res.* 115: D18214. https://doi.org/10.1029/2009JD013635.

Gui, S., Liang, S., Wang, K. et al. (2010). Assessment of three satellite-estimated land surface downwelling shortwave irradiance data sets. *IEEE Geosci. Remote Sens. Lett.* 7 (4): 776–780. https://doi.org/10.1109/LGRS.2010.2048196.

Guo, D. and Wang, H. (2012). The significant climate warming in the northern Tibetan Plateau and its possible causes. *Int. J. Climatol.* 32 (12): 1775–1781. https://doi.org/10.1002/joc.2388.

Guo, D. and Wang, H. (2013). Simulation of permafrost and seasonally frozen ground conditions on the Tibetan Plateau, 1981–2010. *J. Geophys. Res. Atmos.* 118 (11): 5216–5230. https://doi.org/10.1002/jgrd.50457.

Guo, D. and Wang, H. (2014). Simulated change in the near-surface soil freeze/thaw cycle on the Tibetan Plateau from 1981 to 2010. *Chin. Sci. Bull.* 59 (20): 2439–2448. https://doi.org/10.1007/s11434-014-0347-x.

Guo, D., Wang, H., and Wang, A. (2017). Sensitivity of historical simulation of the permafrost to different atmospheric forcing data sets from 1979 to 2009. *J. Geophys. Res. Atmos.* 122: 12269–12284. https://doi.org/10.1002/2017JD027477.

Guo, D.L., Yang, M.X. and Wang, H.J. (2011).Characteristics of land surface heat and water exchange under different soilfreeze/thaw conditions over the central Tibetan Plateau. *Hydrol. Processes*: 25(16): 2531–2541.

Guo, J., Gang, F., Li, Z. et al. (2007). Analyses and numerical modeling of a polar low over the Japan Sea on 19 December 2003. *Atmos. Res.* 85 (3–4): 395–412. https://doi.org/10.1016/j.atmosres.2007.02.007.

Guo, W.Q., Liu, S.Y., Xu, J.L. et al. (2012). Remote sensing monitoring of the glacier movement in the Yulinchuan glacier on the northwest slope of Muztag. *J. Glaciol. Geocryol.* 34 (4): 765–774.

Guo, X.W., Dai, L., Fawei, Z. et al. (2020). Effects of increased precipitation and nitrogen deposition on methane uptake of alpine meadow in Qinghai-Tibet Plateau: in situ experiments. *Pol. J. Ecol.* 68: https://doi.org/10.3161/15052249PJE2020.68.2.001.

Gurney, K.R. et al. (2003). TransCom 3 CO2 inversion intercomparison: 1. Annual mean control results and sensitivity to transport and prior flux information. *Tellus* 55B: 555–579.

Hagg, W. and Braun, L. (2005). The influence of glacier retreat on water yield from high moun tain areas: comparison of Alps and Central Asia. In: *Climate and Hydrology of Mountain Areas*, vol. 18 (ed. C. De Jong, D. Collins and R. Ranzi), 263–275. Chichester: Wiley.

Hahn, D.G. and Manabe, S. (1975). The role of mountains in the South Asian monsoon circulation. *J. Atmos. Sci.* 32: 1515–1541. https://doi.org/10.1175/1520-0469(1975)032<1515:TROMIT>2.0.CO;2.

Haller, T. and Stolp, H. (1985). Quantitative estimation of root exudation of maize plants. *Plant Soil* 86: 207–216. https://doi.org/10.1007/BF02182895.

Hamlet, A.F. and Lettenmaier, D.P. (2007). Effects of 20th century warming and climate variability on flood risk in the western U.S. *Water Resour. Res.* 43: 6. https://doi.org/10.1029/2006WR005099.

Han, B.H., Zhou, B.R., Yan, Y.Q. et al. (2019). Analysis of vegetation coverage change and its driving factors over tibetan plateau from 2000 to 2008. *Acta Agrestia Sin.* 27 (6): 8.

Hanesiak, J.M. and Stewart, R.E. (1995). The mesoscale and microscale structure of a severe ice pellet storm. *Mon. Weather Rev.* 123 (11): 3144–3162. https://doi.org/10.1175/1520-0493(1995)123<3144:TMAMSO>2.0.CO;2.

Hansen, M., DeFries, R., Townshend, J.R.G., and Sohlberg, R. (2000). Global land cover classification at 1 km resolution using a decision tree classifier. *Int. J. Remote Sens.* 21: 1331–1365.

Hansson, K., Simunek, J., Mizoguchi, M. et al. (2004). Water flow and heat transport in frozen soil: numerical solution and freeze-thaw applications. *Vadose Zone* 3: 693–704. https://doi.org/10.2113/3.2.693.

Hao, J., Huang, F., Liu, Y. et al. (2018). Avalanche activity and characteristics of its triggering factors in the western Tianshan Mountains, China. *J. Mt. Sci.* 15 (7): 1397–1411.

Harden, J.W., Sundquist, E.T., Stallard, R.F., and Mark, R.K. (1992). Dynamics of soil carbon during deglaciation of the Laurentide ice sheet. *Science* 258: 1921–1924. https://doi.org/10.1126/science.258.5090.1921.

Hare, F.K. (1961). *The Circulation of the Stratosphere. Publication in Meteorology*, vol. 43. Montreal: Arctic Meteorology Research Group, McGill University.

Harlan, R.L. (1973). Analysis of coupled heat and mass transfer in partial frozen soil. *Water Resour. Res.* 9 (5): 1314–1323. https://doi.org/10.1029/WR009i005p01314.

Harrison, E.F., Minnis, P., Barkstrom, B.R. et al. (1990). Seasonal variation of cloud radiative forcing derived from the earth radiation budget experiment. *J. Geophys. Res. Atmos.* 95: 18687–18703. https://doi.org/10.1029/JD095iD11p18687.

Hartmann, D.L., Wallace, J.M., Limpasvan, V. et al. (2000). Can ozone depletion and global warming interact to produce rapid climate change? *Proc. Natl. Acad. Sci. U. S. A.* 97: 1412–1417. https://doi.org/10.1073/pnas.97.4.1412.

Hayes, D.J., Kicklighter, D.W., McGuire, A.D. et al. (2014). The impacts of recent permafrost thaw on land-atmosphere greenhouse gas exchange. *Environ. Res. Lett.* 9: 045005. https://doi.org/10.1088/1748-9326/9/4/045005.

Haynes, J.M., Vonder Haar, T.H., L'Ecuyer, T., and Henderson, D. (2013). Radiative heating characteristics of earth's cloudy atmosphere from vertically resolved active sensors. *Geophys. Res. Lett.* 40: 624–630. https://doi.org/10.1002/grl.50145.

Heginbottom, J.A., Brown, J., Melnikov, E.S., and Ferrians, O.J. Jr. (1993). Circumarctic map of permafrost and ground ice conditions. In: *Proceedings of the Sixth International Conference on Permafrost*, vol. 2, 1132–1136. Wushan: South China University of Technology Press.

Highwood, E.J., Hoskins, B.J., and Berrisford, P. (2000). Properties of the Arctic tropopause. *Q. J. Roy. Meteorol. Soc.* 226: 1515–1532. https://doi.org/10.1002/qj.49712656515.

Hinkel, K.M. and Nelson, F.E. (2003). Spatial and temporal patterns of active layer thickness at Circumpolar Active Layer Monitoring (CALM) sites in northern Alaska, 1995–2000. *J. Geophys. Res.* 108 (D2): 8168. https://doi.org/10.1029/2001JD000927.

Hinzman, L.D., Goering, D.G., Kinney, T.C., and Li, S. (1997). Numerical simulation of thermokarst formation during disturbance. In: *Disturbance and Recovery in Arctic Lands: an Ecological Perspective* (ed. R.M. Crawford), 191–211. Dordrecht: Kluwer Academic Publishers.

Hinzman, L.D., Bettez, N.D., Bolton, W.R. et al. (2005). Evidence and implications of recent climate change in northern Alaska and other Arctic regions. *Clim. Change* 72: 251–298. https://doi.org/10.1007/s10584-005-5352-2.

Hirota, M., Tang, Y., Hu, Q.-W. et al. (2006). Carbon dioxide dynamics and controls in a deep-water wetland on the Qinghai-Tibetan Plateau. *Ecosystems* 9: 673–688. https://doi.org/10.1007/s10021-006-0029-x.

Hobbie, S.E., Schimel, J.P., Trumbore, S.E. et al. (2002). Controls over carbon storage and turnover in high-latitude soils. *Glob. Chang. Biol.* 6 (S1): 196–210. https://doi.org/10.1046/j.1365-2486.2000.06021.x.

Holton, J.R. (2004). *An Introduction to Dynamic Meteorology*, 4e. San Diego: Elsevier Academic Press.

Hong, J. and Kim, J. (2008). Simulation of surface radiation balance on the Tibetan Plateau. *Geophys. Res. Lett.* 35: L08814. https://doi.org/10.1029/2008GL033613.

Hong, S., Lakshmi, V., and Small, E.E. (2007). Relationship between vegetation biophysical properties and surface temperature using multisensor satellite data. *J. Clim.* 20 (22): 5593–5606. https://doi.org/10.1175/2007JCLI1294.1.

Hong, Y., Liu, G., and Li, J.L. (2016). Assessing the radiative effects of global ice clouds based on CloudSat and CALIPSO measurements. *J. Clim.* 29: 7651–7674. https://doi.org/10.1175/JCLI-D-15-0799.1.

Hoppel, K., Bevilacqua, R., Allen, D. et al. (2003). POAM III observations of the anomalous 2002 Antarctic ozone hole. *Geophys. Res. Lett.* 30: 1394. https://doi.org/10.1029/2003GL016899.

Hoskins, B.J., Mcintyre, M.E., and Robertson, A.W. (1985). On the use and significance of isentropic potential vorticity maps. *Quart. J. Roy. Meteorolog. Soc.* 111: 877–946. https://doi.org/10.1002/qj.49711147002.

Houston, T.G. and Changnon, S.A. (2007). Freezing rain events: a major weather hazard in the conterminous US. *Nat. Hazards* 40 (2): 485–494. https://doi.org/10.1007/s11069-006-9006-0.

Hu, Z.Z., Yang, S., and Wu, R. (2003). Long-term climate variations in China and global warming signals. *J. Geophys. Res.* 11: 1–13. https://doi.org/10.1029/2003JD003651.

Hu, B., Wang, Y., and Liu, G. (2012). Relationship between net radiation and broadband solar radiation in the Tibetan Plateau. *Adv. Atmos. Sci.* 29 (1): 135–143. https://doi.org/10.1007/s00376-011-0221-6.

Hu, G., Zhao, L., Wu, X. et al. (2017). Comparison of the thermal conductivity parameterizations for a freeze-thaw algorithm with a multi-layered soil in permafrost regions. *Catena* 156: 244–251. https://doi.org/10.1016/j.catena.2017.04.011.

Hua, W.J. and Chen, H.S. (2011). Response of land surface processes to global warming and its possible mechanism based on CMIP3 multi-model ensemble. *Chin. J. Atmos. Sci.* 35: 121–133. (in Chinese).

Huang, T.J. (2017). *Spatial-temporal changes of glacier thickness and lake level on the Qinghai-Tibetan Plateau*. Chinese Academy of Sciences: University of Chinese Academy of Sciences (Aerospace Information Research Institute.

Huang, X., Chen, X., Flanner, M. et al. (2018). Improved representation of surface spectral emissivity in a global climate model and its impact on simulated climate. *J. Clim.* 31: 3711–3727.

Huang, X., Chen, C., Yao, B.-Q. et al. (2022). Spatiotemporal dynamics of the carbon budget and the response to grazing in Qinghai grasslands. *Front. Plant Sci.* 12: 775015. https://doi.org/10.3389/fpls.2021.775015.

Hubberten, H.W. and Romanovskii, N.N. (2001). Permafrost response on economic development, environmental security and natural resources. Alfred Wegener Institute for Polar and Marine Research, Potsdam Division, Telegrafenberg A43, D-14473 Potsdam.

Huffman, G.J. and Norman Jr, G.A. (1988). The supercooled warm rain process and the specification of freezing precipitation. *Mon. Weather. Rev.* 116 (11): 2172–2182. DOI:10.1175/1520-0493(1988)116<2172:TSWRPA>

Huntington, T.G., Hodgkins, G.A., Dudley, R.W. et al. (2003). Historical trend in river ice thickness and coherence in hydroclimatological trends in Maine. *Clim. Chang.* 61 (1): 217–236.

Hydrology Bureau, Ministry of Water Resources and Power (1987). *Evaluation of Water Resources in China*. Beijing: Water and Electric Press (in Chinese).

Immerzeel, W., Petersen, L., Ragettli, S., and Pellicciotti, F. (2014). The importance of observed gradients of air temperature and precipitation for modeling runoff from a glacierized watershed in the Nepalese Himalayas. *Water Resour. Res.* 50 (3): 2212–2226. https://doi.org/10.1002/2013WR014506.

Immerzeel, W.W., van Beek, L.P.H., and Bierkens, M.F.P. (2010). Climate change will affect the Asian water towers. *Nat. Geosci.* 328 (5984): 1382–1385. https://doi.org/10.1126/science.1183188.

Immerzeel, W.W., Pellicciotti, F., and Bierkens, M.F.P. (2013). Rising river flows throughout the twenty-first century in two Himalayan glacierized watersheds. *Nat. Geosci.* 6 (8): 1–4. https://doi.org/10.1038/ngeo1896.

Instanes, A., Anisimov, O., Brigham, L. et al. (2005). Infrastructure: Buildings, support systems, and industrial facilities. In: *Arctic Climate Impact Assessment*, 907–944. New York: Cambridge University Press https://doi.org/10.1109/EICCCC.2006.277269.

IPCC, Intergovernmental Panel for Climate Change (2013). Working Group I Contribution to the IPCC Fifth Assessment Report Climate Change 2013: The Physical Science Basis Summary for Policymakers.

Ives, J.D., Barry, R.G., and Alford, D.K. (1974). Arctic and alpine environments. *J. Anim. Ecol.* 45 (2): 608.

Jacob, T., Wahr, J., Pfeffer, W.T., and Swenson, S. (2012). Recent contributions of glaciers and ice caps to sea level rise. *Nature* 482: 514–518. https://doi.org/10.1038/nature10847.

Jafarov, E.E., Romanovsky, V.E., Genet, H. et al. (2013). The effects of fire on the thermal stability of permafrost in lowland and upland black spruce forests of interior Alaska in a changing climate. *Environ. Res. Lett.* 8 (3): 035030. https://doi.org/10.1088/1748-9326/8/3/035030.

Jakobsson, M. (2002). Hypsometry and volume of the Arctic Ocean and its constituent seas. *Geochem. Geophys. Geosyst.* 3: 1028. https://doi.org/10.1029/2001GC000302.

Jeffries, M.K. (1992). Arctic ice shelves and ice islands: origin, growth and disintegration, physical characteristics, structural stratigraphic variability, and dynamics. *Rev. Geophys.* 30: 245–267. https://doi.org/10.1029/92RG00956.

Ji, L., Senay, G.B., and Verdin, J.P. (2014). Evaluation of the global land data assimilation system (GLDAS) air temperature data products. In: *American Geophysical Union Fall Meeting Abstracts*, 0818. https://doi.org/10.1175/JHM-D-14-0230.1.

Ji, Z.M. and Kang, S.C. (2012). Projection of snow cover changes over China under RCPs scenarios. *Clim. Dyn.* 41: 589–600.

Jia, G.J., Epstein, H.E., and Walker, D.A. (2003). Greening of Arctic Alaska, 1981–2001. *Geophys. Res. Lett.* 30 (20): 2067. https://doi.org/10.1029/2003GL018268.

Jia, H., Chen, F., and Pan, D. (2019). Disaster chain analysis of avalanche and landslide and the river blocking dam of the Yarlung Zangbo River in Milin County of Tibet on 17 and 29 October 2018. *Int. J. Environ. Res. Public Health* 16 (23): 4707. doi: 10.3390/ijerph16234707.

Jin, H., Li, S., Cheng, G. et al. (2000). Permafrost and climatic change in China. *Glob. Planet. Chang* 26: 387–404. https://doi.org/10.1016/S0921-8181(00)00051-5 (in Chinese).

Jin, H.J., He, R.X., Cheng, G.D. et al. (2009). Change in frozen ground and co-environmental impacts in the source area of the Yellow River on northeastern Qinghai-Tibet Plateau, China. *Environ. Res. Lett.* 4 (4): 045206.

Jin, Z., Zhuang, Q., He, J.-S. et al. (2015). Net exchanges of methane and carbon dioxide on the Qinghai- Tibetan Plateau from 1979 to 2100. *Environ. Res. Lett.* 10: https://doi.org/10.1088/1748-9326/10/8/085007.

Jobbagy, E.G. and Jackson, R.B. (2000). The vertical distribution of soil organic carbon and its relation to climate and vegetation. *Ecol. Appl.* 10: 423–436. https://doi.org/10.1890/1051-0761(2000)010[0423:TVDOSO]2.0.CO;2.

Johannessen, O.M., Bengtsson, L., Miles, M.W. et al. (2004). Arctic climate change: observed and modelled temperature and sea-ice variability. *Tellus* 56A: 328–341. https://doi.org/10.3402/tellusa.v56i4.14418.

Johansson, T., Malmer, N., Crill, P.M. et al. (2006). Decadal vegetation changes in a northern peatland, greenhouse gas fluxes and net radiative forcing. *Glob. Chang. Biol.* 12: 2352–2369. https://doi.org/10.1111/j.1365-2486.2006.01267.x.

Jones, D.L. and Darrah, P.R. (1993). Re-sorption of organic compounds by roots of *Zea mays L* and its consequences in the rhizosphere. 2. Experimental and model evidence for simultaneous exudation and re-sorption of soluble C compounds. *Plant Soil* 153: 47–59. https://doi.org/10.1007/BF00010543.

Jones, P.D. and Mogberg, A. (2003). Hemispheric and large scale air temperature variations: an extensive revision and an update to 2001. *J. Clim.* 16: 206–223. https://doi.org/10.1175/1520-0442(2003)016<0206:HALSSA>2.0.CO;2.

Jonsson, A. and Karlsson, M. (2003). Sources of carbon dioxide supersaturation in Clearwater and humic lakes in northern Sweden. *Ecosystems* 6: 224–235.

Jorgenson, M.T. and Brown, J. (2005). Classification of the Alaskan Beaufort Sea coast and estimation of carbon and sediment inputs from coastal erosion. *Geo-Mar. Lett.* 25: 69–80.

Jorgenson, M.T., Racine, C.H., Walters, J.C., and Osterkamp, T.E. (2001). Permafrost degradation and ecological changes associated with a warming climate in central Alaska. *Clim. Chang.* 48: 551–579.

Julien, Y., Sobrino, J.A., and Verhoef, W. (2006). Changes in land surface temperatures and NDVI values over Europe between 1982 and 1999. *Remote Sens. Environ.* 103 (1): 43–55. https://doi.org/10.1016/j.rse.2006.03.011.

Kääb, A., Berthier, E., Nuth, C. et al. (2012). Contrasting patterns of early twenty-first-century glacier mass change in the Himalayas. *Nat. Lett.* 488: 495–498. https://doi.org/10.1038/nature1132.

Kääb, A., Leinss, S., Gilbert, A. et al. (2018). Massive collapse of two glaciers in western Tibet in 2016 after surge-like instability. *Nat. Geosci.* 11 (2): 114–120. https://doi.org/10.1038/s41561-017-0039-7.

Kajikawa, M., Sakurai, K.I., and Kikuchi, K. (1988). Characteristic features of supercooled raindrops in the mid-winter season of Arctic Canada. *J. Meteorol. Soc. Jpn. Ser. II* 66 (2): 393–398. https://doi.org/10.2151/jmsj1965.66.2_393.

Kane, D.L., Hinzman, L.D., and Zarling, J.P. (1991). Thermal response of the active layer to climate warming in a permafrost environment. *Cold Reg. Sci. Technol.* 19: 111–122. https://doi.org/10.1016/0165-232X(91)90002-X.

Kane, D.L., Hinkel, K.M., and Goering, D.J. (2001). Nonconductive heat transfer associated with frozen soils. *Glob. Planet. Chang.* 29: 275–292. https://doi.org/10.1016/S0921-8181(01)00095-9.

Kane, D.L., Yoshikawa, K., and McNamara, J.P. (2013). Regional groundwater flow in an area mapped as continuous permafrost, NE Alaska (USA). *Hydrgeol. J.* 21 (1): 41–52. https://doi.org/10.1007/s10040-012-0937-0.

Kang, E.S. (2000). Review and prospect of hydrological research in cold and arid regions of China. *J. Glaciol. Geocryol.* 22 (002): 178–188. (in Chinese).

Kang, H.W., Gu, X.Q., Zhu, C.W., and Whitfield, P.H. (2004). Precipitation recycling in Southern and Central China Chin. *J. Atmos. Sci.* 28 (6): 892–900. (in Chinese).

Kaplan, J.O., Bigelow, N.H., Prentice, I.C. et al. (2003). Climate change and Arctic ecosystems II. Modeling, paleodata-model comparisons, and future projections. *J. Geophys. Res.* 108 (8171): https://doi.org/10.1029/2002JD002559.

Karlsson, K.-G., Anttila, K., Trentmann, J. et al. (2017). Clara-a2: the second edition of the CM SAF cloud and radiation data record from 34 years of global AVHRR data. *Atmos. Chem. Phys.* 17: 5809–5828.

Kaštovská, E. and Šantrùèková, H. (2007). Fate and dynamics of recently fixed C in pasture plant-soil system under field conditions. *Plant Soil* 300: 61–69. https://doi.org/10.1007/s11104-007-9388-0.

Kato, T., Tang, Y., Gu, S. et al. (2004). Carbon dioxide exchange between the atmosphere and an alpine meadow ecosystem on the Qinghai-Tibetan Plateau, China. *Agric. For. Meteorol.* 124: 121–134. https://doi.org/10.1016/j.agrformet.2003.12.008.

Kay, A.L. and Crooks, S.M. (2014). An investigation of the effect of transient climate change on snowmelt, flood frequency and timing in northern Britain. *Int. J. Climatol.* 34 (12): 3368–3381. https://doi.org/10.1002/joc.3913.

Kayastha, R.B., Takehuci, Y., Nakawo, M., and Ageta, Y. (2000). Practical prediction of ice melting beneath various thickness of debris cover on Khumbu Glacier, Nepal, using a positive degree-day factor. In: *Debris-Covered Glaciers: Proceedings of an International Workshop Held at the University of Washington in Seattle* (13–15 September 2000), (264), 71. Washington, USA.

Ke, C., Li, X., Xie, H. et al. (2016). Variability in snow cover phenology in China from 1952 to 2010. *Hydrol. Earth Syst. Sci.* 20 (2): 755–770.

Kidston, J., Scaife, A.A., Hardiman, S.C. et al. (2015). Stratospheric influence on tropospheric jet streams, storm tracks and surface weather. *Nat. Geosci.* 8: 433–440. https://doi.org/10.1038/ngeo2424.

Kim, Y.S. (1994). Analysis and comparison of diurnal-variations of cloud radiative forcing - earth radiation budget experiment and international satellite cloud climatology project results. *J. Geophys. Res.-Atmos.* 99: 20847–20862.

King, A.D., Butler, A.H., Jucker, M. et al. (2019). Observed relationships between suddenstratospheric warmings and European climate extremes. *J. Geophys. Res. Atmos.* 124: 13943–13961. https://doi.org/10.1029/2019JD030480.

Kirchner, I., Stenchikov, G.L., Graf, H.-F. et al. (1999). Climate model simulation of winter warming and summer cooling following the 1991 mount Pinatubo volcanic eruption. *J. Geophys. Res.* 104: 19039–19055. https://doi.org/10.1029/1999JD900213.

Kirillov, S., Dmitrenko, I., Rysgaard, S. et al. (2018). The inferred formation of a subice platelet layer below the multiyear landfast sea ice in the Wandel Sea (NE Greenland) induced by meltwater drainage. *J. Geophys. Res. Oceans* 123 (5): 3489–3506. https://doi.org/10.1029/2017JC013672.

Klein, T. and Heinemann, G. (2002). Interaction of katabatic winds and mesocyclones near the eastern coast of Greenland. *Meteorol. Appl.* 9: 407–422. https://doi.org/10.1017/S1350482702004036.

Kling, G.W., Kipphut, G.W., and Miller, M.C. (1991). Arctic Lakes and streams as gas conduits to the atmosphere: implications for tundra carbon budgets. *Science* 251: 298–301. https://doi.org/10.1126/science.251.4991.298.

Kodera, K. and Chiba, M. (1995). Tropospheric circulation changes associated with stratospheric sudden warming: a case study. *J. Geophys. Res.* 100 (D6): 11055–11068. https://doi.org/10.1029/95JD00771.

Koichi, U. (1986). Conductive heat flux through active layer in tundra plain. *J. Fac. Sci.*, Hokkaido University, Series 7, Geophysics 8 (1): 29–35.

Kondragunta, S., Flynn, L.E., Neuendorffer, A. et al. (2005). Vertical structure of the anomalous 2002 Antarctic ozone hole. *J. Atmos. Sci.* 62: 801–811. https://doi.org/10.1175/JAS-3324.1.

Kong, Y. and Wang, C.H. (2017). Responses and changes in the permafrost and snow water equivalent in the Northern Hemisphere under a scenario of 1.5 C warming. *Adv. Clim. Change Res.* 8 (4): 235–244. https://doi.org/10.1016/j.accre.2017.07.002.

Kong, Y. and Wang, C.-H. (2017). Responses and changes in the permafrost and snow water equivalent in the Northern Hemisphere under a scenario of 1.5 °C warming. *Adv. Clim. Chang. Res.* https://doi.org/10.1016/j.accre.2017.07.002.

Köppen, W. (1936). *Das geographische System der Klimate*, 1–44. Berlin: Gebrüder Borntraeger.

Koren, V., Schaake, J., Mitchell, K. et al. (1999). A parameterization of snowpack and frozen ground intended for NCEP weather and climate models. *J. Geophys. Res.* 104: 19569–19585. https://doi.org/10.1029/1999JD900232.

Koryakin, V.S. (1990). Ledniki Novoi Zemliiklimat. In: *Glaciers of Novaya Zemlya and climate*, 23–29. Priroda (n.p. 1).

Kossin, J.P., Schubert, W.H., and Montgomery, M.T. (2000). Unstable interactions between a hurricane's primary eyewall and a secondary ring of enhanced vorticity. *J. Atmos. Sci.* 57: 3893–3917. https://doi.org/10.1175/1520-0469(2001)058<3893:UIBAHS>2.0.CO;2.

Koster, R.D. and Suarez, M.J. (2001). Soil moisture memory in climate models. *J. Hydrometeorol.* 2: 558–570. https://doi.org/10.1175/1525-7541(2001)002<0558:SMMICM>2.0.CO;2.

Koster, R.D., Dirmeyer, P.A., Hahmann, A.N. et al. (2002). Comparing the degree of land-atmosphere interaction in four atmospheric general circulation models. *J. Hydrometeorol.* 3 (3): 363–375. https://doi.org/10.1175/1525-7541(2002)003<0363.CTDOLA>2.0.CO;2.

Koster, R.D., Dirmeyer, P.A., Guo, Z. et al. (2004). Regions of strong coupling between soil moisture and precipitation. *Science* 305: 1138–1140. https://doi.org/10.1126/science.1100217.

Koster, R.D., Sud, Y.C., Guo, Z. et al. (2006). GLACE: the global land–atmosphere coupling experiment. Part I: overview. *J. Hydrometeorol.* 7 (4): 590–610. https://doi.org/10.1175/JHM510.1.

Kotlyakov, V.M., Osipova, G.B., and Tsvetkov, D.G. (1997). Fluctuation of unstable: mountain glaciers: scale and character. *Ann. Glaciol.* 24: 338–343. https://doi.org/10.3189/S0260305500012416.

Kretschmer, M., Cohen, J., Matthias, V. et al. (2018). The different stratospheric influence on cold-extremes in Eurasia and North America. *Nat. Clim. Atmos. Sci.* 1: 44. https://doi.org/10.1038/s41612-018-0054-4.

Krishnamurti, T.N., Daggupaty, S.M., Fein, J. et al. (1973). Tibetan high and upper tropospheric tropical circulations during northern summer. *Bull. Am. Meteorol. Soc.* 54: 1234–1249. https://doi.org/10.1175/1520-0477-54.12.1234.

Krüger, K., Naujokat, B., Labitzke, K. et al. (2005). The unusual midwinter warming in the southern hemisphere stratosphere 2002: a comparison to northern hemisphere phenomena. *J. Atmos. Sci.* 62 (3): 603–613. https://doi.org/10.1175/JAS-3316.1.

Kudryavtsev, V.A., Garagulya, L.S., Kondrat, yeva, K.A., and Melamed, V.G. (1974). *Fundamentals of Frost Forecasting in Geological Engineering Investigations*. Moscow: Nauka (English translation appears as U.S. Army Cold Regions Research and Engineering Laboratory Draft Translation 606) (in Russian).

Kug, J.-S., Jeong, J.-H., Jang, Y.-S. et al. (2015). Two distinct influences of Arctic warming on cold winters over North America and East Asia. *Nat. Geosci.* 8: 759–762. https://doi.org/10.1038/ngeo2517.

Kukulies, J., Chen, D., and Wang, M. (2019). Temporal and spatial variations of convection and precipitation over the Tibetan Plateau based on recent satellite observations. Part I: cloud climatology derived from CloudSat and CALIPSO. *Int. J. Climatol.* 39: 5396–5412. https://doi.org/10.1002/joc.6162.

Kumari, B.P. and Goswami, B.N. (2010). Seminal role of clouds on solar dimming over the Indian monsoon region. *Geophys. Res. Lett.* 37: L06703. https://doi.org/10.1029/2009GL042133.

Kumari, B.P., Londhe, A.L., Daniel, S. et al. (2007). Observational evidence of solar dimming: offsetting surface warming over India. *Geophys. Res. Lett.* 34: L21810. https://doi.org/10.1029/2007GL031133.

Kurylyk, B.L., MacQuarrie, K.T.B., and McKenzie, J.M. (2014). Climate change impacts on groundwater and soil temperatures in cold and temperate regions: implications, mathematical theory, and emerging simulation tools. *Earth Sci. Rev.* 138: 313–334. https://doi.org/10.1016/j.earscirev.2014.06.006.

Kuzyakov, Y., Ehrensberger, H., and Stahr, K. (2001). Carbon partitioning and below-ground translocation by lolium perenne. *Soil Biol. Biochem.* 33: 61–74. https://doi.org/10.1016/S0038-0717(00)00115-2.

Labitzke, K. (1968). Midwinter warmings in the upper stratosphere in 1966. *Q. J. R. Meteotol. Soc.* 94: 279–291. https://doi.org/10.1002/qj.49709440105.

Labitzke, K. (1982). On interannual variability of the middle stratosphere during northern winters. *J. Meteorol. Soc. Jpn.* 60: 124–139. https://doi.org/10.2151/jmsj1965.60.1_124.

Lachlan-Cope, T. (2010). Antarctic clouds. *Polar Res.* 29: 150–158.

Lackmann, G.M. (2001). Analysis of a surprise western New York snowstorm. *Weather Forecasting* 16 (1): 99–116. https://doi.org/10.1175/1520-0434(2001)016<0099:AOASWN>2.0.CO;2.

Lal, R.P. and Ram, S. (2009). Climatology of blizzards over Schirmacher Oasis, east Antarctica. *Mausam* 60 (1): 39–50. https://doi.org/10.54302/mausam.v60i1.960.

Lamb, P.J. et al. (2016). Synoptic pattern analysis and climatology of ice and snowstorms in the southern Great Plains, 1993–2011. *Weather Forecasting* 31 (4): 1109–1136. https://doi.org/10.1175/WAF-D-15-0172.1.

Laternser, M. and Schneebeli, M. (2002). Temporal trend and spatial distribution of avalanche activity during the last 50 years in Switzerland. *Nat. Hazards* 27 (3): 201–230. https://doi.org/10.1023/A:1020327312719.

Lavigne, A., Eckert, N., Bel, L., and Parent, E. (2015). Adding expert contributions to the spatiotemporal modelling of avalanche activity under different climatic influences. *J. R. Stat. Soc. Ser. C. Appl. Stat.* 64 (4): 651–671. https://doi.org/10.1111/rssc.12095.

Lawrence, D.M. and Slater, A.G. (2008). Incorporating organic soil into a global climate model. *Clim. Dyn.* 30 (2–3): 145–160. https://doi.org/10.1007/s00382-007-0278-1.

Lawrence, D.M., Oleson, K.W., Flanner, M.G. et al. (2011). Parameterization improvements and functional and structural advances in version 4 of the community land model. *J. Adv. Model. Earth Syst.* 3: 365–375. https://doi.org/10.1029/2011MS00045.

Le Roux, E., Evin, G., Eckert, N. et al. (2021). Elevation-dependent trends in extreme snowfall in the French Alps from 1959 to 2019. *Cryosphere* 15 (9): 4335–4356. https://doi.org/10.5194/tc-15-4335-2021.

Leake, J.R., Ostle, N.J., Rangel-Castro, J.I., and Johnson, D. (2006). Carbon fluxes from plants through soil organisms determined by field 13CO2 pulse-labelling in an upland grassland. *Appl. Soil Ecol.* 33: 152–175. https://doi.org/10.1016/j.apsoil.2006.03.001.

Lemke, P., Ren, J., Alley, R.B. et al. (2007). *Observations: Changes in Snow, Ice and Frozen Ground, Contribution of Working Group I to the Fourth Assessment Report of the Intergovernmental Panel on Climate Change*. Cambridge: Cambridge University Press.

Leopold, L.B. (1972). River channel change with time: an example. *Geol. Soc. Am. Bull.*, 1973, 84 (6): 1845–1860.

Leung, L. R., Qian, Y. , and Bian, X. (2003). Hydro climate of the western United States based on observations and regional climate simulation of 1981–2000. Part I: Seasonal statistics. *J. Climate*: 16, 1892–1911.

Lhasa Group of Tibetan Plateau Meteorology Research (1981). *Research of 500-mb Vortex and Shear Lines over the Tibetan Plateau in Summer*, 341–348. Beijing: Science Press (in Chinese).

Li, G., Lu, J., Jin, B., and Bu, N. (2001). The effects of anomalous snow cover of the Tibetan Plateau on the surface heating. *Adv. Atmos. Sci.* 18: 1207–1214. https://doi.org/10.1007/s00376-001-0034-0.

Li, G., Zhao, F., Huang, C. et al. (2014). Analysis of 30-year climatology of the Tibetan Plateau vortex in summer with NCEP reanalysis data. *Chin. J. Atmos. Sci.* 38: 756–769. (in Chinese)

Li, G.P. and Lu, J.H. (1996). Some possible solutions of nonlinear internal inertial gravity wave equations in the atmosphere. *Adv. Atmos. Sci.* 13 (2): 244–252. https://doi.org/10.1007/BF02656866.

Li, G.P. and Xu, Q. (2005). Effect of dynamic pumping in the boundary layer on the Tibetan Plateau vortices. *Atmos.* 6: 965–972. https://doi.org/10.3878/j.issn.1006-9895.2005.06.12. (in Chinese).

Li, G.P., Lu, H.G., Huang, C.H. et al. (2016). A climatology of the surface heat source on the Tibetan plateau in summer and its impacts on the formation of the Tibetan Plateau vortex. *Chin. J. Atmos. Sci.* 40 (1): 131–141. (in Chinese). https://doi.org/10.1007/s10409-004-0010-x.

Li, J. and Zeng, Q. (2001). A unified monsoon index. *Geophys. Res. Lett.* 29: 1274. https://doi.org/10.1029/2001GL013874.

Li, J., Huang, J., Stamnes, K. et al. (2015). A global survey of cloud overlap based on CALIPSO and CloudSat measurements. *Atmos. Chem. Phys.* 15: 519–536. https://doi.org/10.5194/acp-15-519-2015.

Li, J.P. and Zeng, Q.C. (2005). A new monsoon index, its interannual variability and relation with monsoon precipitation. *Clim. Environ. Res.* 10 (3): 351–364. (in Chinese).

Li, L., Zhang, R., Wen, M., and Liu, L. (2014). Effect of the atmospheric heat source on the development and eastward movement of the Tibetan Plateau vortices. *Tellus A* 66 (1): 24451. https://doi.org/10.3402/tellusa.v66.24451.

Li, R., Zhao, L., Wu, T. et al. (2014). The impact of surface energy exchange on the thawing process of active layer over the northern Qinghai-Xizang Plateau, China. *Environ. Earth Sci.* 72: 2091–2099. https://doi.org/10.1007/s12665-014-3117-9.

Li, S. and Cheng, G. (1996). *Frozen Ground Map of the Qinghai-Tibet Plateau*. Lanzhou: Gansu Culture Press (in Chinese).

Li, S., Yang, K., and Wang, C. (2018). Bias characteristics of land surface model (CLM4.5) over Tibetan Plateau during soil freezing-thawing period and its causes. *J. Glaciol. Geocryol.* 40: 322–334.

Li, X., Cheng, G.D., Jin, H.J. et al. (2008). Cryospheric change in China. *Glob. Planet. Chang.* 62: 210–218. https://doi.org/10.1016/j.gloplacha.2008.02.001 (in Chinese).

Li, X., Jin, R., Pan, X.D. et al. (2012). Changes in the near-surface soil freeze-thaw cycle on the Qinghai-Tibetan Plateau. *Int. J. Appl. Earth Obs. Geoinf.* 17: 33–42. https://doi.org/10.1016/j.jag.2011.12.002.

Li, Z., Kurz, W.A., Apps, M.J., and Beukema, S.J. (2003). Belowground biomass dynamics in the carbon budget model of the Canadian forest sector: recent improvements and implications for the estimation of NPP and NEP. *Can. J. For. Res.* 33: 126–136. https://doi.org/10.1139/x02-165.

Liang, F. and Xia, X.A. (2005). Long-term trends in solar radiation and the associated climatic factors over China for 1961–2000. *Ann. Geophys.* 23 (7): 2425–2432. https://doi.org/10.5194/angeo-23-2425-2005.

Liang, X., Liu, Y., and Wu, G. (2005). Effect of Tibetan Plateau on the site of onset and intensity of the Asian summer monsoon. *Acta. Meteor. Sin.* 63: 799–805.

Liang, H., Liu, J.M., Zhang, J.C. et al. (2006). Research oil retrieval of the amount of atmospheric water vapor over Qinghai-Xizang Plateau. *Plateau Meteorol.* 25 (6): 1055–1063. (in Chinese).

Liang, S., Wang, K., Zhang, X., and Wild, M. (2010). Review on estimation of land surface radiation and energy budgets from ground measurement, remote sensing and model simulations. *IEEE J. Sel. Top. Appl. Earth Obs. Remote Sens.* 3 (3): 225–240. https://doi.org/10.1109/JSTARS.2010.2048556.

Liang, S. et al. (2013). *A Long-Term Global LAnd Surface Satellite (GLASS) Dataset for Environmental Studies, Int. J. Digital Earth*, https://doi.org/10.1080/17538947.2013.805262.

Liang, X., Lettenmaier, D.P., and Wood, E.F. (1994). A simple hydrologically based model of land surface water and energy fluxes for general circulation models. *J. Geophys. Res.* 99 (D7): 14415–14428. https://doi.org/10.1029/94JD00483.

Liang, X., Wood, E.F., and Lettenmaier, D.P. (1996). Surface soil moisture parameterization of the VIC-2 l model: evaluation and modification. *Global Planet. Change* 13: 195–206. https://doi.org/10.1016/0921-8181(95)00046-1.

Liang, X. -Z., Xu, M., Kunkel, K. E. et al. (2007). Regional climate model simulation of U.S.–Mexico summer precipitation using the optimal ensemble of two cumulus parameterizations. *J. Climate*: 20, 5201–5207.

Limpasuvan, V., Thompson, D.W.J., and Hartmann, D.L. (2004). The lifecycle of the Northern Hemisphere sudden stratospheric warmings. *J. Clim.* 17: 2584–2596. https://doi.org/10.1175/1520-0442(2004)017<2584:TLCOTN>2.0.CO;2.

Lin, Z. (2015). Analysis of tibetan plateau vortex activities using era-interim data for the period 1979–2013. *J. Meteorol. Res.* 29: 720–734. https://doi.org/10.1007/s13351-015-4273-x.

Lin, Z., Zhou, Z., and Jia, L. (2013). Objective identifying method of Qinghai-Xizang Plateau vortex using NCEP/NCAR reanalysis dataset. *Plateau Meteorol.* 32: 1580–1588. (in Chinese).

Lin, Z.Y. and Wu, X.D. (1990). A preliminary analysis about the tracks of moisture transportation on the Qinghai–Xizang Plateau. *Geogr. Res.* 9 (3): 33–40.

Lin, Z.Y. and Zhao, X.Y. (1996). Spatial characteristics of changes in temperature and precipitation of the Qinghai-Xizang (Tibet) Plateau. *Sci. China Ser. D.* 39: 442–448.

Lindenschmidt, K.E. and Rokaya, P. (2019). A stochastic hydraulic modelling approach to determining the probable maximum staging of ice-jam floods. *J. Environ. Inf.* 34 (1): 45–54. https://doi.org/10.3808/jei.201900416.

Lindenschmidt, K.E., Rokaya, P., Das, A. et al. (2019). A novel stochastic modelling approach for operational real-time ice-jam flood forecasting. *J. Hydrol.* 575: 381–394. https://doi.org/10.1016/j.jhydrol.2019.05.048.

Liou, K.N., Lee, W.-L., and Hall, A. (2007). Radiative transfer in mountains: application to the Tibetan Plateau. *Geophys. Res. Lett.* 34: L23809. https://doi.org/10.1029/2007GL031762.

Liu, R.Z. (1986). Dynamic analysis of stability of polar high. *Chin. J. Atmos. Sci.* 10 (003): 88–95. https://doi.org/10.3878/j.issn.1006-9895.1986.03.11 (in Chinese).

Liu, X.D. and Chen, B.D. (2000). Climatic warming in the Tibetan Plateau during recent decades. *Int. J. Climatol.* 20: 1729–1742. https://doi.org/10.1002/1097-0088(20001130)20:14<1729::AID-JOC556>3.0.CO;2-Y.

Liu, X. and Li, G.P. (2007). Analytical solutions for the thermal forcing vortices in the boundary layer and its applications. *Appl. Math. Mech.* 28 (4): 429–439. (in Chinese).

Liu, X.R. and Li, G.P. (2014). Numerical simulation and potential vorticity diagnosis of an eastward moving southwest vortex. *Plateau Meteorol.* 33 (05): 1204–1216. https://doi.org/10.7522/j.issn.1000-0534.2013.00151 (in Chinese).

Liu, Y.F. and Li, G.P. (2016). Climatic characteristics of atmospheric heat source over the Tibetan Plateau and its possible relationship with the generation of the Tibetan Plateau vortex in the summer. *Chin. J. Atmos. Sci.* 40 (4): 864–876. https://doi.org/10.3878/j.issn.1006-9895.1512.15184 (in Chinese).

Liu, X. and Yanai, M. (2002). Influence of Eurasian spring snow cover on Asian summer rainfall. *Int. J. Climatol.* 22: 1075–1089. https://doi.org/10.1002/joc.784.

Liu, X. and Yin, Z.-Y. (2001). Spatial and temporal variation of summer precipitation over the eastern Tibetan Plateau and the North Atlantic Oscillation. *J. Clim.* 14: 2896–2909. https://doi.org/10.1175/1520-0442(2001)014,2896:SATVOS.2.0.CO;2.

Liu, Z.X., Su, Z., Yao, T.D. et al. (2000). Glacier resources and their distribution characteristics on the Tibet Plateau. *Resour. Sci.* 22 (05): 49–52.

Liu, J., Hayakawab, N., Lub, M. et al. (2003). Hydrological and geocryological response of winter streamflow to climate warming in Northeast China. *Cold Reg. Sci. Technol.* 37: 15–24. https://doi.org/10.1016/S0165-232X(03)00012-0.

Liu, B.H., Xu, M., Henderson, M. et al. (2004). Taking China's temperature: daily range, warming trends, and regional variations, 1955–2000. *J. Clim.* 17 (22): 4453–4462. https://doi.org/10.1175/3230.1.

Liu, Y., Hoskins, B., and Blackburn, M. (2007). Impact of Tibetan orography and heating on the summer flow over Asia. *J. Meteorol. Soc. Jpn.* 85B: 1–19. https://doi.org/10.2151/jmsj.85B.1.

Liu, S.L., Liu, R.G., and Liu, Y. (2010). Spatial and temporal variation of global LAI during 1981–2006. *J. Geogr. Sci.* 20 (3): 323–332.

Liu, W., Chen, S., Qin, X. et al. (2012). Storage, patterns, and control of soil organic carbon and nitrogen in the northeastern margin of the Qinghai–Tibetan Plateau. *Environ. Res. Lett.* 7: 035401. https://doi.org/10.1088/1748-9326/7/3/035401.

Liu, B., Sun, Y., Wang, Y., and Zhang, Y. (2013). Monitoring and assessment of vegetation variation in North China based on SPOT/NDVI. *J. Arid Land Resour. Environ.* 27 (9): 98–103.

Liu, J.H., Gao, J.X., and Wang, W.J. (2013). Variations of vegetation coverage and its relations to global climate changes on the Tibetan Plateau during 1981–2005. *J. Mt. Sci.* 2013 (2): 9. https://doi.org/10.1016/0090-2616(81)90039-5.

Liu, F., Mao, X., Zhang, Y. et al. (2014). Risk analysis of snow disaster in the pastoral areas of the Qinghai-Tibet Plateau. *J. Geog. Sci.* 24 (3): 411–426. https://doi.org/10.1007/s11442-014-1097-z.

Liu, S., Zamanian, K., Schleuss, P.-M. et al. (2018). Degradation of Tibetan grasslands: consequences for carbon and nutrient cycles. *Agric. Ecosyst. Environ.* 252: 93–104. https://doi.org/10.1016/j.agee.2017.10.011.

Liu, Y., Li, D., Chen, L., and Ming, F. (2020). Study on the mechanical criterion of ice lens formation based on pore size distribution. *Appl. Sci.* 10: 8981. https://doi.org/10.3390/app10248981.

Liu, Q., Chen, G., Wang, L. et al. (2021). Southward cold airmass flux associated with the East Asian winter monsoon: diversity and impacts. *J. Clim.* 34 (8): 3239–3254. https://doi.org/10.1175/JCLI-D-20-0319.1.

Lolli, S., Campbell, J.R., Lewis, J.R. et al. (2017). Daytime top-of-the-atmosphere cirrus cloud radiative forcing properties at Singapore. *J. Appl. Meteorol. Clim.* 56: 1249–1257. https://doi.org/10.1175/JAMC-D-16-0262.1.

López-Moreno, J.I. and García-Ruiz, J.M. (2004). Influence of snow accumulation and snowmelt on streamflow in the central Spanish Pyrenees / Influence de l'accumulation et de la fonte de la neige sur les écoulements dans les Pyrénées centrales espagnoles. *Hydrol. Sci. J.* 49: 5, null-802.

López-Moreno, J.I., Goyette, S., and Beniston, M. (2009). Impact of climate change on snowpack in the Pyrenees: horizontal spatial variability and vertical gradients. *J. Hydrol.* 374: 384–396. https://doi.org/10.1016/j.jhydrol.2009.06.049.

Lorenz, R., Davin, E.L., and Seneviratne, S.I. (2012). Modeling land-climate coupling in Europe: Impact of land surface representation on climate variability and extremes. *J. Geophys. Res.* 117: D20109. https://doi.org/10.1029/2012JD017755.

Lu, C.H. and Ding, Y.H. (2013). Observational responses of stratospheric sudden warming to blocking highs and its feedbacks on the troposphere. *Chin. Sci. Bull.* 58: 1374–1384. https://doi.org/10.1007/s11434-012-5505-4.

Lu, H., Wu, N., Guo, Z. et al. (2004). Distribution of carbon isotope composition of modern soils on the Qinghai-Tibetan Plateau. *Biogeochemistry.* 70: 275–299. https://doi.org/10.1023/B:BIOG.0000049343.48087.ac.

Luo, Y., Zhang, R., Qian, W. et al. (2011). Intercomparison of deep convection over the Tibetan Plateau-Asian monsoon region and subtropical North America in boreal summer using CloudSat/CALIPSO data. *J. Climate*. 24 (8): 2164-2177.

Luo, S., Qian, Z., and Wang, Q. (1982). Synoptic and climatic studies on relationships between the Tibetan high and the summer floods and droughts in East China. *Plateau Meteorol.* 1 (2): 1–10. https://doi.org/10.1007/s00376-007-0136-4 (in Chinese).

Luo, S., Lü, S., Zhang, Y. et al. (2008). Simulation analysis on land surface process of BJ site of central Tibetan Plateau using CoLM. *Plateau Meteorol.* 27: 259–271. https://doi.org/10.3724/SP.J.1047.2008.00014.

Luo, D., Xiao, Y., Yao, Y. et al. (2016). Impact of ural blocking on winter warm arctic–cold eurasian anomalies. Part i: blocking-induced amplification. *J. Clim.* 29 (11): 3949–3971. https://doi.org/10.1175/JCLI-D-15-0612.1.

Luo, S., Fang, X., Lyu, S. et al. (2017). Improving CLM4.5 simulations of land-atmosphere exchange during freeze-thaw processes on the Tibetan Plateau. *J. Meteorol. Res.* 31: 916. https://doi.org/10.1007/s13351-017-6063-0.

Luo, D., Chen, X., Overland, J. et al. (2019). Weakened potential vorticity barrier linked to Arctic Sea-ice loss and increased mid-latitude cold extremes. *J. Clim.* 32: 4235–4261. https://doi.org/10.1175/JCLI-D-18-0449.1.

Lv, C., Gong, H., and Wang, B. (2010). Senow disaster risk assessment based on fuzzy comprehensive evaluation. In: *[C]2010 18th International Conference on Geoinformatics*, 1–6. Beijing, China: IEEE https://doi.org/10.1109/GEOINFORMATICS.2010.5567707.

Ma, X.B. and Hu, Z.Y. (2005). Precipitation variation characteristics and abrupt change over Qinghai-Xizang plateau in the last 40 years. *J. Desert Res.* 25 (1): 137–139. (in Chinese).

Ma, Y., Kang, S., Zhu, L. et al. (2008). Roof of the World: Tibetan observation and research platform. *Bull. Am. Meteorol. Soc.* 89: 1487–1492. https://doi.org/10.1175/2008BAMS2545.1.

Ma, Y., Zhong, L., Wang, B. et al. (2011). Determination of land surface heat fluxes over heterogeneous landscape of the Tibetan Plateau by using the MODIS and in-situ data. *Atmos. Chem. Phys. Discuss.* 11 (7): 19,617–19,638. https://doi.org/10.5194/acp-11-10461-2011.

Ma, Z.G., Wei, H.L., and Fu, Z.B. (2015). Progress in the research in the relationship between soil moisture and climate change. *Adv. Earth Science* 14 (3): 88–94. (in Chinese).

Mackay, J.R. (1995). Active layer changes (1968 to 1993) following the forest-tundra fire near Inuvik, NWT. *Canada Arct. Alp. Res.* 27: 323–336.

Makhover, A.P. (1983). *The Heat Budget of Arctic Ice in Winter*. St. Petersburg: Gidrometeoizda (in Russian).

Manabe, S. and Brocoli, A.J. (1990). Mountains and arid climate of middle latitude. *Science* 247: 192–195. https://doi.org/10.1126/science.247.4939.192.

Manabe, S. and Wetherald, R.T. (1987). Large-scale changes of soil wetness induced by an increase in atmospheric carbon dioxide. *J. Atmos. Sci.* 44: 1211–1235. https://doi.org/10.1175/1520-0469(1987)044,1211:LSCOSW.2.0.CO;2.

Manney, G.L. and Sabutis, J.L. (2000). Development of the polar vortex in the 1999–2000 Arctic winter stratosphere. *Geophys. Res. Lett.* 27: 2589–2592. https://doi.org/10.1029/2000GL011703.

Manney, G.L., Sabutis, J.L., and Swinbank, R. (2001). A unique stratospheric warming event in November 2000. *Geophys. Res. Lett.* 28 (13): 2629–2632. https://doi.org/10.1029/2001GL012973.

Manney, G.L., Schwartz, M.J., Krüger, K. et al. (2010). Aura microwave limb sounder observations of dynamics and transport during the record-breaking 2009 Arctic stratospheric major warming. *Geophys. Res. Lett.* 36: L12815. https://doi.org/10.1029/2009GL038586.

Mao, F., Tang, S.H., Sun, H., and Zhang, J.H. (2008). A study of dynamic change of dry and wet climate regions in the Tibetan Plateau over the last 46 years. *Chin. J. Atmos. Sci.* 32 (3): 499–507. (in Chinese).

Marks, D., Kimball, J., Tingey, D. et al. (1998). The sensitivity of snowmelt processes to climate conditions and forest cover during rain-on-snow: a case study of the 1996 Pacific northwest flood. *Hydrol. Process.* 12 (10–11): 1569–1587. https://doi.org/10.1002/(SICI)1099-1085(199808/09)12:10/11<1569::AID-P682>3.0.CO;2-L.

Mason, R.M. and Andson, C.E. (1958). The development and decay of the 100mb summertime anticyclone over Southern Asia. *Mon. Weather Rev.* 91: 3–12. https://doi.org/10.1175/1520-0493(1963)091<0003:TDADOT>2.3.CO;2.

Masson, V., Moigne, P.L., Martin, E. et al. (2013). The SURFEXv7.2 land and ocean surface platform for coupled or offline simulation of Earth surface variables and fluxes. *Geosci. Model Dev.* 6 (4): 929–960. https://doi.org/10.5194/gmd-6-929-2013.

Matthias, V., Dörnbrack, A., and Stober, G. (2016). The extraordinarily strong and cold polar vortex in the early northern winter 2015/2016. *Geophys. Res. Lett.* 43: 12287–12294. https://doi.org/10.1002/2016GL071676.

Matus, A.V. and L'Ecuyer, T.S. (2017). The role of cloud phase in Earth's radiation budget. *J. Geophys. Res. Atmos.* 122: 2559–2578. https://doi.org/10.1002/2016JD025951.

Maussion, F., Scherer, D., Mölg, T. et al. (2013). Precipitation seasonality and variability over the Tibetan Plateau as resolved by the high Asia reanalysis. *J. Clim.* 27: 1910–1927. https://doi.org/10.1175/JCLI-D-13-00282.1.

Maykut, G.A. (1986). The surface heat and mass balance. In: *The Geophysics of Sea Ice*. NATO ASI Ser., Ser. B Phys., vol. 146 (ed. N. Untersteiner), 395–464. New York: Plenum.

Mazhitova, G.G. (2008). Soil temperature regimes in the discontinuous permafrost zone in the east European Russian Arctic. *Eurasian Soil Sci.* 41: 48–62.

McCabe, G.J. and Wolock, D.M. (2009). Recent declines in Western U.S. snowpack in the context of twentieth-century climate variability. *Earth Interact.* 13: 1–15.

McCabe, G.J. and Wolock, D.M. (2010). Long-term variability in Northern Hemisphere snow cover and associations with warmer winters. *Clim. Change* 99: 141–153. https://doi.org/10.1007/s10584-009-9675-2.

Mccarthy, N., Bentsen, N.S., Willoughby, I., and Balandier, P. (2011). The state of forest vegetation management in europe in the 21st century. *Eur. J. For. Res.* 130 (1): 7–16. https://doi.org/10.1007/s10342-010-0429-5.

McCauley, C.A., White, D.M., Lilly, M.R., and Nyman, D.M. (2002). A comparison of hydraulic conductivities, permeabilities and infiltration rates in frozen and unfrozen soils. *Cold Reg. Sci. Technol.* 34: 117–125. https://doi.org/10.1016/S0165-232X(01)00064-7.

McGuire, A.D., Melillo, J.M., Kicklighter, D.W. et al. (1997). Equilibrium responses of global net primary production and carbon storage to doubled atmospheric carbon dioxide: sensitivity to changes in vegetation nitrogen concentration. *Glob. Biogeochem. Cycles* 11: 173–189. https://doi.org/10.1029/97GB00059.

McGuire, A.D., Clein, J., Melillo, J.M. et al. (2000). Modeling carbon responses of tundra ecosystems to historical and projected climate: the sensitivity of pan-arctic carbon storage to temporal and spatial variation in climate. *Glob. Chang. Biol.* 6: S141–S159. https://doi.org/10.1046/j.1365-2486.2000.06017.x.

McGuire, A.D., Anderson, L.G., Christensen, T.R. et al. (2009). Sensitivity of the carbon cycle in the Arctic to climate change. *Ecol. Monogr.* 79: 523–555. https://doi.org/10.1890/08-2025.1.

McNamara, J.P., Kane, D.L., and Hinzman, L.D. (1998). An analysis of streamflow hydrology in the Kuparuk River basin, Arctic Alaska: a nested watershed approach. *J. Hydrol.* 206: 39–57. https://doi.org/10.1016/S0022-1694(98)00083-3.

McNoldy, B.D. (2004). Triple eyewall in hurricane Juliette. *Bull. Am. Meteorol. Soc.* 85: 1663–1666. http:!//www.jstor.org/stable/26221211.

Melillo, J.M., Callaghan, T.V., Woodward, F.I. et al. (1990). Climate change: effects on ecosystems. In: *Climate Change—The IPCC Scientific Assessment* (ed. G.J. Jenkins and J.J. Ephraums), 282–310. New York: Cambridge University Press.

Meybeck, M., Green, P., and Vorosmarty, C. (2001). New typology for mountains and other relief classes: an application to global continental water resources and population distribution. *Mt. Res. Dev.* 21: 34–45. https://doi.org/10.1016/S0165-232X(01)00064-7.

Michel, F.A. and van Everdingen, R.O. (1994). Changes in hydrogeologic regimes in permafrost regions due to climatic change. *Permafrost Periglacial Process.* 5: 191–195. https://doi.org/10.1002/ppp.3430050308.

Migliavacca, F., Confortola, G., Soncini, A. et al. (2015). Hydrology and potential climate changes in the Rio Maipo(Chile). *Geogr. Fis. Din. Quat.* 38: 155–168.

Mihalcea, C., Brock, B.W., Diolaiuti, G. et al. (2008). Using ASTER satellite and ground-based surface tempera ture measurements to derive supraglacial debris cover and thickness patterns on Miage Glacier (Mont Blanc Massif, Italy). *Cold Reg. Sci. Technol.* 52: 341–354. https://doi.org/10.1016/j.coldregions.2007.03.004.

Mihalcea, C., Mayer, C., Diolaiuti, G. et al. (2008). Spatial distribution of debris thickness and melting from remote sensing and meteorological data, at debris-covered Baltoro glacier, Karakoram, Pakistan. *Ann. Glaciol.* 48: 49–57. https://doi.org/10.3189/172756408784700680.

Min, M., Wang, P.C., Campbell, J.R. et al. (2010). Midlatitude cirrus cloud radiative forcing over China. *J. Geophys. Res. Atmos.* 115: D20. https://doi.org/10.1029/2010JD014161.

Mo, T. and Lou, Z. (2020). Numerical simulation of frost heave of concrete lining Trapezoidal channel under an open system. *Water* 12: 335. https://doi.org/10.3390/w12020335.

Mock, C.J., Carter, K.C., and Birkeland, K.W. (2017). Some perspectives on avalanche climatology. *Ann. Am. Assoc. Geogr.* 107 (2): 299–308. https://doi.org/10.1080/24694452.2016.1203285.

Movchan, D. (2019). Spatio-temporal trend analysis of vegetation productivity in Europe using MODIS data sets. *Ukr. J. Remote Sens.* 22: 4–11.

Muller, S.W. (1943). Permafrost or permanently frozen ground and related engineering problems. Strategic Engineering Study, Special Report, Vol. 62. Washington, D.C.: Office of Chief of Engineering, U.S. Army (reprinted by J. W. Edwards, Ann Arbor, MI, 1947).

Myhre, G., Samset, B.H., Schultz, M. et al. (2013). Radiative forcing of the direct aerosol effect from AeroCom phase II simulations. *Atmos. Chem. Phys.* 13 (4): 1853–1877. https://doi.org/10.5194/acp-13-1853-2013.

Myneni, R.B., Keeling, C.D., Tucker, C.J. et al. (1997). Increased plant growth in the northern high latitudes from 1981 to 1991. *Nature* 386 (6626): 698–702. https://doi.org/10.1038/386698a0.

Nagurn, A.P. (1998). Climatic characteristics of the tropopause over the Arctic Basin. *Annu. Geophys.* 16: 110–115. https://doi.org/10.1007/s00585-997-0110-6.

Nakajima, T., Yoon, S.C., Ramanathan, V. et al. (2007). Overview of the atmospheric brown cloud east Asian regional experiment 2005 and a study of the aerosol direct radiative forcing in East Asia. *J. Geophys. Res. Atmos.* 112: D24S91. https://doi.org/10.1029/2007JD009009.

Nan, S., Zhao, P., Yang, S., and Chen, J. (2009). Springtime tropospheric temperature over the Tibetan Plateau and evolution of the tropical Pacific SST. *J. Geophys. Res.* 114: D10104. https://doi.org/10.1029/2008JD011559.

National Oceanic and Atmospheric Administration. Flooding_factsheet. https://training.weather.gov/nwstc/Hydrology/flooding_factsheet.pdf

Nelson, F.E. and Anisimov, O.A. (1993). Permafrost zonation in Russia under anthropogenic climatic change. *Permafr. Periglac. Process.* 4: 137–148. https://doi.org/10.1002/ppp.3430040206.

Nelson, F.E., Mueller, G.R., and Shiklomanov, N.I. (1997a). Sample design for assessing variability of active-layer thickness. In: *Abstracts of the 93rd Annual Meeting of the Association of American Geographers*, 191.

Nelson, F.E., Shijlomanov, N.I., Mueller, G. et al. (1997b). Estimating active-layer thickness over a larger region: Kupaaruk River basin, Alaska. *Arct. Alp. Res.* 29: 367–378.

Ni, C.C., Li, G.P., and Xiong, X.Z. (2017). Analysis of a vortex precipitation event over Southwest China using AIRS and in situ measurements. *Adv. Atmos. Sci.* 34 (4): 559–570. https://doi.org/10.1007/s00376-016-5262-4.

Ni, J. (2002). Carbon storage in grasslands of China. *J. Arid Environ.* 50: 205–218. https://doi.org/10.1006/jare.2001.0902.

Ni-Meister, W. (2008). Recent advances on soil moisture data assimilation. *Phys. Geogr.* 29: 19–37.

Nicolsky, D.J., Romanovsky, V.E., Alexeev, V.A., and Lawrence, D.M. (2007). Improved modeling of permafrost dynamics in a GCM land-surface scheme. *Geophys. Res. Lett.* 34: L08501. https://doi.org/10.1029/2007GL029525.

Nijssen, B., Schnur, R., and Lettenmaier, D.P. (2001). Global retrospective estimation of soil moisture using the Variable Infiltration Capacity land surface model. *J. Clim.* 14: 1790–1808.

Ni-Meister, W., Houser, P. R. and Walker, J. P. (2006). Soil moisture initialization for climate prediction: Assimilation of scanning multi frequency microwave radiometer soil moisture data into a land surface model. *J. Geophys.Res.* 111: D20102.

Ninomiya, K. (1989). Polar/comma-cloud lows over the Japan Sea and the Northwestern Pacific in winter. *J. Meteorol. Soc. Jpn.* 67: 83–97. https://doi.org/10.2151/jmsj1965.67.1_83.

Ninomiya, K. (1991). Polar low development over the east coast of the Asian continent on 9–11 December 1985. *J. Meteorol. Soc. Jpn.* 69: 669–685. https://doi.org/10.2151/jmsj1965.69.6_669.

Ninomiya, K. (1994). A meso-scale low family formed over the northeastern Japan Sea in the northwestern part of parent polar low. *J. Meteorol. Soc. Jpn.* 72: 589–603. https://doi.org/10.2151/jmsj1965.72.4_589.

Niu, Y.G. (1997). Ice flood hazards and protective measures. Yellow River. (In Chinese).

Niu, G.Y. and Yang, Z.L. (2006). Effects of frozen soil on snowmelt runoff and soil water storage at a continental scale. *J. Hydrometeorol.* 7 (5): 937–952. https://doi.org/10.1175/JHM538.1.

Niu, J., Yang, Z.L., Mitchell, K.E. et al. (2011). The community Noah land surface model with multiparameterization options (Noah-MP), 2. Evaluation over global river basins. *J. Geophys. Res.* 116: D12110. https://doi.org/10.1029/2010JD015140;Nordeng, T.E. and Rasmussen, E.A. (2010). A most beautiful polar low. A case study of a polar low development in the bear island region. *Tellus A* 44 (2): 81–99. https://doi.org/10.1034/j.1600-0870.1992.00001.x.

Nobre, P., Moura, A. D., and Sun, L. (2001). Dynamical downscaling of seasonal climate prediction over Nordeste Brazil with ECHAM3 and NCEP's regional spectral models at IRI. *Bull. Amer. Meteor. Soc*: 82: 2787–2796.

Nordstroem, C., Soegaard, H., Christensen, T. et al. (2001). Seasonal carbon dioxide balance and respiration of a high-arctic fen ecosystem in NE-Greenland. *Theor. Appl. Climatol.* 70: 149–166. https://doi.org/10.1007/s007040170012.

Norris, J.R. and Wild, M. (2009). Trends in aerosol radiative effects over China and Japan inferred from observed cloud cover, solar "dimming," and solar "brightening.". *J. Geophys. Res.* 114: D00D15. https://doi.org/10.1029/2008JD011378.

North, G.R., Pyle, J.A., and Zhang, F. (ed.) (2014). *Encyclopedia of Atmospheric Sciences*, vol. 1. Amsterdam: Elsevier.

Novak, M. (2010). Dynamics of the near-surface evaporation zone and corresponding effects on the surface energy balance of a drying bare soil. Agricultural & *Forest Meteorology*. 150(10): 1358–1365.

Oberman, N.G. (2012). Long-term temperature regime of the Northeast European permafrost region during contemporary climate warming. In: *Proceedings of the 10th International Conference on Permafrost* (ed. V.P. Melnikov, D.S. Drozdov and V.E. Romanovsky), 287e291. Salekhard: The Northern Publisher.

Oku, Y., Ishikawa, H., Haginoya, S., and Ma, Y.M. (2006). Recent trends in land surface temperature on the Tibetan Plateau. *J. Clim.* 19: 2995–3003. https://doi.org/10.1175/JCLI3811.1.

Oleson, K.W., Lawrence, D.M., Gordon, B., et al. (2010). Technical description of version 4.0 of the Community Land Model (CLM). NCAR Tech. Note NCAR/TN-4781STR, 257. https://doi.org/10.5065/D6FB50WZ.

Oleson, K.W., Lawrence, D.M., Bonan, G.B., et al. (2013). Technical Description of Version 4.5 of the Community Land Model (CLM).

O'Neill, A. and Taylor, B.F. (1979). A study of the major stratospheric warming of 1976/77. *Quart. J. Roy. Meteorolog. Soc.* 105: 71–92. https://doi.org/10.1002/qj.49710544306.

Ose, T. (1996). The comparison of the simulated response to the regional snow mass anomalies over Tibet, Eastern Europe, and Siberia. *J. Meteor. Soc. Japan*. 74: 845–866.

Overduin, P.P., Kane, D.L., and Loon, W.K.P.v. (2006). Measuring thermal conductivity in freezing and thawing soil using the soil temperature response to heating. *Cold Reg. Sci. Technol.* 45 (1): 8–22. https://doi.org/10.1016/j.coldregions.2005.12.003.

Overland, J.E. and Wang, M. (2019). Impact of the winter polar vortex on greater North America. *Int. J. Climatol.* 39: 5815–5821. https://doi.org/10.1002/joc.6174.

Overland, J.E., Hall, R., Hanna, E. et al. (2020). The polar vortex and extreme weather: the beast from the east in winter 2018. *Atmos.* 11 (6): 664. https://doi.org/10.3390/atmos11060664.

Overland, J.E., Ballinger, T.J., Cohen, J. et al. (2021). How do intermittency and simultaneous processes obfuscate the Arctic influence on midlatitude winter extreme weather events? *Environ. Res. Lett.* 16 (4): 043002. https://doi.org/10.1088/1748-9326/abdb5d.

Overland, J.E., Oppenheimer, M., and Yohe, G. (2021). Rare events in the Arctic. *Clim. Change* 168 (3): 1–13. https://doi.org/10.1007/s10584-021-03238-2.

Pagani, M., Huber, M., Liu, Z. et al. (2011). The role of carbon dioxide during the onset of Antarctic glaciation. *Science* 334 (6060): 1261–1264. https://doi.org/10.1126/science.1203909.

Palazzoli, I., Maskey, S., Uhlenbrook, S. et al. (2015). Impact of prospective climate change on water resources and crop yields in the Indrawati basin, Nepal. *Agric. Syst.* 133: 143–157. https://doi.org/10.1016/j.agsy.2014.10.016.

Pan, G.X. (1999). Pedogenic carbonates in aridic soils of China and the significance interrestrial carbon transfer. *J. Nanjing Agric. Univ.* 22: 51–57. (in Chinese).

Park, H., Walsh, J., Fedorov, A.N. et al. (2013). The influence of climate and hydrological variables on opposite anomaly in active-layer thickness between Eurasian and North American watersheds. *Cryosphere* 7: 631–645. https://doi.org/10.5194/tc-7-631-2013.

Park, H., Alexander, F., Mikhail, N.Z. et al. (2015). Effect of snow cover on pan-Arctic permafrost thermal regimes. *Clim. Dyn.* 44: 2873e2895. https://doi.org/10.1007/s00382-014-2356-5.

Paul, E.A. and Clark, F.E. (1989). *Soil Microbiology and Biochemistry*. New York: Academic Press.

Pavolonis, M.J. and Key, J.R. (2003). Antarctic cloud radiative forcing at the surface estimated from the AVHRR polar pathfinder and ISCCP D1 datasets, 1985–93. *J. Appl. Meteorol.* 42: 827–840. https://doi.org/10.1175/1520-0450(2003)042<0827:ACRFAT>2.0.CO;2.

Pawson, S. and Kubitz, T. (1996). Climatology of planetary waves in the northern stratosphere. *J. Geophys. Res.* 101 (D12): 16987–16996. https://doi.org/10.1029/96JD01226.

Peings, Y. and Douville, H. (2010). Influence of the Eurasian snow cover on the Indian summer monsoon variability in observed climatologies and CMIP3 simulations. *Clim. Dyn.* 34: 643–660. https://doi.org/10.1007/s00382-009-0565-0.

Pei, L. (2019). The Preliminary Study of Characteristics and Types of Ice Avalanche Disaster in the Tibetan Plateau. China University of Geosciences (Beijing). (In Chinese)

Peng, H., Ma, W., Mu, Y. et al. (2015). Degradation characteristics of permafrost under the effect of climate warming and engineering disturbance along the Qinghai–Tibet Highway. *Nat. Hazards* 75 (3): https://doi.org/10.1007/s11069-014-1444-5.

Penn, S. (1957). The prediction of snow vs. rain. Forecasting Guide No. 2, U.S. Weather Bureau.

Perlwitz, J. and Harnik, N. (2003). Observational evidence of a stratospheric influence on the troposphere by planetary wave refraction. *J. Clim.* 16: 3011–3026. https://doi.org/10.1175/1520-0442(2003)016<3011:OEOASI>2.0.CO;2.

Peterson, B.J., Holmes, R.M., McClelland, J.W. et al. (2002). Increasing river discharge to the Arctic Ocean. *Science* 298: 2171–2173. https://doi.org/10.1126/science.1077445.

Phelps, A.R., Peterson, K.M., and Jeffries, M.O. (1998). Methane efflux from high-latitude lakes during spring ice melt. *J. Geophys. Res. Atmos.* 103 (D22): 29029–29036. https://doi.org/10.1029/98JD00044.

Piao, S., Fang, J., Zhou, L. et al. (2003). Interannual variations of monthly and seasonal normalized difference vegetation index (NDVI) in China from 1982 to 1999. *J. Geophys. Res.* 108 (D14): 4401. https://doi.org/10.1029/2002JD002848.

Pinker, R.T., Zhang, B., and Dutton, E.G. (2005). Do satellites detect trends in surface solar radiation? *Science* 308 (5723): 850–854. https://doi.org/10.1126/science.1103159.

Pleim, J.E. and Gilliam, R. (2009). An indirect data assimilation scheme for deep soil temperature in the Pleim-Xiu land surface model. *J. Appl. Meteorol. Climatol.* 48: 1362–1376. https://doi.org/10.1175/1520-0450(2003)042<1811:DOALSM>2.0.CO;2.

Pleim, J.E. and Xiu, A. (2003). Development of a land surface model. Part II: Data assimilation. *J. Appl. Meteorol.* 42: 1811–1822. https://doi.org/10.1175/1520-0450(2003)042<1811:DOALSM>2.0.CO;2.

Pogliotti, P., Guglielmin, M., Cremonese, E. et al. (2015). Warming permafrost and active layer variability at Cime Bianche, Western European Alps. *Cryosphere* 9: 647–661. https://doi.org/10.5194/tc-9-647-2015.

Politovich, M.K. (1989). Aircraft icing caused by large, supercooled droplets. *J. Appl. Meteorol.* 1988–2005 (28): 856–868. https://doi.org/10.1175/1520-0450(1989)0282.0.CO;2.

Polvani, L.M. and Waugh, D.W. (2004). Upward wave activity flux as a precursor to extreme stratospheric events and subsequent anomalous surface weather regimes. *J. Clim.* 17 (18): 3548–3554. https://doi.org/10.1175/1520-0442(2004)017<3548:UWAFAA>2.0.CO;2.

Potter, C., Boriah, S., Steinbach, M. et al. (2008). Terrestrial vegetation dynamics and global climate controls in North America: 2001–2005. *Earth Interact.* 12 (8): 1. https://doi.org/10.1175/2008EI249.1.

Prentice, I.C. et al. (2001). The carbon cycle and atmospheric carbon dioxide. In: *Climate Change 2001: The Scientific Basis*, 183–238. New York: Cambridge University Press.

Prowse, T., Alfredsen, K., Beltaos, S. et al. (2011). Past and future changes in Arctic lake and river ice. *Ambio* 40 (1): 53–62. https://doi.org/10.1007/s13280-011-0216-7.

Prowse, T.D. (2001). River-ice ecology. I: hydrologic, geomorphic, and water-quality aspects. *J. Cold Reg. Eng.* 15: 1–16. https://doi.org/10.1061/(ASCE)0887-381X(2001)15:1(1).

Prowse, T.D. and Beltaos, S. (2002). Climatic control of river-ice hydrology: a review. *Hydrol. Process.* 16 (4): 805–822. https://doi.org/10.1002/hyp.369.

Pu, Z. X., and W.-K. Tao (2004). Mesoscale assimilation of TMI rainfall data with 4DVAR: Sensitivity studies. *J. Meteor. Soc.* 82: 1389–1397. Japan.

Pu, Z. and Xu, L. (2009). MODIS/Terra observed snow cover over the Tibet Plateau: distribution, variation and possible connection with the East Asian summer monsoon (EASM). *Theor. Appl. Climatol.* 97 (3–4): 265–278. https://doi.org/10.1007/s00704-008-0074-9.

Pudasaini, S.P. and Hutter, K. (2007). *Avalanche Dynamics: Dynamics of Rapid Flows of Dense Granular Avalanches*. Springer Science & Business Media.

Pyrina, M., Hatzianastassiou, N., Matsoukas, C. et al. (2015). Cloud effects on the solar and thermal radiation budgets of the Mediterranean basin. *Atmos. Res.* 152: 14–28. https://doi.org/10.1016/j.atmosres.2013.11.009.

Qi, W., Feng, L., Liu, J. et al. (2020). Snow as an important natural reservoir for runoff and soil moisture in northeast China. *J. Geophys. Res. Atmos.* 125: e2020JD033086. https://doi.org/10.1029/2020JD033086.

Qian, Y., Zhang, Q., Yao, Y., and Zhang, X. (2002). Seasonal variation and heat preference of the south asia high. *Adv. Atmos. Sci.* 19 (5): 821–836. https://doi.org/10.1007/s00376-002-0047-3.

Qian, Y., Zhang, Y., and Zheng, Y. Q. (2003). Impacts of the Tibetan Plateau snow anomaly in winter and spring on precipitation in China in spring and summer. *J. Arid Meteor.* 21: 1–7.

Qian, Y., Zheng, Y., Zhang, Y., and Miao, M. (2003). Responses of China's summer monsoon climate to snow anomaly over the Tibetan Plateau. *Int. J. Climatol.* 23: 593–613. https://doi.org/10.1002/joc.901.

Qian, Y., Flanner, M.G., Leung, L.R., and Wang, W. (2011). Sensitivity studies on the impacts of Tibetan Plateau snowpack pollution on the Asian hydrological cycle and monsoon climate. *Atmos. Chem. Phys.* 11: 1929–1948. https://doi.org/10.5194/acp-11-1929-2011.

Qin, D., Liu, S.Y., and Li, P.J. (2006). Snow cover distribution, variability, and response to climate change in western China. *J. Clim.* 19: 1820–1833. https://doi.org/10.1175/JCLI3694.1.

Qin, J., Yang, K., Liang, S., and Guo, X. (2009). The altitudinal dependence of recent rapid warming over the Tibetan Plateau. *Clim. Change* 97 (1): 321–327. https://doi.org/10.1007/s10584-009-9733-9.

Qin, J., Yang, K., Liang, S. et al. (2011). Evaluation of surface albedo from GEWEX-SRB and ISCCP-FD data against validated MODIS product over the Tibetan Plateau. *J. Geophys. Res.* 116: D24116. https://doi.org/10.1029/2011JD015823.

Qin, D.H., Yao, T.D., and Ding, Y.J. (2014). *Cryosphere Dictionary*. Beijing: China Meteorological Press.

Qin, D., Yao, T., Ding, Y., and Ren, J. (2021). *Introduction to Cryospheric Science*. Springer Nature.

Qiu, J. (2008). The third pole. *Nature* 454: 393–396. https://doi.org/10.1038/454393a.

Quinn, P.K., Bates, T.S., Baum, E. et al. (2008). Short-lived pollutants in the Arctic: their climate impact and possible mitigation strategies. *Atmos. Chem. Phys.* 8 (6): 1723–1735. https://doi.org/10.5194/acp-8-1723-2008.

Rafiq, M. and Mishra, A. (2016). Investigating changes in Himalayan glacier in warming environment: a case study of Kolahoi glacier. *Environ. Earth Sci.* 75 (1469): https://doi.org/10.1007/s12665-016-6282-1.

Ragettli, S., Pellicciotti, F., Immerzeel, W.W. et al. (2015). Unraveling the hydrology of a Himalayan catchment through integration of high resolution in situ data and remote sensing with an advanced simulation model. *Adv. Water Resour.* 78: 94–111. https://doi.org/10.1016/j.advwatres.2015.01.013.

Räisänen, J. (2008). Warmer climate: less or more snow? *Clim. Dyn.* 30: 307–319.

Ramanathan, V. (1987). The role of earth radiation budget studies in climate and general circulation research. *J. Geophys. Res. Atmos.* 92 (D4): 4075–4095. https://doi.org/10.1029/JD092iD04p04075.

Ramanathan, V., Ramana, M.V., Roberts, G. et al. (2007). Warming trends in Asia amplified by brown cloud solar absorption. *Nature* 448: 575–578. https://doi.org/10.1038/nature06019.

Rannie, W.F. (1983). Breakup and freezeup of the Red River at Winnipeg, Manitoba Canada in the 19th century and some climatic implications. *Clim. Change* 5 (3): 283–296. https://doi.org/10.1007/BF02423523.

Rasmussen, E. (1979). The polar low as an extratropical CISK distribution. *Q. J. Roy. Meteorol. Soc.* 105: 531–549. https://doi.org/10.1002/qj.49710544504.

Rasmussen, E. and Turner, J. (2002). *Polar Lows: Mesoscale Weather Systems in the Polar Regions.* Cambridge: Cambridge University Press https://doi.org/10.1029/2004EO120011.

Rasmussen, R.M., Bernstein, B.C., Murakami, M. et al. (1995). The 1990 Valentine's Day Arctic outbreak. Part I: mesoscale and microscale structure and evolution of a Colorado front range shallow upslope cloud. *J. Appl. Meteorol. Climatol.* 34: 1481–1511. https://doi.org/10.1175/1520-0450-34.7.1481.

Rauber, R.M., Olthoff, L.S., Ramamurthy, M.K. et al. (2000). The relative importance of warm rain and melting processes in freezing precipitation events. *J. Appl. Meteorol. Climatol.* 39 (7): 1185–1195. https://doi.org/10.1175/1520-0450-34.7.1481.

Rauscher, S.A., Pal, J.S., Diffenbaugh, N.S., and Benedetti, M.M. (2008). Future changes in snowmelt-driven runoff timing over the western US. *Geophys. Res. Lett.* 35 (16): https://doi.org/10.1029/2008GL034424.

Raynolds, M.K., Walker, D.A., and Maier, H.A. (2006). NDVI patterns and phytomass distribution in the circumpolar Arctic. *Remote Sens. Environ.* 102 (3–4): 271–281. https://doi.org/10.1016/j.rse.2006.02.016.

Raynolds, M.K., Comiso, J.C., Walker, D.A., and Verbyla, D. (2008). Relationship between satellite-derived land surface temperatures, arctic vegetation types, and NDVI. *Remote Sens. Environ.* 112 (4): 1884–1894. https://doi.org/10.1016/j.rse.2007.09.008.

Reichle, R., Koster, H., R. D., Dong, J., and A. A. Berg (2004). Global soil moisture from satellite observations, land surface models, and ground data: Implications for data assimilation. *J. Hydrometeor*: 5, 430–442.

Reichle, R. H., Koster, R. D., Liu, P. et al. (2007). Comparison and assimilation of global soil moisture retrievals from the Advanced Microwave Scanning Radiometer for the Earth Observing System (AMSR-E) and the Scanning Multichannel Microwave Radiometer (SMMR). *J. Geophys. Res*: 112, D09108.

Reichler, T., Kushner, P.J., and Polvani, L.M. (2005). The coupled stratosphere–troposphere response to impulsive forcing from the troposphere. *J. Atmos. Sci.* 62 (9): 3337–3352. https://doi.org/10.1175/JAS3527.1.

Renfrew, I. (2003). Polar lows. *Encycl. Atmos. Sci.* 67 (12): 1761–1768.

Ricaud, P., Lefèvre, F., Berthet, G. et al. (2005). Polar vortex evolution during the 2002 Antarctic major warming as observed by the Odin satellite. *J. Geophys. Res. Atmos.* 110 (D5): https://doi.org/10.1029/2004JD005018.

Richard, A.C. and Hering, W.S. (1959). The stratospheric warming of January–February. *J. Meteorol.* 16: 91–108. https://doi.org/10.1175/1520-0469(1959)016<0091:TSWOJF>2.0.CO;2.

Riggs, G.A., Hall, D.K., and Salomonson, V.V. (2006). MODIS Snow products user guide to collection 5. http://modis-snow-ice.gsfc.nasa.gov/sugkc2.html.

Robbins, C.C. and Cortinas, J.V. (2002). Local and synoptic environments associated with freezing rain in the contiguous United States. *Weather Forecast.* 17: 47–65. https://doi.org/10.1175/1520-0434(2002)017<0047:LASEAW>2.0.CO;2.

Robinson, C.H. (2002). Controls on decomposition and soil nitrogen availability at high latitudes. *Plant Soil* 242 (1): 65–81. https://doi.org/10.1023/A:1019681606112.

Rodell, M., Houser, P.R., Jambor, U. et al. (2004). The global land data assimilation system. *Bull. Am. Meteorol. Soc.* 85: 381–394. https://doi.org/10.1175/BAMS-85-3-381.

Rohrer, M.B., Braun, L.N., and Lang, H. (1994). Long-term records of snow cover water equivalent in the Swiss Alps: 1. Analysis. *Hydrol. Res.* 25 (1–2): 53–64. https://doi.org/10.2166/nh.1994.0019.

Rokaya, P., Lindenschmidt, K.E., Pietroniro, A., and Clark, M. (2022). Modelling of ice jam floods under past and future climates: a review. *J. Hydrol. X* 15: https://doi.org/10.1016/j.hydroa.2022.100120.

Romanovsky, V.E., Drozdov, D.S., Oberman, N.G. et al. (2010). Thermal state of permafrost in Russia. *Permafrost Periglacial Processes* 21 (2): 136–155. https://doi.org/10.1002/ppp.683.

Ropelewski, C.F. and Halpert, M.S. (1987). Global and regional scale precipitation patterns associated with the El Niño/Southern Oscillation. *Mon. Weather Rev.* 115 (8): 1606–1626. https://doi.org/10.1175/1520-0493(1987)115<1606:GARSPP>2.0.CO;2.

Rosenfeld, D., Sherwood, S., Wood, R., and Donner, L. (2014). Climate effects of aerosol-cloud interactions. *Science* 343 (6169): 379–380. https://doi.org/10.1126/science.1247490.

Ross, R.J. and Elliott, W.P. (1996). Tropospheric water vapor climatology and trends over North America: 1973–93. *J. Clim.* 9: 3561–3574. https://doi.org/10.1175/1520-0442(1996)009<3561:TWVCAT>2.0.CO;2.

Rosso, R. (1984). Nash model relation to Horton order ratios. *Water Resour. Res.* 20 (7): 914–920. https://doi.org/10.1029/WR020i007p00914.

Rothrock, D.A., Percival, D.B., Wensnahan, M. et al. (2008). The decline in Arctic sea-ice thickness: separating the spatial, annual, and interannual variability in a quarter century of submarine data. *J. Geophys. Res.* 113: C05003. https://doi.org/10.1029/2007JC004252.

Salama, M., van der Velde, R., Zhong, L. et al. (2012). Decadal variations of land surface temperature anomalies observed over the Tibetan Plateau by the Special Sensor Microwave Imager (SSM/I) from 1987 to 2008. *Clim. Change* 114 (3): 769–781. https://doi.org/10.1007/s10584-012-0427-3.

Sarmiento, J.L. and Gruber, N. (2006). *Ocean Biogeochemical Dynamics*. Princeton, NJ: Princeton University Press.

Satheesh, S.K., Ramanathan, V., Li-Jones, X. et al. (1999). A model for the natural and anthropogenic aerosols over the tropical Indian Ocean derived from Indian Ocean experiment data. *J. Geophys. Res. Atmos.* 104 (D22): 27421–27440. https://doi.org/10.1029/1999JD900478.

Sato, K., Inoue, J., and Watanabe, M. (2014). Influence of the Gulf stream on the Barents Sea ice retreat and Eurasian coldness during early winter. *Environ. Res. Lett.* 9 (8): 084009. https://doi.org/10.1088/1748-9326/9/8/084009.

Sayers, P., Li, Y., Gerald, G. et al. (2013). Flood risk management: A strategic approach. Asian Development Bank. 1–206.

Sazonova, T.S., Romanovsky, V.E., Walsh, J.E., and Sergueev, D.O. (2004). Permafrost dynamics in the 20th and 21st centuries along the East Siberian transect. *J. Geophys. Res.* 109: D01108. https://doi.org/10.1029/2003JD003680.

Scherhag, R. (1952). Die explosionsartige stratospharener warmungen desspatwinters 1951–1952. *Berichte des Deutschen Wetterdienstes* 52: 173–118.

Scherhag, R. (1960). Stratospheric temperature changes and the associated changes in pressure distribution. *J. Atmos. Sci.* 17 (6): 575–583. https://doi.org/10.1175/1520-0469(1960)017<0575:STCATA>2.0.CO;2.

Scherler, D., Bookhagen, B., and Strecker, M.R. (2011). Spatially variable response of Himalayan glaciers to climate change affected by debris cover. *Nat. Geosci.* 4 (3): 156–159. https://doi.org/10.1038/NGEO1068.

Scherrer, S.C., Appenzeller, C., and Laternser, M. (2004). Trends in Swiss alpine snow days – the role of local and large scale climate variability. *Geophys. Res. Lett.* 31: L13215. https://doi.org/10.1029/2004GL020255.

Schimanke, S., Körper, J., Spangehl, T., and Cubasch, U. (2011). Multi-decadal variability of sudden stratospheric warmings in an AOGCM. *Geophys. Res. Lett.* 38 (1): https://doi.org/10.1029/2010GL045756.

Schlesinger, W.H. and Bernhardt, E.S. (1998). Biogeochemistry: an analysis of global change. *Q. Rev. Biol.* 73 (1): 353–423. https://doi.org/10.1093/obo/9780199830060-0111.

Schulze, E.-D. et al. (1999). Productivity of forests in the Eurosiberian boreal region and their potential to act as a carbon sink: a synthesis. *Glob. Chang. Biol.* 5: 703–722. https://doi.org/10.1046/j.1365-2486.1999.00266.x.

Schuur, E.A., Bockheim, J., Canadell, J.G. et al. (2008). Vulnerability of permafrost carbon to climate change: implications for the global carbon cycle. *BioScience* 58 (8): 701–714. https://doi.org/10.1641/B580807.

Scott, R.C., Lubin, D., Vogelmann, A.M., and Kato, S. (2017). West Antarctic ice sheet cloud cover and surface radiation budget from NASA A-train satellites. *J. Clim.* 30: 6151–6170. https://doi.org/10.1175/JCLI-D-16-0644.1.

Seiki, T. and Nakajima, T. (2014). Aerosol effects of the condensation process on a convective cloud simulation. *J. Atmos. Sci.* 71 (2): 833–853. https://doi.org/10.1175/JAS-D-12-0195.1.

Sekiguchi, M., Nakajima, T., Suzuki, K. et al. (2003). A study of the direct and indirect effects of aerosols using global satellite data sets of aerosol and cloud parameters. *J. Geophys. Res. Atmos.* 108 (D22): https://doi.org/10.1029/2002JD003359.

Seneviratne, S.I., Koster, R.D., Guo, Z. et al. (2006). Soil moisture memory in AGCM simulations: analysis of global land–atmosphere coupling experiment (GLACE) data. *J. Hydrometeorol.* 7 (5): 1090–1112. https://doi.org/10.1175/JHM533.1.

Seneviratne, S.I., Lüthi, D., Litschi, M., and Schär, C. (2006). Land–atmosphere coupling and climate change in Europe. *Nature* 443 (7108): 205–209.

Seneviratne, S.I., Corti, T., Davin, E.L. et al. (2010). Investigating soil moisture–climate interactions in a changing climate: a review. *Earth-Sci. Rev.* 99 (3–4): 125–161. https://doi.org/10.1016/j.earscirev.2010.02.004.

Seneviratne, S.I., Wilhelm, M., Stanelle, T. et al. (2013). Impact of soil moisture-climate feedbacks on CMIP5 projections: first results from the GLACE-CMIP5 experiment. *Geophys. Res. Lett.* 40 (19): 5212–5217. https://doi.org/10.1002/grl.50956.

Serreze, M. and Barry, R. (2005). *The Arctic Climate System (Cambridge Atmospheric and Space Science Series)*. Cambridge: Cambridge University Press. doi:10.1017/CBO9780511535888.

Seyfried, M.S. and Murdock, M.D. (1997). Use of air permeability to estimate infiltrability of frozen soil. *J. Hydrol.* 202 (1–4): 95–107. https://doi.org/10.1016/S0022-1694(97)00061-9.

Shang, D. and Wang, C. (2006). The effect of the frozen-thaw process in Tibetan Plateau on summer monsoon over Eastern Asia. *J. Arid Meteor.* 24: 19–22. (in Chinese).

Shapiro, M.A., Fedor, L.S., and Hampel, T. (1987). Research aircraft measurements of a polar low over the Norwegian Sea. *Tellus A* 39 (4): 272–306. https://doi.org/10.3402/tellusa.v39i4.11761.

Shapiro, M.A., Hampel, T., and Krueger, A.J. (1987). The Arctic tropopause fold. *Mon. Weather Rev.* 115 (2): 444–454. https://doi.org/10.1175/1520-0493(1987)115<0444:TATF>2.0.CO;2.

Shea, J.M., Immerzeel, W.W., Wagnon, P. et al. (2015). Modelling glacier change in the Everest region, Nepal Himalaya. *Cryosphere* 9 (3): 1105–1128. https://doi.org/10.5194/tc-9-1105-2015.

Sheffield, J., Ziegler, A.D., Wood, E.F., and Chen, Y. (2004). Correction of the high-latitude rain day anomaly in the NCEP–NCAR reanalysis for land surface hydrological modeling. *J. Clim.* 17 (19): 3814–3828. https://doi.org/10.1175/1520-0442(2004)017<3814:COTHRD>2.0.CO;2.

Shen, H.D. (2010). *Study on River Ice*. Beijing: Yellow River Water Conservancy Press.

Shen, J., Li, K., Cui, Z. et al. (2022). Improvement of summer precipitation simulation in China by assimilating spring soil moisture over the Tibetan Plateau. *Theor. Appl. Climatol.* 147: 785–801. https://doi.org/10.1007/s00704-021-03840-5.

Sherhag, R. (1952). Die explosionsartigen Stratospharenerwarmungen des Spatwinters 1951–1952. *Ber. Deut. Wetterd.* 6: 51–63.

Shi, G.Y., Hayasaka, T., Ohmura, A. et al. (2008). Data quality assessment and the long-term trend of ground solar radiation in China. *J. Appl. Meteorol. Climatol.* 47 (4): 1006–1016. https://doi.org/10.1175/2007JAMC1493.1.

Shi, H.X. and Wang, C.H. (2015). Projected 21st century changes in snow water equivalent over Northern Hemisphere landmasses from the CMIP5 model ensemble. *Cryosphere* 9 (5): 1943–1953. https://doi.org/10.5194/tc-9-1943-2015.

Shi, Q. and Liang, S. (2013). Characterizing the surface radiation budget over the Tibetan Plateau with ground-measured, reanalysis, and remote sensing data sets: 2. Spatiotemporal analysis. *J. Geophys. Res. Atmos.* 118 (16): 8921–8934. https://doi.org/10.1002/jgrd.50719.

Shijin, W., Lanyue, Z., and Yanqiang, W. (2019). Integrated risk assessment of snow disaster over the Qinghai-Tibet Plateau. *Geomat. Nat. Haz. Risk* 10: 740–757. https://doi.org/10.1080/19475705.2018.1543211.

Shindell, D. (2007). Local and remote contributions to Arctic warming. *Geophys. Res. Lett.* 34: L14704. https://doi.org/10.1029/2007GL030221.

Shindell, D. and Faluvegi, G. (2009). Climate response to regional radiative forcing during the twentieth century. *Nat. Geosci.* 2 (4): 294–300. https://doi.org/10.1038/ngeo473.

Shindell, D.T., Faluvegi, G., Koch, D.M. et al. (2009). Improved attribution of climate forcing to emissions. *Science* 326: 716–718. https://doi.org/10.1126/science.1174760.

Shiyan, T. and Fukang, Z. (1964). Pattern change at 100mb and its relation to advance and retreat of the subtropical Pacific high. *Acta Meteorol. Sin.* 34: 385–395. (in Chinese).

Shoop, S.A. and Bigl, S.R. (1997). Moisture migration during freeze and thaw of unsaturated soils: modeling and large scale experiments. *Cold Reg. Sci. Technol.* 25: 33–45. https://doi.org/10.1016/S0165-232X(96)00015-8.

Shukla, J. and Mintz, Y. (1982). Influence of land-surface evapotranspiration on the earth's climate. *Science* 215 (4539): 1498–1501. https://doi.org/10.1126/science.215.4539.1498.

Shupe, M.D. (2011). Clouds at Arctic atmospheric observatories. part II: thermodynamic phase characteristics. *J. Appl. Meteorol. Climatol.* 50: 645–661. https://doi.org/10.1175/2010JAMC2468.1.

Shupe, M.D., Walden, V.P., Eloranta, E. et al. (2011). Clouds at Arctic atmospheric observatories. Part I: occurrence and macrophysical properties. *J. Appl. Meteorol. Climatol.* 50: 626–644. https://doi.org/10.1175/2010JAMC2467.1.

Siegmund, P.C. and Velthoven, P.V. (2005). To what extent can the 2002 Antarctic major stratospheric warming be explained by horizontal advection of the vortex from the pole? *Meteorol. Atmos. Phys.* 90 (3–4): 65–177. https://doi.org/10.1007/s00703-004-0085-8.

Siliang, L., Ronggao, L., and Yang, L. (2010). Spatial and temporal variation of global lai during 1981–2006. *J. Geog. Sci.* 20: 323–332. https://doi.org/10.1007/s11442-010-0323-6.

Singh, P. and Kumar, N. (1997). Impact assessment of climate change on the hydrological response of a snow and glacier melt runoff dominated Himalayan river. *J. Hydrol.* 193 (1–4): 316–350. https://doi.org/10.1016/S0022-1694(96)03142-3.

Sitch, S., Mcguire, A.D., Kimball, J. et al. (2007). Assessing the circumpolar carbon balance of arctic tundra with remote sensing and process-based modeling approaches. *Ecol. Appl.* 17: 213–234. https://doi.org/10.1890/1051-0761(2007)017[0213:ATCBOC]2.0.CO;2.

Slater, A.G. and Lawrence, D.M. (2013). Diagnosing present and future permafrost from climate models. *J. Clim.* 26: 5608–5623. https://doi.org/10.1175/JCLI-D-12-00341.1.

Slater, A.G., Pitman, A.J., and Desborough, C.E. (1998). Simulation of freeze-thaw cycles in a general circulation model land surface scheme. *J. Geophys. Res.* 103 (D10): 11303–11312. https://doi.org/10.1029/97JD03630.

Slater, M.E., Williams, S.R., Harris, P. et al. (2007). Preliminary evaluation of infraclavicular catheters inserted using ultrasound guidance: through-the-catheter anesthesia is not inferior to through-the-needle blocks. *Reg. Anesth. Pain Med.* 32: 296–302. https://doi.org/10.1016/j.rapm.2007.02.005.

Small, E. E., Giorgi, F., and Sloan, L. C. (1999). Regional climate model simulation of precipitation in central Asia: Mean and inter annual variability. *J. Geophys. Res.* 104: 6563–6582.

Smith, L.C., Sheng, Y., MacDonald, G.M., and Hinzman, L.D. (2005). Disappearing Arctic lakes. *Science* 308: 1429. https://doi.org/10.1126/science.1108142.

Smith, S.L., Romanovsky, V.E., Lewkowicz, A.G. et al. (2010). Thermal state of permafrost in North America: a contribution to the International Polar Year. *Permafr. Periglac. Process.* 21: 117–135. https://doi.org/10.1002/ppp.690.

Sobota, I. and Nowak, M. (2014). Changes in the dynamics and thermal regime of the permafrost and active layer of the High Arctic coastal area in north-west Spitsbergen, Svalbard. *Geogr. Ann. Ser. B* 96: 227–240. https://doi.org/10.1111/geoa.12045.

Solomon, A., Shupe, M.D., Persson, O. et al. (2014). The sensitivity of springtime Arctic mixed-phase stratocumulus clouds to surface-layer and cloud-top inversion-layer moisture sources. *J. Atmos. Sci.* 71: 574–595. https://doi.org/10.1175/JAS-D-13-0179.1.

Solomon, S., Qin, D., Manning, M. et al. (2007). Climate change 2007: The physical science basis. Contribution of Working Group I to the Fourth Assessment Report of the Intergovernmental Panel on Climate Change. Cambridge University Press, Cambridge, UK.

Soncini, A., Bocchiola, D., Confortola, G. et al. (2015). Future hydrological regimes in the upper Indus basin: a case study from a high-altitude glacierized catchment. *J. Hydrometeorol.* 16: 306–326. https://doi.org/10.1175/JHM-D-14-0043.1.

Soncini, A., Bocchiola, D., Confortola, G. et al. (2016). Future hydrological regimes and glacier cover in the everest region: the case study of the upper dudh koshi basin. *Sci. Total Environ.* 565: 1084–1101. https://doi.org/10.1016/j.scitotenv.2016.05.138.

Spaans, E.J.A. and Baker, J.M. (1996). The soil freezing characteristic: its measurement and similarity to the soil moisture characteristic. *Soil Sci. Soc. Am. J.* 60: 13–19. https://doi.org/10.2136/sssaj1996.03615995006000010005x.

Srivastava, P.K., Han, D., Rico-Ramirez, M.A. et al. (2015). Performance evaluation of WRF-Noah land surface model estimated soil moisture for hydrological application: Synergistic evaluation using SMOS retrieved soil moisture. *J. Hydrol.* 529: 200–212.

St. Jacques, J.-M. and Sauchyn, D.J. (2009). Increasing winter baseflow and mean annual streamflow from possible permafrost thawing in the Northwest Territories, Canada. *Geophys. Res. Lett.* 36: L01401. https://doi.org/10.1029/2008GL035822.

Stamnes, K., Ellingson, R.G., Curry, J.A. et al. (1999). Review of science issues, deployment strategy, and status for the ARM North Slope of Alaska–Adjacent Arctic Ocean climate research site. *J. Clim.* 12 (1): 46–63. https://doi.org/10.1175/1520-0442(1999)012<0046:ROSIDS>2.0.CO;2.

Stephens, B.B., Gurney, K.R., Tans, P.P. et al. (2007). Weak northern and strong tropical land carbon uptake from vertical profiles of atmospheric CO_2. *Science* 316: 1732–1735. https://doi.org/10.1126/science.1137004.

Stephens, G.L. (2005). Cloud feedbacks in the climate system: a critical review. *J. Clim.* 18: 237–273. https://doi.org/10.1175/JCLI-3243.1.

Stevens, B., Giorgetta, M., Esch, M. et al. (2013). Atmo spheric component of the MPI-M earth system model: ECHAM6. *J. Adv. Model. Earth Syst.* 5: 1–27. https://doi.org/10.1002/jame.20015.

Stewart, R.E. and King, P. (1987). Freezing precipitation in winter storms. *Mon. Weather Rev.* 115 (7): 1270–1280. https://doi.org/10.1175/1520-0493(1987)115<1270:FPIWS>2.0.CO;2.

Stewart, I.T. (2009). Changes in snowpack and snowmelt runoff for key mountain regions. *Hydrol. Process.* 23 (1): 78–94. https://doi.org/10.1002/hyp.7128.

Stewart, I.T., Cayan, D.R., and Dettinger, M.D. (2004). Changes in snowmelt runoff timing in western North America under a "business as usual" climate change scenario. *Clim. Change* 62: 217–232. https://doi.org/10.1023/B:CLIM.0000013702.22656.e8.

Stocker, T. (ed.) (2014). *Climate Change 2013: The Physical Science Basis: Working Group I Contribution to the Fifth Assessment Report of the Intergovernmental Panel on Climate Change*. Cambridge: Cambridge University Press.

Stocker, T.F., Qin, D.H., Plattner, G.K. et al. (2013). *The Scientific Basis, Working Group I Contribution to the Fifth Assessment Report of the Intergovernmental Panel on Climate Change, Summary for Policymakers*. New York: Cambridge University Press.

Stoffel, M. and Corona, C. (2018). Future winters glimpsed in the Alps. *Nat. Geosci.* 11 (7): 458–460. https://doi.org/10.1038/s41561-018-0177-6.

Stolarski, R.S., McPeters, R.D., and Newman, P.A. (2005). The ozone hole of 2002 as measured by TOMS. *J. Atmos. Sci.* 62: 716–720. https://doi.org/10.1175/JAS-3338.1.

Stone, R. (2010). Earth-observation summit endorses global data sharing. *Science* 330 (6006): 902. https://doi.org/10.1126/science.330.6006.902.

Stow, D., Daeschner, S., Hope, A. et al. (2003). Variability of seasonally integrated normalized difference vegetation index across the North slope of Alaska in the 1990s. *Int. J. Remote Sens.* 24 (5): 1111–1117. https://doi.org/10.1080/0143116021000020144.

Strapazzon, G., Schweizer, J., Chiambretti, I. et al. (2021). Effects of climate change on avalanche accidents and survival. *Front. Physiol.* 12: 450. https://doi.org/10.3389/fphys.2021.639433.

Stroeve, J., Holland, M.M., Meier, W. et al. (2007). Arctic sea ice decline: faster than forecast. *Geophys. Res. Lett.* 34: L09501. https://doi.org/10.1029/2007GL029703.

Sturm, M., Holmgren, J., Liston, G. et al. (1995). A seasonal snow cover classification system for local to global applications. *J. Clim.* 8: 1261–1283. https://doi.org/10.1175/1520-0442(1995)008<1261:ASSCCS>2.0.CO;2.

Su, F., Adam, J.C., Trenberth, K.E., and Lettenmaier, D.P. (2006). Evaluation of surface water fluxes of the pan-Arctic land region with a land surface model and ERA-40 reanalysis. *J. Geophys. Res.* 111: D05110. https://doi.org/10.1029/2005JD006387.

Su, F., Duan, X., Chen, D. et al. (2013). Evaluation of the global climate models in the CMIP5 over the Tibetan Plateau. *J. Clim.* 26: 3187–3208. https://doi.org/10.1175/JCLI-D-12-00321.1.

Sud, Y.C., Chao, W.C., and Walker, G.K. (1993). Dependence of rainfall on vegetation: theoretical considerations, simulation experiments, observations, and inferences from simulated atmospheric soundings. *J. Arid Environ.* 25 (1): 5–18.

Sun, M., Zhou, S., Yao, X. et al. (2022). Surging glaciers in High Mountain Asia between 1986 and 2021. *EGUsphere* 1–30.

Sun, G. and Song, Z. (1987). *Formation of the South Asia High and Its Relation to Circulation Change and Rain Belt of China. Ifluences (if the Tibetan Plateau on the Summertime Weather in China)*, 93–100. Beijing: Science Press (in Chinese).

Sun, S.F., Zhang, X., and Guo, A. (2003). A simplified version of the coupled heat and moisture transport model. *Global Planet. Change* 37 (3–4): 265–276. https://doi.org/10.1016/S0921-8181(02)00210-2.

Surabi, M., Denman, K.L., Guy, B. et al. (2007). Couplings between changes in the climate system and biogeochemistry. In: *Climate Change 2007: The Physical Science Basis. Contribution of Working Group I to the Fourth Assessment Report of the Intergovernmental Panel on Climate Change* (ed. S. Solomon, D. Qin, M. Manning, et al.), 499–587. Cambridge: Cambridge University Press.

Swenson, S.C. and Lawrence, D.M. (2012). A new fractional snow-covered area parameterization for the Community Land Model and its effect on the surface energy balance. *J. Geophys. Res.* 117: D21107. https://doi.org/10.1029/2012JD018178.

Swenson, S.C., Lawrence, D.M., and Lee, H. (2012). Improved simulation of the terrestrial hydrological cycle in permafrost regions by the Community Land Model. *J. Adv. Model. Earth Syst.* 4: M08002. https://doi.org/10.1029/2012MS000165.

Tachibana, Y., Komatsu, K.K., Alexeev, V.A. et al. (2019). Warm hole in Pacific Arctic sea ice cover forced mid-latitude Northern Hemisphere cooling during winter 2017–18. *Sci. Rep.* 9: 5567. https://doi.org/10.1038/s41598-019-41682-4.

Takamura, T., Sugimoto, N., Shimizu, A. et al. (2007). Aerosol radiative characteristics at Gosan, Korea, during the Atmospheric Brown Cloud East Asian Regional Experiment 2005. *J. Geophys. Res. Atmos.* 112: D22S36. https://doi.org/10.1029/2007JD008506.

Takata, K. and Kimoto, M. (2000). A numerical study on the impact of soil freezing on the continental-scale cycle. *J. Meteorol. Soc. Jpn.* 78 (3): 199–221. https://doi.org/10.2151/jmsj1965.78.3_199.

Tan, I., Storelvmo, T., and Zelinka, M.D. (2016). Observational constraints on mixed-phase clouds imply higher climate sensitivity. *Science* 352: 224–227. https://doi.org/10.1126/science.aad5300.

Tanaka, K., Ishikawa, H., Hayashi, T. et al. (2001). Surface energy budget at Amdo on the Tibetan Plateau using GAME/Tibet IOP98 data. *J. Meteorol. Soc. Jpn.* 79: 505–517. https://doi.org/10.2151/jmsj.79.505.

Tang, W.J., Yang, K., Qin, J. et al. (2011). Solar radiation trend across China in recent decades: a revisit with quality-controlled data. *Atmos. Chem. Phys.* 11 (1): 393–406. https://doi.org/10.5194/acp-11-393-2011.

Tarnocai, C. (2006). The effect of climate change on carbon in Canadian peatlands. *Global Planet. Change* 53: 222–232. https://doi.org/10.1016/j.gloplacha.2006.03.012.

Tarnocai, C., Canadell, J.G., Mazhitova, G. et al. (2009). Soil organic carbon pools in the northern circumpolar permafrost region. *Global Biogeochem. Cycles* 23: GB2023. https://doi.org/10.1029/2008GB003327.

Tedesco, M., Mote, T., Fettweis, X. et al. (2016). Arctic cut-off high drives the poleward shift of a new Greenland melting record. *Nat. Commun.* 7: 11723. https://doi.org/10.1038/ncomms11723.

Teuling, A.J. and Seneviratne, S.I. (2008). Contrasting spectral changes limit albedo impact on land-atmosphere coupling during the 2003 European heat wave. *Geophys. Res. Lett.* 35: L03401. https://doi.org/10.1029/2007gl032778.

Tian, H.J., Wang, L., and Shi, J.H. (2020). Vegetation dynamics of the north Chinese grassland in the past 20 years. *Pratacult. Sci.* 37 (11): 2165–2174.

Todhunter, P.E. (2001). A hydroclimatological analysis of the red Rwer of the North snowmelt flood catastrophe of 1997. *JAWRA J. Am. Water Resour. Assoc.* 37 (5): 1263–1278. https://doi.org/10.1111/j.1752-1688.2001.tb03637.x.

Tong, L.Q., Tu, J.N., Pei, L.X. et al. (2018). Preliminary discussion of the frequently debris flow events in the Sedongpu basin at Gyalaperi peak, Yarlung Zangbo river. *J. Eng. Geol.* 26: 1552–1561. https://doi.org/10.13544/j.cnki.jeg.2018-401 (in Chinese).

Torrence, C. and Webster, P. (1998). The annual cycle of persistence in the El Niño -southern oscillation. *Q. J. Roy. Meteorol. Soc.* 124 (550): 1985–2004. https://doi.org/10.1002/qj.49712455010.

Trenberth, K.E. (1998). Atmospheric moisture residence time and cycling: implication for rainfall rates and climate change. *Climate Change* 39: 667–694. https://doi.org/10.1023/A:1005319109110.

Trenberth, K.E. (1999). Atmospheric moisture recycling: role of advection and local evaporation. *J. Clim.* 12: 1368–1381. https://doi.org/10.1175/1520-0442(1999)012<1368:AMRROA>2.0.CO;2.

Trenberth, K.E. and Stepaniak, D.P. (2003a). Co-variability of components of poleward atmospheric energy transports on seasonal and interannual timescales. *J. Clim.* 16: 3690–3704. https://doi.org/10.1175/1520-0442(2003)016<3691:COCOPA>2.0.CO;2.

Trenberth, K.E. and Stepaniak, D.P. (2003b). Seamless poleward atmospheric energy transports and implications for the Hadley circulation. *J. Clim.* 16: 3705–3721. https://doi.org/10.1175/1520-0442(2003)016<3706:SPAETA>2.0.CO;2.

Trenberth, K.E., Dai, A., Rasmussen, R.M., and Parsons, D.B. (2003). The changing character of precipitation. *Bull. Am. Meteorol. Soc.* 84: 1205–1217. https://doi.org/10.1175/BAMS-84-9-1205.

Trenberth, K.E., Fasullo, J., and Smith, L. (2005). Trends and variability in column-integrated atmospheric water vapor. *Clim. Dyn.* 24: 741–758. https://doi.org/10.1007/s00382-005-0017-4.

Trenberth, K.E., Fasullo, J.T., and Kiehl, J. (2009). Earth's global energy budget. *Bull. Am. Meteorol. Soc.* 90: 311–323. https://doi.org/10.1175/2008BAMS2634.1.

Troy, T.J. (2010). The hydrology of northern Eurasia: uncertainty and change in the terrestrial water and energy budgets. PhD dissertation, Princeton University.

Troy, T.J., Sheffield, J., and Wood, E.F. (2011). Estimation of the terrestrial water budget over northern Eurasia through the use of multiple data sources. *J. Clim.* 24 (13): 3272–3293. https://doi.org/10.1175/2011JCLI3936.1.

Tsuboki, K. and Asai, T. (2004). The multi-scale structure and development mechanism of mesoscale cyclones over the sea of Japan in winter. *J. Meteorol. Soc. Jpn.* 82: 597–621. https://doi.org/10.2151/jmsj.2004.597.

Tucker, C.J. (1979). Red and photographic infrared linear combinations for monitoring vegetation. *Remote Sens. Environ.* 8 (2): 127–150. https://doi.org/10.1016/0034-4257(79)90013-0.

Tuniel, N. (2002). *The Physical Geography of Northern Eurasia*, 149–168. Oxford: Oxford University Press.

Turunen, J., Tomppo, E., Tolonen, K., and Reinikainen, A. (2002). Estimating carbon accumulation rates of undrained mires in Finland: application to boreal and subarctic regions. *Holocene* 12: 69–80. https://doi.org/10.1191/0959683602hl522rp.

Twomey, S. (1977). The influence of pollution on the shortwave albedo of clouds. *J. Atmos. Sci.* 34: 1149–1152. https://doi.org/10.1175/1520-0469(1977)0342.0.CO;2.

Tzeng, R.-Y. and Lee, Y.-H. (2001). The effects of land-surface characteristics on the East Asian summer monsoon. *Clim. Dyn.* 17: 317–326. https://doi.org/10.1007/s003820000115.

Ullah, S., Liu, S., Adnan, M. et al. (2020). Risks of Glaciers Lakes Outburst Flood along China Pakistan Economic Corridor. 10.5772/intechopen.93459.

United States Geological Survey (2019). Snowmelt Runoff and the Water Cycle. www.usgs.gov/special-topics/water-science-school/science/snowmelt-runoff-and-water-cycle.

UCAR Center for Science Education. (2022). Winter storms. https://scied.ucar.edu/learning-zone/storms/winter-storms

Vaughan, D.G., Comiso, J.C., Allison, I. et al. (2013). Observations: cryosphere. In: *Climate Change: The Physical Science Basis. Contribution of Working Group I to the Fifth Assessment Report of the Intergovernmental Panel on Climate Change* (ed. T.F. Stocker, D. Qin, G.-K. Plattner, et al.), 319–382. Cambridge: Cambridge University Press.

Verburg, P.S.J., VanLoon, W.K.P., and Lukewille, A. (1999). The CLIMEX soil-heating experiment: soil response after 2 years of treatment. *Biol. Fertil. Soils* 28: 271–276. https!!://doi.org/10.1007/s003740050493.

Viganò, G., Confortola, G., Fornaroli, R. et al. (2015). Future climate change may affect habitat in Alpine streams: a case study in Italy. *ASCE J. Hydrol. Eng.* 21 (2): 220–231.

Vigdorchik, M.E. (1980). *Arctic Pleistocene History and the Development of Submarine Permafrost.* Boulder, CO: Westview Press.

Villamil-Otero, G.A., Zhang, J., He, J., and Zhang, X. (2018). Role of extratropical cyclones in the recently observed increase in poleward moisture transport into the Arctic Ocean. *Adv. Atmos. Sci.* 35: 85–94. https://doi.org/10.1007/s00376-017-7116-0.

Viviroli, D., Dürr, H.H., Messerli, B. et al. (2007). Mountains of the world, water towers for humanity: typology, mapping, and global significance. *Water Resour. Res.* 43: W07447. https://doi.org/10.1029/2006WR005653.

Vojtek, M., Faško, P., and Šťastný, P. (2003). Some selected snow climate trends in Slovakia with respect to altitude. *Acta Math. Univ. Comenianae* 32: 17–27.

Vormoor, K., Lawrence, D., Schlichting, L. et al. (2016). Evidence for changes in the magnitude and frequency of observed rainfall vs. snowmelt driven floods in Norway. *J. Hydrol.* 538: 33–48. https://doi.org/10.1016/j.jhydrol.2016.03.066.

Vries, D.A. (1963). Thermal properties of soils. In: *Physics of the Plant Environment* (ed. W.R. van Wijk). Amsterdam: North-Holland.

Walker, G. (1910). On the meteorological evidence for supposed changes of climate in India. General Government Branch Press. *Mem. India Meteor. Dept.* 21: 1–21.

Walker, G. (2007). A world melting from the top down. *Nature* 446: 718–721.

Walter, B.P., Heimann, M., and Matthews, E. (2001a). Modeling modern methane emissions from natural wetlands. 1. Model description and results. *J. Geophys. Res.* 106: 34189–34206. https://doi.org/10.1029/2001JD900165.

Walter, B.P., Heimann, M., and Matthews, E. (2001b). Modeling modern methane emissions from natural wetlands. 2. Interannual variations 1982–1993. *J. Geophys. Res.* 106: 34207–34219. https://doi.org/10.1029/2001JD900164.

Walter, K.M., Zimov, S.A., Chanton, J.P. et al. (2006). Methane bubbling from Siberian thaw lakes as a positive feedback to climate warming. *Nature* 443: 71–75.

Walter, K.M., Smith, L.C., and Chapin, F.S. III (2007). Methane bubbling from northern lakes: present and future contributions to the methane budget. *Phil. Trans. R. Soc. A* 365: 1657–1676. https://doi.org/10.1098/rsta.2007.2036.

Ward and Paul (2001). Antarctica Fact File. http://www.coolantarctica.com/Antarctica%20fact%20file/antarctica%20fact%20file%20index.htm.

Wang, A. and Zeng, X. (2012). Evaluation of multireanalysis products with in situ observations over the Tibetan Plateau. *J. Geophys. Res.* 117: D05102. https://doi.org/10.1029/2011JD016553.

Wang, B., Wu, R., and Fu, X. (2000). Pacific–East Asian teleconnection: how does ENSO affect East Asian climate? *J. Clim.* 13: 1517–1536. https://doi.org/10.1175/1520-0442(2000)013<1517:PEATHD>2.0.CO;2.

Wang, B., Bao, Q., Hoskins, B. et al. (2008). Tibetan plateau warming and precipitation changes in East Asia. *Geophys. Res. Lett.* 35: L14702. https://doi.org/10.1029/2008GL034330.

Wang, C. and Cui, Y. (2011). Characteristics of the difference of temperature between surface and atmosphere over the Tibetan plateau in the early stage of east Asian summer monsoon onset. *Clim. Environ. Res.* 16: 586–596. https://doi.org/10.3724/SP.J.1146.2006.01085.

Wang, C., & Cui, Z. (2018). Improvement of short-term climate prediction with indirect soil variables assimilation in China, *J. Climate.* 31(4): 1399–1412.

Wang, C. and Guo, Y. (2012). Precipitable water conversion rates over the Qinghai-Xizang (Tibet) Plateau: changing characteristics with global warming. *Hydrol. Process.* 26: 1509–1516. https://doi.org/10.1002/hyp.8246.

Wang, C. and Shang, D.C. (2007). Effect of the variation of the soil temperature and moisture in the transition from dry-season to wet-season over Northern Tibet Plateau. *Plateau Meteorol.* 26: 677–685. https://doi.org/10.1002/jrs.1570.

Wang, C. and Shi, R. (2007). Simulation of the land surface processes in the Western Tibetan Plateau in summer. *J. Glaciol. Geocryol.* 29: 73–81. https://doi.org/10.1002/jrs.1570.

Wang, C., Dong, W., and Wei, Z. (2001). The feature of seasonal frozen soil in Qinghai Tibet Plateau. *Acta Geog. Sin.* 56: 523–531. https://doi.org/10.3321/j.issn:0375-5444.2001.05.003.

Wang, C., Dong, W., and Wei, Z. (2003). A study on relationship between freezing–thawing processes of the Qinghai-Xizang Plateau and the atmospheric circulation over East Asia. *Chin. J. Geophys.* 46: 438–448. https://doi.org/10.1002/cjg2.3361.

Wang, C., Dong, A., Wang, S. et al. (2000). The correlation between precipitation in northwest China during spring and snowdepth in the Tibetan Plateau. *J. Glaciol. Geocryol.* 22: 340–346

Wang C., Yang, K., Zhang, F. (2020). Impacts of soil freeze-thaw process and snow melting over Tibetan Plateau on Asian summer monsoon system: a review and perspective. *Frontiers in Earth Science.* 8: 133.

Wang C., Yang, K., Zhang, F. et al. (2021). Climate effects of soil freeze-thaw process over Qinghai-Xizang plateau: progress and perspectives. *Plateau Meteorology.* 40(6): 1318–1336.

Wang, C., Cheng, G., Deng, A., and Dong, W. (2008). Numerical simulation on climate effects of freezing-thawing processes using CCM3. *Sci. Cold Arid. Reg.* 1: 68–79.

Wang, C., Wang, Z., and Cui, Y. (2009a). Snow cover of China during the last 40 years: spatial distribution and interannual variation. *J. Glaciol. Geocryol.* 31: 301–310. http:!//ir.casnw.net/handle/362004/20228.

Wang, C., Li, Y., and Wang, Y. (2015). The impacts of interdecadal variability of circulation and winter monsoon on winter snowfall over the Tibetan Plateau. *Climatic Environ. Res.* 20: 421–432. CNKI:SUN:QHYH.0.2015-04-005 (in Chinese).

Wang C., Wu, D., and Zhang, F. (2019). Modification of the convective adjustment time scale in the KainFritsch Eta scheme for the case of weakly forced deep convection over the Tibetan Plateau region. *Q. J. Roy. Meteorol. Soc*: 1–18.

Wang, C., Wu, D., Kong, Y. et al. (2017). Changes of soil thermal and hydraulic regimes in northern hemisphere permafrost regions over the 21st century. *Arct. Antarct. Alp. Res.* 49 (2): 305–319. https://doi.org/10.1657/AAAR0016-026.

Wang, C., Yang, K., Li, Y. et al. (2017). Impacts of spatiotemporal anomalies of Tibetan Plateau snow cover on summer precipitation in eastern China. *J. Clim.* 30 (1–3): 885–903. https://doi.org/10.1175/JCLI-D-16-0041.1.

Wang, C., Wang, Z., Kong, Y. et al. (2019). Most of the northern hemisphere permafrost remains under climate change. *Sci. Rep.* 9: 3295. https://doi.org/10.1038/s41598-019-39942-4.

Wang, C., Zhao, W., and Cui, Y. (2020). Changes in the seasonally frozen ground over the eastern Qinghai-Tibet Plateau in the past 60 years. *Front. Earth Sci.* 8: 1–12. https://doi.org/10.3389/feart.2020.00270.

Wang, C.H. and Yang, K. (2018). A new scheme for considering soil water-heat transport coupling based on Community Land Model: model description and preliminary validation. *J. Adv. Model. Earth Syst.* 10: 927–950. https://doi.org/10.1002/2017MS001148.

Wang, C.H., Shi, R., and Zuo, H.C. (2008). Analysis on simulation of characteristic of land surface in western Qinghai-Xizang Pleteau during frozen and thawing. *Plateau Meteorol.* 27 (2): 239–248. (In Chinese). https://doi.org/10.3724/SP.J.1047.2008.00014.

Wang, C.H., Wang, Z.L., and Cui, Y. (2009b). Spatial distributions and interannual variations of snow cover over China in the last 40 years. *Sci. Clod Arid Reg.* 1: 509–518.

Wang, C.H., Jin, S.L., and Shi, H.X. (2014). Area change of the frozen ground in China in the next 50 years. *J. Glaciol. Geocryol.* 36: 1–8. https://doi.org/10.7522/j.issn.10000240.2014.000 (in Chinese).

Wang, G. and Eltahir, E.A.B. (1999). Use of ENSO information in medium-and long-range forecasting of the Nile floods. *J. Clim.* 12: 1726–1737. https://doi.org/10.1175/1520-0442(1999)012<1726:UOEIIM>2.0.CO;2.

Wang, G., Ju, Q., Cheng, G., and Yuanmin, L. (2002). Soil organic carbon pool of grassland soils on the Qinghai-Tibetan Plateau and its global implication. *Sci. Total Environ.* 291: 207–217. https://doi.org/10.1016/S0048-9697(01)01100-7.

Wang, G., Hu, H., and Li, T. (2009). The influence of freeze-thaw cycles of active soil layer on surface runoff in a permafrost watershed. *J. Hydrol.* 375: 438–449. https://doi.org/10.1016/j.jhydrol.2009.06.046.

Wang, H.J. (2001). The weakening of the Asia monsoon circulation after the end of 1970's. *Adv. Atmos. Sci.* 3: 376–385. https://doi.org/10.1007/BF02919316.

Wang, H.S., Jia, G.S., Feng, J.M., and Ma, Z.G. (2010). Spatial distribution and seasonality of gross primary production in northern China. *Chin. J. Atmos. Sci.* 34 (5): 882–890. https://doi.org/10.3724/SP.J.1037.2010.00186.

Wang, J., Shen, Y.P., Xin, L.A. et al. (2001). Impact of climate change on snowmelt runoff in the mountainous regions of Northwest China. *J. Glaciol. Geocryol.* 23: 28–33.

Wang, K. and Liang, S. (2009). Global atmospheric downward longwave radiation over land surface under all-sky conditions from 1973 to 2008. *J. Geophys. Res.* 114: D19101. https://doi.org/10.1029/2009JD011800.

Wang, K., Wang, P., Li, Z. et al. (2007). A simple method to estimate actual evapotranspiration from a combination of net radiation, vegetation index, and temperature. *J. Geophys. Res.* 112: D15107. https://doi.org/10.1029/2006JD008351.

Wang, L., Zhuo, J., Qi, J. et al. (2017). Development of a land surface model with coupled snow and frozen soil physics. *Water Resour. Res.* 53: 5085–5103. https://doi.org/10.1002/2017WR020451.

Wang, L., Zeng, F., Zhang, Y.J. et al. (2019a). Progress in the study of soil carbon stocks and their influencing factors on the Qinghai-Tibet plateau. *J. Ecol.* 38 (11): 10.

Wang, L., Zeng, H., Zhang, Y. et al. (2019b). A review of research on soil carbon storage and its influencing factors in the Tibetan Plateau. *Chin. J. Ecol.* 38 (11): https://doi.org/10.13292/j.1000-4890.201911.009.

Wang, M., Zhou, S., and Duan, A. (2012). Trend in the atmospheric heat source over the central and eastern Tibetan Plateau during recent decades: comparison of observations and reanalysis data. *Chin. Sci. Bull.* 57 (5): 548–557. https://doi.org/10.1007/s11434-011-4838-8.

Wang, Q., Jacob, D.J., Fisher, J.A. et al. (2011). Sources of carbonaceous aerosols and deposited black carbon in the Arctic in winter-spring: implications for radiative forcing. *Atmos. Chem. Phys.* 11 (23): 12453–12473. https://doi.org/10.5194/acp-11-12453-2011.

Wang, W., Anderson, B.T., Phillips, N. et al. (2006). Feedbacks of vegetation on summertime climate variability over the North American grasslands. Part I: statistical analysis. *Earth Interact.* 10 (17): https://doi.org/10.1175/EI196.1.

Wang, W., Rinke, A., Moore, J.C. et al. (2016). Diagnostic and model dependent uncertainty of simulated Tibetan permafrost area. *Cryosphere* 10: 287–306. https://doi.org/10.5194/tc-10-287-2016.

Wang, X. and Key, J.R. (2005a). Arctic surface, cloud, and radiation properties based on the AVHRR polar pathfinder dataset. Part I: spatial and temporal characteristics. *J. Clim.* 18 (14): 2558–2574. https://doi.org/10.1175/JCLI3438.1.

Wang, X. and Key, J.R. (2005b). Arctic surface, cloud, and radiation properties based on the AVHRR polar pathfinder dataset. Part II: recent trends. *J. Clim.* 18 (14): 2575–2593. https://doi.org/10.1175/JCLI3439.1.

Wang, X., Li, Y., Yu, S. et al. (2009). Statistical study on the plateau low vortex activities. *Plateau Meteorol.* 28: 64–71. https://doi.org/10.1016/S1003-6326(09)60084-4 (in Chinese).

Wang, Y. and Wang, C. (2015). Features of clouds and convection during the pre- and post-onset periods of the Asian summer monsoon. *Theor. Appl. Climatol.* 123: 551–564. https://doi.org/10.1007/s00704-015-1372-7.

Wang, Z.L. and Wang, C.H. (2012). Predicting the snow water equivalent over China in the next 40 years based on climate models from IPCC AR4. *J. Glaciol. Geocryol.* 34: 1273–1283.

Ward, P. (2001). *Antarctica Fact File*. www.coolantarctica.com/Antarctica%20fact%20file/antarctica%20 fact%20file%20index.htm.

Warembourg, F.R. and Estelrich, H.D. (2000). Towards a better understanding of carbon flow in the rhizosphere: a time dependent approach using carbon-14. *Biol. Fertil. Soils* 30: 528–534. https!!://doi.org/10.1007/s003740050032.

Warren, S.G., Brandt, R.E., and O'Rawe Hinton, P. (1998). Effect of surface roughness on bidirectional reflectance of Antarctic snow. *J. Geophys. Res.: Planets* 103 (E11): 25789–25807. https://doi.org/10.1029/98JE01898.

Waugh, D.W. (1997). Elliptical diagnostics of stratospheric polar vortices. *Q. J. Roy. Meteorol. Soc.* 123: 1725–1748. https://doi.org/10.1002/qj.49712354213.

Waugh, D.W. and Randel, W.J. (1999). Climatology of Arctic and Antarctic polar vortices using elliptical diagnostics. *J. Atmos. Sci.* 56: 1594–1613. https://doi.org/10.1175/1520-0469(1999)056<1594:COAAAP>2.0.CO;2.

Waugh, D.W., Randel, W.J., Pawson, S. et al. (1999). Persistence of the lower stratospheric polar vortices. *J. Geophys. Res.* 104 (22): 27191–27201. https://doi.org/10.1029/1999JD900795.

Waugh, D.W., Sobel, A.H., and Polvani, L.M. (2016). What is the polar vortex, and how does it influence weather? *Bull. Am. Meteorol. Soc.* 98 (1): BAMS-D-15-00212.1. https://doi.org/10.1175/BAMS-D-15-00212.1.

Webster, P.J. (1995). The annual cycle and the predictability of the tropical coupled ocean-atmosphere system. *Meteorol. Atmos. Phys.* 56: 33–55. https://doi.org/10.1007/BF01022520.

Webster, P.J. and Yang, S. (1992). Monsoon and Enso: selectively interactive systems. *Q. J. R. Meteorol. Soc.* 118: 877–926. https://doi.org/10.1002/qj.49711850705.

Webster, M., Forest, C., Reilly, J. et al. (2003). Uncertainty analysis of climate change and policy response. *Clim. Change* 61: 295–320. https://doi.org/10.1023/B:CLIM.0000004564.09961.9f.

Wei, J.J. and Li, G.P. (2016). Characteristic of moist potential vorticity and second order moist potential vorticity of heavy rainfall over Gangxi cause by southwest vortex of southeast path. *J. Chendu Univ. Inf. Technol.* 31 (6): 592–598. (in Chinese).

Wei, L. and Li, D.L. (2003). Evaluation of NCEP/DOE surface flux data over Qinghai-Xizang Plateau. *Plateau Meteorol.* 22: 478–487. https://doi.org/10.3321/j.issn:1000-0534.2003.05.008 (in Chinese).

Wei, Z.G., Huang, R.H., and Dong, W.J. (2003). Interannual and interdecadal variations of air temperature and precipitation over the Tibetan Plateau. *Atmos. Sci.* 27: 157–170. https://doi.org/10.3878/j.issn.1006-9895.2003.02.03 (in Chinese).

Weller, G., Cubley, S., Parker, S. et al. (1972). The tundra microclimate during snow melt at Barrow, Alaska. *Arctic* 25: 29–300. https://doi.org/10.2307/40508069.

White, A., Cannell, M.G.R., and Friend, A.D. (2000). The high-latitude terrestrial carbon sink: a model analysis. *Glob. Chang. Biol.* 6: 227–245. https://doi.org/10.1046/j.1365-2486.2000.00302.x.

Wild, M. (2012). Enlightening global dimming and brightening. *Bull. Am. Meteorol. Soc.* 93 (1): 27–37. https://doi.org/10.1175/bams-d-11-00074.1.

Wild, M., Trussel, B., Ohmura, A. et al. (2009). Global dimming and brightening: an update beyond 2000. *J. Geophys. Res.* 114: D00D13. https://doi.org/10.1029/2008jd011382.

Williams, R.S. and Ferrringo, J.G. (2002). Satellite Image Atlas of Glaciers of the World-North America. Washington, DC: U.S. Geological Survey, Professional paper, 1386-J.

Willoughby, H.E. (1990). Temporal changes of the primary circulation in tropical cyclones. *J. Atmos. Sci.* 47: 242–264. https://doi.org/10.1175/1520-0469(1990)0472.0.CO;2.

Willoughby, H.E., Clos, J.A., and Shoreibah, M.G. (1982). Concentric eye walls, secondary wind maxima, and the evolution of the hurricane vortex. *J. Atmos. Sci.* 39: 395–411. https://doi.org/10.1175/1520-0469(1982)0392.0.CO;2.

Wilson, C. (1967). Cold Regions Science and Engineering Monograph 1, Section A3b: Climatology of the Cold Regions: Northern Hemisphere.

Wilson, C.V. and Godson, W.L. (1962). *The Stratospheric Temperature Field at High Latitude. Arctic Meteorology Research Group Publication in Meteorology No.46*. Montreal: McGill University.

Wilson, D., Hisdal, H., and Lawrence, D. (2010). Has streamflow changed in the Nordic countries? Recent trends and comparisons to hydrological projections. *J. Hydrol.* 394: 334–346. https://doi.org/10.1016/j.jhydrol.2010.09.010.

van der Woerd, J., Owen, L.A., Tapponnier, P. et al. (2004). Giant – M8 earthquake-triggered ice avalanches in the eastern Kunlun Shan, northern Tibet: characteristics, nature and dynamics. *Geol. Soc. Am. Bull.* 116 (3–4): 394–406. https://doi.org/10.1130/B25317.1.

Woo, M., Kane, D.L., Carey, S.K., and Yang, D. (2008). Progress in permafrost hydrology in the new millennium. *Permafrost Periglacial Process.* 19: 237–254. https://doi.org/10.1002/ppp.613.

Woollings, T., Barnes, E., Hoskins, B. et al. (2018). Daily to decadal modulation of jet variability. *J. Clim.* 31: 1297–1314. https://doi.org/10.1175/JCLI-D-17-0286.1.

Wu, G.X., Liu, Y.M., Liu, X. et al. (2005). How the heating over the Tibetan Plateau affects the Asian climate in summer. *Chin. J. Atmos. Sci.* 29: 47–56. (in Chinese).

Wu, G.X., Liu, Y., Zhang, Q. et al. (2007). The influence of mechanical and thermal forcing by the Tibetan Plateau on Asian climate. *J. Hydrometeorol.* 8: 770–789. https://doi.org/10.1175/JHM609.1.

Wu, G.X., Liu, Y., Zhu, X. et al. (2009). Multi-scale forcing and the formation of subtropical desert and monsoon. *Ann. Geophys.* 27: 3631–3644. https://doi.org/10.5194/angeo-27-3631-2009.

Wu, G.X., Liu, Y., He, B. et al. (2011). Thermal controls on the Asian summer monsoon. *Sci. Rep.* 2 (5): 65–65. https://doi.org/10.1038/srep00404.

Wu, G.X., Liu, Y., Dong, B. et al. (2012a). Revisiting Asian monsoon formation and change associated with Tibetan Plateau forcing: I. Formation. *Clim. Dyn.* 39: 1169–1181. https://doi.org/10.1007/s00382-012-1334-z.

Wu, G.X., Liu, Y., He, B. et al. (2012b). Thermal controls on the Asian summer monsoon. *Sci. Rep.* 2: 65–65. https://doi.org/10.1038/srep00404.

Wu, G.X., Duan, A.M., Liu, Y.M. et al. (2014). Tibetan Plateau climate dynamics: recent research progress and outlook. *Natl. Sci. Rev.* 2: 100–116. https://doi.org/10.1093/nsr/nwu045.

Wu, Q., Shen, Y., and Shi, B. (2003). Relationship between frozen soil together with water-heat process and ecological environment in the Tibetan Plateau. *J. Glaciol. Geocryol.* 25: 250–254. (in Chinese).

Wu, Q., Zhang, T., and Liu, Y. (2012). Thermal state of the active layer and permafrost along the Qinghai-Xiang (Tibet) Railway from 2006 to 2010. *Cryosphere* 6: 607–612. https://doi.org/10.5194/tc-6-607-2012.

Wu, Q.B. and Zhang, T.J. (2010). Changes in active layer thickness over the Qinghai-Tibetan Plateau from 1995 to 2007. *J. Geophys. Res.* 115: D09107. https://doi.org/10.1029/2009JD012974.

Wu, S.H., Yin, Y.H., Zhen, D., and Yang, Q.Y. (2005). Climate changes in the Tibetan Plateau during the last three decades. *Acta Geograph. Sin.* 60: 3–11. https://doi.org/10.3321/j.issn:0375-5444.2005.01.001 (in Chinese).

Wu, T.W. and Qian, Z.A. (2003). The relation between the Tibetan winter snow and the Asian summer monsoon and rainfall: an observational investigation. *J. Clim.* 16: 2038–2051. https://doi.org/10.1175/1520-0442(2003)016<2038:TRBTTW>2.0.CO;2.

Wu, Y., Tan, H., Deng, Y. et al. (2010). Partitioning pattern of carbon flux in a Kobresia grassland on the Qinghai-Tibetan Plateau revealed by field 13C pulse-labeling. *Glob. Chang. Biol.* 16: 2322–2333. https://doi.org/10.1111/j.1365-2486.2009.02069.x.

Wu, X., Fang, H., Zhao, Y. et al. (2017). A conceptual model of the controlling factors of soil organic carbon and nitrogen densities in a permafrost-affected region on the eastern Qinghai-Tibetan Plateau. *J. Geophys. Res. Biogeosci.* 122: 1705–1717. https://doi.org/10.1002/2016JG003641.

Wu, X.J., Wang, X.M., Liu, S.W. et al. (2021). Snow cover loss compounding the future economic vulnerability of western China. *Sci. Total Environ.* 755: 143025. https://doi.org/10.1016/j.scitotenv.2020.143025.

Xia, Y.M., Ek, B., Wu, Y.H. et al. (2015). Comparison of NLDAS-2 simulated and NASMD observed daily soil moisture. Part I: comparison and analysis. *J. Hydrometeorol.* 16: 1962–1980. https://doi.org/10.1175/JHM-D-14-0096.1.

Xia, Y.Z. (2012). *The Research of Distribution and Formation for Debris Flow due to Break-Off Glacier Lake in the Ranwu-Peilong Section of Sichuan-Tibet*. Chongqing: Jiaotong University (in Chinese).

Xiao, Y., Zhao, L., Dai, Y. et al. (2013). Representing permafrost properties in CoLM for the Qinghai-Xizang (Tibetan) Plateau. *Cold Reg. Sci. Technol.* 87: 68–77. https://doi.org/10.1016/j.coldregions.2012.12.004Get.

Xie, C.W., Zhao, L., Wu, T.H., and Dong, X.C. (2012). Changes in the thermal and moisture regime within the active layer in the Qinghai-Tibet Plateau. *J. Mt. Sci.* 9: 483–491. https://doi.org/10.1007/s11629-012-2352-3.

Xie, Z.C., Wang, X., Kang, S. Er et al. (2006). Glacial runoff in China: an evaluation and prediction for the future 50 years. *Glaciers and Permafrost* https://doi.org/10.1007/s11442-006-0415-5 (in Chinese).

Xu, B., Wang, M., Joswiak, D.R. et al. (2009). Deposition of anthropogenic aerosols in a southeastern Tibetan glacier. *J. Geophys. Res.* 114: D17209. https://doi.org/10.1029/2008JD011510.

Xu, C., Wan, J., and Lv, F. (2004). Statistic and synoptic analysis on blowing snow and snowstorm in Zhongshan Station, Antarctica. *Chin. J. Polar Res.* 16 (1): 39–45. (in Chinese).

Xu, G. and Li, S. (1994). The influence of the abnormal snow cover over the Qinghai-Tibet Plateau on precipitation. *Quart. J. Appl. Meteor.* 5: 62–67. (in Chinese).

Xu, J., Zhang, B.P., Tan, J. et al. (2009). Spatial relationship between altitudinal vegetation belts and climatic factors in the Qinghai-tibetan Plateau. *J. Mt. Sci.* 27 (6): 663–670. https://doi.org/10.3969/j.issn.1008-2786.2009.06.004.

Xu, L. and Dirmeyer, P. (2011). Snow-atmosphere coupling strength in a global atmospheric model. *Geophys. Res. Lett.* 38: L13401. https://doi.org/10.1029/2011GL048049.

Xu, L. and Dirmeyer, P. (2012). Snow–atmosphere coupling strength. Part II: albedo effect versus hydrological effect. *J. Hydrometeorol.* 14: 404–418. https://doi.org/10.1175/JHM-D-11-0103.1.

Xu, L. and Dirmeyer, P. (2013). Snow–atmosphere coupling strength. Part I: effect of model biases. *J. Hydrometeorol.* 14: 389–403. https://doi.org/10.1175/JHM-D-11-0102.1.

Xu, X.D., Tao, S.Y., and Wang, J.Z. (2002). The relationship between water vapor transport features of Tibetan Plateau: monsoon 'large triangle' affecting region and drought-flood abnormality of China. *Acta. Meteor. Sin.* 60: 259–265. https://doi.org/10.3321/j.issn:0577-6619.2002.03.001 (in Chinese).

Xu, X.N. and Guo, X.R. (1978). The structures of the polar anticyclone and their circulation characteristics. *Chinese J. Atmospheric Sci.* 2: 258-262. 10.3878/j.issn.1006-9895.1978.03.10. (in Chinese).

Xu, X.Z., Wang, J.C., and Zhang, L.X. (2001). *Physics of Frozen Soils*. Beijing: Science Press (in Chinese).

Yan, Y., Liu, X., Liu, Y., and Lu, J. (2020). Comparison of mixed-phase clouds over the Arctic and the Tibetan Plateau: seasonality and vertical structure of cloud radiative effects. *Clim. Dyn.* 54: 4811–4822. https://doi.org/10.1007/s00382-020-05257-8.

Yanai, M., Li, C., and Song, Z. (1992). Seasonal heating of the Tibetan Plateau and its effects on the evolution of the summer monsoon. *J. Meteorol. Soc. Jpn.* 70: 319–351. https://doi.org/10.2151/jmsj1965.70.1B_319.

Yanase, W., Niino, H., and Saito, K. (2002). High-resolution numerical simulation of a polar low. *Geophys. Res. Lett.* 29: https://doi.org/10.1029/2002GL014736.

Yanase, W., Fu, G., Niino, H., and Kato, T. (2004). A polar low over the Japan Sea on 21 January 1997. Part II: a numerical study. *Mon. Weather Rev.* 132: 1552–1574. https://doi.org/10.1175/1520-0493(2004)132<1552:APLOTJ>2.0.CO;2.

Yang, C. (1991). Glacier water resources in China. *Resour. Sci.* 1: 46–55.

Yang, D., Kane, D., Hinzman, L. et al. (2002). Siberian Lena River hydrologic regime and recent change. *J. Geophys. Res.* 107 (D23): 4694. https://doi.org/10.1029/2002JD002542.

Yang, K. and Wang, C. (2019a). Seasonal persistence of soil moisture anomalies related to freeze–thaw over the Tibetan Plateau and prediction signal of Summer precipitation in eastern China. *Clim. Dyn.* 53: 2411–2424.

Yang, K. and Wang, C. (2019b). Water storage effect of soil freeze-thaw process and its impacts on soil hydro-thermal regime variations. *Agric. For. Meteorol.* 265: 280–294. https://doi.org/10.1016/j.agrformet.2018.11.011.

Yang, K. and Wang, C. (2022). East–west reverse coupling between spring soil moisture and summer precipitation and its possible responsibility for wet bias in GCMs over Tibetan Plateau. *J. Geophys. Res. Atmos.* 127: e2021JD036286. https://doi.org/10.1029/2021JD036286.

Yang, K., Hu, T., and Wang, C. (2017). A numerical study on the relationship between the spring–winter snow cover anomalies over the northern and southern Tibetan Plateau and summer precipitation in East China. *Chin. J. Atmos. Sci.* 41: 345–356. doi: 10.3878/j.issn.1006-9895.1604.16119.

Yang, K., Koike, T., Stackhouse, P. et al. (2006). An assessment of satellite surface radiation products for highlands with Tibet instrumental data. *Geophys. Res. Lett.* 33: L22403. https://doi.org/10.1029/2006GL027640.

Yang, K., Ma, Y., Pinker, R.T. et al. (2008). Evaluation of satellite estimates of downward shortwave radiation over the Tibetan Plateau. *J. Geophys. Res.* 113: D17204. https://doi.org/10.1029/2007JD009736.

Yang, K., He, J., Tang, W. et al. (2010). On downward shortwave and longwave radiations over high altitude regions: observation and modeling in the Tibetan Plateau. *Agric. Forest. Meteorol.* 150: 38–46. https://doi.org/10.1016/j.agrformet.2009.08.004.

Yang, K., Ye, B., Zhou, D. et al. (2011). Response of hydrological cycle to recent climate changes in the Tibetan Plateau. *Clim. Change* 109 (3–4): 517–534. https://doi.org/10.1007/s10584-011-0099-4.

Yang, K., Ding, B., Qin, J. et al. (2012). Can aerosol loading explain the solar dimming over the Tibetan Plateau? *Geophys. Res. Lett.* 39: L20710. https://doi.org/10.1029/2012GL053733.

Yang, K., Wang, C., and Bao, H. (2016). Contribution of soil moisture variability to summer precipitation in the Northern Hemisphere. *J. Geophys. Res. Atmos.* 121 (D12): 108–124. https://doi.org/10.1002/2016JD025644.

Yang, K., Wang, C., and Li, S. (2018). Improved simulation of frozen-thawing process in land surface model (CLM4.5). *J. Geophys. Res. Atmos.* 123: 13238–13258. https://doi.org/10.1029/2017JD028260.

Yang, M., Wang, X., Pang, G. et al. (2019). The Tibetan Plateau cryosphere: observations and model simulations for current status and recent changes. *Earth Sci. Rev.* 190: 353–369.

Yang, M., Yao, T., He, Y. et al. (2002). The water cycles between land surface and atmosphere in northern part of Tibetan Plateau. *Sci. Geog. Sin.* 22: 29–33. https://doi.org/10.1007/s11769-002-0038-4.

Yang, M., Yao, T., Gou, X. et al. (2003). The soil moisture distribution, thawing-freezing processes and their effects on the seasonal transition on the Qinghai-Xizang (Tibetan) Plateau. *J. Asian Earth Sci.* 21: 457–465. https://doi.org/10.1016/S1367-9120(02)00069-X.

Yang, M., Wang, S., Yao, T. et al. (2004). Desertification and its relationship with permafrost degradation in Qinghai-Xizang (Tibet) plateau. *Cold Reg. Sci.Technol.* 39: 47–53. https://doi.org/10.1016/j.coldregions.2004.01.002.

Yang, M., Yao, T., Gou, X. et al. (2007). Comparison analysis of the summer monsoon precipitation between northern and southern slopes of Tanggula Mountains, Qinghai-Xizang (Tibetan) Plateau: a case study in summer 1998. *Hydrol. Process.* 21 (14): 1841–1847.

Yang, M., Nelson, F.E., Shiklomanov, N.I. et al. (2010). Permafrost degradation and its environmental effects on the Tibetan Plateau: a review of recent research. *Earth Sci. Rev.* 103: 31–44. https://doi.org/10.1016/j.earscirev.2010.07.002.

Yang, W., Yao, T.D., Xu, B.Q. et al. (2008). Quick ice mass loss and abrupt retreat of the maritime glaciers in the Kangri Karpo Mountains, southeast Tibetan Plateau. *Chi. Sci. Bull.* 53 (16): 2547–2551.

Yang, X., Pavelsky, T.M., and Allen, G.H. (2020). The past and future of global river ice. *Nature* 577 (7788): 69–73. https://doi.org/10.1038/s41586-019-1848-1.

Yang, Y., Chen, R., Liu, G. et al. (2022). Trends and variability in snowmelt in China under climate change. *Hydrol. Earth Syst. Sci.* 26: 305–329. https://doi.org/10.5194/hess-26-305-2022.

Yang, Y.C. and Gao, D.Y. (1987). Studies on the moisture passage on the lower reaches of the Brahmaputra. *China Sci.* 8: 893–902. (in Chinese).

Yang, Y.C., Li, B.S., and Gao, D.Y. (1987). A preliminary investigation to the water vapor pass in the lower reaches of Yalung Zangbo River Valley. *Sci. China B* 32: 580–593. https://doi.org/10.1351/pac198961050973 (in Chinese).

Yang, Z. (1997). Research on cold regions hydrology in China. In: *The Sixth Proceedings of the Symposium on Hydrology*, 50–56. Beijing: Science Press.

Yao, T., Thompson, L., Yang, W. et al. (2012). Different glacier status with atmospheric circulations in Tibetan Plateau and surroundings. *Nat. Clim. Change* 2: 663–667. https://doi.org/10.1038/nclimate1580.

Yao, X., Sun, J., Kang, L. et al. (2014). Advances on research of shear convergence line over Qinghai-Xizang Plateau. *Plateau Meteorol.* 33 (1): 294–300. doi: 10.7522/j.issn.1000-0534. 2013.00164. (in Chinese).

Yao, Z., Xiangming, X., Cui, J. et al. (2016). Consistency between sun-induced chlorophyll fluorescence and gross primary production of vegetation in North America. *Remote Sens. Environ.* 183: 154–169. https://doi.org/10.1016/j.rse.2016.05.015.

Yasunari, T., Kitoh, A., and Tokioka, T. (1991). Local and remote responses to excessive snow mass over Eurasia appearing in the Northern spring and summer climate-a study with the MRI GCM. *J. Meteorol. Soc. Jpn.* 69: 473–487. https://doi.org/10.2151/jmsj1965.69.4_473.

Ye, D.Z. and Gao, Y.X. (1979). *Meteorology of the Qinghai-Xizang Plateau*. Beijing: Chinese Science Press.

Ye, H. and Bao, Z.H. (2001). Lagged teleconnections between snow depth in northern Eurasia, rainfall in Southeast Asia and sea-surface temperatures over the tropical Pacific Ocean. *Int. J. Climatol.* 21: 1607–1621. https://doi.org/10.1002/joc.695.

Ye, J., Li, F., Sun, G., and Guo, A. (2010). Solar dimming and its impact on estimating solar radiation from diurnal temperature range in China, 1961–2007. *Theor. Appl. Climatol.* 101 (1–2): 137–142. https://doi.org/10.1007/s00704-009-0213-y.

Yeh, T.C., Wetherald, R.T., and Manabe, S. (1984). The effect of soil moisture on the short-term climate and hydrology change – a numerical experiment. *Mon. Weather Rev.* 112: 474–490.

Yi, S., Woo, M.-K., and Arain, A.M. (2007). Impacts of peat and vegetation on permafrost degradation under climate warming. *Geophys. Res. Lett.* 34: L16504. https://doi.org/10.1029/2007/GL030550.

Yi, S., McGuire, A.D., Harden, J. et al. (2009). Interactions between soil thermal and hydrological dynamics in the response of Alaska ecosystems to fire disturbance. *J. Geophys. Res.* 114: G02015. https://doi.org/10.1029/2008JG000841.

You, Q., Sanchez-Lorenzo, A., Wild, M. et al. (2013). Decadal variation of surface solar radiation in the Tibetan plateau from observations, reanalysis and model simulations. *Clim. Dyn.* 40: 2073–2086. https://doi.org/10.1007/s00382-012-1383-3.

You, Q., Wu, T., Shen, L. et al. (2020). Review of snow cover variation over the Tibetan Plateau and its influence on the broad climate system. *Earth Sci. Rev.* 201: 103043. https://doi.org/10.1016/j.earscirev.2019.103043.

Yu, F., Qi, H.L., Yao, X.L., and Liu, Y.Z. (2015). Comparison of permafrost degradation under natural ground surfaces and embankments of the Qinghai-Tibet Highway. *Cold Reg. Sci. Technol.* 114: 1–8. https://doi.org/10.1016/j.coldregions.2015.02.002.

Yu, L. and Wang, C. (2012). The simulation and revision of sil moisture over Tibetan Plateau. *J. Glaciol. Geocryol.* 61 (2): 531–538. (in Chinese). https://doi.org/10.7498/aps.61.029201.

Yu, S. (2008). New research advances of the Tibetan Plateau vortex in summer. *Torrential Rain Disasters* 27: 367–372. (in Chinese).

Yuan, X., and Liang, X.-Z. (2011). Improving cold season precipitation prediction by the nested CWRF-CFS system. *Geophys. Res. Lett.* 38: L02706.

Yun, H., Wu, Q., Zhuang, Q. et al. (2018). Consumption of atmospheric methane by the Qinghai–Tibet Plateau alpine steppe ecosystem. *Cryosphere* 12: 2803–2819. https://doi.org/10.5194/tc-12-2803-2018.

Zarnani, A., Musilek, P., Shi, X.Y. et al. (2012). Learning to predict ice accretion on electric power lines. *Eng. Appl. Artif. Intell.* 25: 609–617. https://doi.org/10.1016/j.engappai.2011.11.004.

Zeng, X.B. and Decker, M. (2009). Improving the numerical solution of soil moisture-based Richards equation for land models with a deep or shallow water table. *J. Hydrometeorol.* 10 (1): 308–319. https://doi.org/10.1175/2008JHM1011.1.

Zerr, R.J. (1997). Freezing rain: an observational and theoretical study. *J. Appl. Meteorol. Climatol.* 36 (12): 1647–1661. https://doi.org/10.1175/1520-0450(1997)036<1647:FRAOAT>2.0.CO;2.

Zhai, P., Zhang, X.B., Wang, H., and Pan, X.H. (2004). Trends in total precipitation and frequency of daily precipitation extremes over China. *J. Clim.* 18: 1096–1108. https://doi.org/10.1175/JCLI-3318.1.

Zhan, Y., Lu, Z., Yao, H., and Xian, S.H. (2018). A coupled thermo-hydromechanical model of soil slope in seasonally frozen regions under freeze-thaw action. *Adv. Civ. Eng.* e7219826.

Zhang, D., Cong, Z., Ni, G. et al. (2015). Effects of snow ratio on annual runoff within the Budyko framework. *Hydrol. Earth Syst. Sci.* 19: 1977–1992. https://doi.org/10.5194/hess-19-1977-2015.

Zhang, J., Wu, L., and Dong, W. (2011). Land-atmosphere coupling and summer climate variability over East Asia. *J. Geophys. Res. Atmos.* 116 (D5): 420–424. https://doi.org/10.1029/2010JD014714.

Zhang, J., Zhang, C., Xiao, J., and Jiang, J. (2019). A PZT-based electromechanical impedance method for monitoring the soil freeze–thaw process. *Sensors* 19: 1107. https://doi.org/10.3390/s19051107.

Zhang, J.J. and Peng, Y.Q. (1983). Characteristics of the South Asia high in time and frequency. *Acta. Meteor. Sin.* 41 (3): 348–353. (in Chinese).

Zhang, L., Xia, X., Liu, S. et al. (2020). Significant methane ebullition from alpine permafrost rivers on the East Qinghai–Tibet Plateau. *Nat. Geosci.* 13: 1–6. https://doi.org/10.1038/s41561-020-0571-8.

Zhang, P.C., Tang, Y.H., Hirota, M. et al. (2009). Use of a regression method to partition sources of ecosystem respiration in an alpine meadow. *Soil Biol. Biochem.* 41: 663–670. https://doi.org/10.1016/j.soilbio.2008.12.026.

Zhang, P.Z. (1991). Energetic analysis during the February 1981 stratospheric warming. *Acta Meteorol. Sin.* 49 (3): 369–374. (in Chinese).

Zhang, Q., Qian, Y.F., and Zhang, X.H. (2000). Interannual and inlerdecadal wirialions of the South Asia high. *Sci. Atmos. Sin.* 24 (1): 67–78. (in Chinese).

Zhang, R.H., Sumi, A., and Kimoto, M. (1999). A diagnostic study of the impact of El Niño on the precipitation in China. *Adv. Atmos. Sci.* 16: 229–241. https://doi.org/10.1007/BF02973084.

Zhang, S. and Tao, S. (2001). Influences of snow cover over the Tibetan Plateau on Asian summer monsoon. *Chin. J. Atmos. Sci.* 25: 372–390.

Zhang, T., Osterkamp, T.E., and Stamnes, K. (1997). Effects of climate on the active layer and permafrost on the North Slope of Alaska, USA. *Permafr. Periglac. Process.* 8: 45–67. https://doi.org/10.1002/(SICI)1099-1530(199701)8:1<45::AID-PPP240>3.0.CO;2-K.

Zhang, T.J., Oliver, W., Frauenfeld, M. et al. (2005). Spatial and temporal variability in active layer thickness over the Russian Arctic drainage basin. *J. Geophys. Res.* 110: D16101. https://doi.org/10.1029/2004JD005642.

Zhang, T.R., Chai, X.M., and Li, Z.Z. (2010). Characteristics of vegetation coverage in northern China and its relationship with sandstorm. *Plateau Meteorol.* 29 (1): 137–145. https://doi.org/10.3788/gzxb20103906.0998.

Zhang, W.J. (1983). The throbbing glacier of mount Nanga Bawa. *J. Glaciol. Geocryol.* 05 (04): 75–76.

Zhang, X., Sun, S., and Xue, Y. (2007). Development and testing of a frozen soil parameterization for cold region studies. *J. Hydrometeorol.* 8 (4): 852–861. https://doi.org/10.1175/JHM605.1.

Zhang, X.Y., Friedl, M.A., Schaaf, C.B., and Strahler, A.H. (2004). Climate controls on vegetation phenological patterns in northern mid- and high latitudes inferred from modis data. *Glob. Chang. Biol.* 10 (7): 1133–1145.

Zhang, Y., Li, T., and Wang, B. (2004). Decadal change of the spring snow depth over the Tibetan Plateau: the associated circulation and influence on the East Asian summer monsoon. *J. Clim.* 17: 2780–2793. https://doi.org/10.1175/1520-0442(2004)017<2780:DCOTSS>2.0.CO;2.

Zhang, Y., Tang, Y., Jiang, J., and Yang, Y.H. (2007). Characterizing the dynamics of soil organic carbon in grasslands on the Qinghai-Tibetan Plateau. *Sci. China, Ser. D: Earth Sci.* 50: 113–120. https://doi.org/10.1007/s11430-007-2032-2.

Zhang, Y.S., Ohata, T., and Kadota, T. (2003). Land-surface hydrological processes in the permafrost region of the eastern Tibetan Plateau. *J. Hydrol.* 283: 41–56. https://doi.org/10.1016/S0022-1694(03)00240-3.

Zhang, Z.Q. and Wu, Q.B. (2012). Predicting changes of active layer thickness on the climate warming. *J. Glaciol. Grocryol.* 34 (03): 505e511.

Zhao, H. and Moore, G.W.K. (2004). On the relationship between Tibetan snow cover, the Tibetan plateau monsoon and the Indian summer monsoon. *Geophys. Res. Lett.* 31: L14204. https://doi.org/10.1029/2004GL020040.

Zhao, X. (2009). *Alpine Meadow Ecosystem and Global Change*. Beijing: Science Press (in Chinese).

Zhao, L., Li, J., Xu, S. et al. (2010). Seasonal variations in carbon dioxide exchange in an alpine wetland meadow on the Qinghai-Tibetan Plateau. *Biogeosciences* 7: https://doi.org/10.5194/bg-7-1207-2010.

Zhao, Y. and Qian, Y.F. (2007). Relationship between the surface thermal anomalies in the Tibetan Plateau and the rainfall in the Jianghuai area in summer. *Chin. J. Atmos. Sci.* 1: 145–154. (in Chinese).

Zheng, D., van der Velde, R., Su, Z. et al. (2017). Evaluation of Noah frozen soil parameterization for application to a Tibetan meadow ecosystem. *J. Hydrometeorol.* 18: 1749–1763. https://doi.org/10.1175/JHM-D-16-0199.1.

Zhong, L., Su, Z.B., Ma, Y.M. et al. (2011). Accelerated changes of environmental conditions on the Tibetan Plateau caused by climate change. *J. Clim.* 24: 6540–6550. https://doi.org/10.1175/JCLI-D-10-05000.1.

Zhou, D., Fan, G., Huang, R. et al. (2007). Interannual variability of the normalized difference vegetation index on the Tibetan Plateau and its relationship with climate change. *Adv. Atmos. Sci.* 24 (3): 474–484. https://doi.org/10.1007/s00376-007-0474-2.

Zhou, J.F. (1991). Numerical experiments of south-north movement of the ridgeline of the South Asia high. *Acta. Meteor. Sin.* 49: 96–99. (in Chinese).

Zhou, Y., Yue, Y., Gao, Z., and Zhou, Y. (2017). Climatic characteristics and determination method for freezing rain in China. *Adv. Meteorol.* 2017: e4635280. https://doi.org/10.1155/2017/4635280.

Zhu, F., Lu, L., Chen, X., and Zhao, W. (1980). *The South Asia High*, 16–17. Beijing: Science Press (in Chinese).

Zhu, L., Huang, G., Fan, G. et al. (2018). Elevation-dependent sensible heat flux trend over the Tibetan Plateau and its possible causes. *Clim. Dyn.* 52: 3997–4009. https://doi.org/10.1007/s00382-018-4360-7.

Zhu, T., Zhang, D., and Weng, F. (2004). Numerical simulation of Hurricane Bonnie (1998). Part I: eyewall evolution and intensity changes. *Mon. Weather Rev.* 132: 225–241. https://doi.org/10.1175/1520-0493(2004)132<0225:NSOHBP>2.0.CO;2.

Zhu, Y. and Ding, Y. (2007). Influences of snow cover over Tibetan Plateau on weather and climate: advances and problems. *Meteorol. Sci. Technol.* 35: 1–8. (in Chinese).

Zhu, Y.X., Ding, Y., and Xu, H. (2007). Decadal relationship between atmospheric heat source and winter–spring snow cover over the Tibetan Plateau and rainfall in east China. *Acta. Meteor. Sin.* 65: 946–958. https://doi.org/10.1029/2007JD008874 (in Chinese).

Zhuang, Q., Melillo, J.M., Kicklighter, D.W. et al. (2004). Methane fluxes between terrestrial ecosystems and the atmosphere at northern high latitudes during the past century: a retrospective analysis with a process-based biogeochemistry model. *Glob. Biogeochem. Cycles* 18: GB3010. https://doi.org/10.1029/2004GB002239.

Zhuang, Q., Melillo, J.M., Sarofim, M.C. et al. (2006). CO2 and CH4 exchanges between land ecosystems and the atmosphere in northern high latitudes over the 21st century. *Geophys. Res. Lett.* 33: L17403. https://doi.org/10.1029/2006GL026972.

Zhuang, Q., Chen, M., Xu, K. et al. (2013). Response of global soil consumption of atmospheric methane to changes in atmospheric climate and nitrogen deposition. *Global Biogeochem. Cycles* 27: 650–663. https://doi.org/10.1002/gbc.20057.

Zimov, S.A., Davydov, S.P., Zimova, G.M. et al. (2006). Permafrost carbon: stock and decomposability of a globallysignificant carbon pool. *Geophys. Res. Lett.* 33: L20502. https://doi.org/10.1029/2006GL027484.

Zimov, S.A., Schuur, E.A.G., and Chapin, F.S. III (2006). Permafrost and the global carbon budget. *Science* 312: 1612–1613. https://doi.org/10.1126/science.1128908.

Zuev, V.V. and Savelieva, E. (2019). The cause of the strengthening of the Antarctic polar vortex during October–November periods. *J. Atmos. Sol.-Terr. Phys.* 190: 1–5. https://doi.org/10.1016/j.jastp.2019.04.016.

Zuev, V.V. and Savelieva, E. (2020). Arctic polar vortex dynamics during winter 2006/2007. *Pol. Sci.* 25: 100532. https://doi.org/10.1016/j.polar.2020.100532.

Zuev, V.V., Savelieva, E.S., Borovko, I.V., and Krupchatnikov, V.N. (2020). Antarctic polar vortex weakening due to a temperature decrease in the lower subtropical stratosphere. In: *26th International Symposium on Atmospheric and Ocean Optics, Atmospheric Physics*. http:!//dx.doi.org/10.1117/12.2574652.

Zuo, Q., Gao, S., and Sun, X. (2016). Effects of the upstream temperature anomaly on freezing rain and snowstorms over Southern China in early 2008. *J. Meteorol. Res.* 30 (5): 694–705. https://doi.org/10.1007/s13351-016-5253-5.

Zwiers, F.W. (1993). Simulation of the Asian summer monsoon with the CCC GCM-1. *J. Clim.* 6: 469–486. https://doi.org/10.1175/1520-0442(1993)006,0469:SOTASM.2.0.CO;2.

Index

Page references in *italics* refer to Figures; those in **bold** refer to Tables

Climatology in Cold Regions, First Edition. Chenghai Wang.

Wiley The manufacturer's authorized representative according to the EU General Product Safety Regulation is Wiley-VCH GmbH, Boschstr. 12, 69469 Weinheim, Germany, e-mail: Product_Safety@wiley.com.

Printed and bound by CPI Group (UK) Ltd, Croydon, CR0 4YY
19/08/2026
02201090-0001